AF564856

Aquatic Ecosystem Management

NIPA® GENX ELECTRONIC RESOURCES & SOLUTIONS P. LTD.
New Delhi-110 034

About the Editors

Dr. Adnan Abubakr is working as Professor of Limnology and Head in the Division of Aquatic Environmental Management, Faculty of Fisheries, Sher-e-Kashmir University of Agricultural Sciences and Technology, Kashmir, SKUAST-K. He has earned M.Sc. degree & Ph.D. degree in Limnology from Department of Limnology, Barkatullah University Bhopal, and have specialized in Water and Wastewater examination, analytical and bacteriological techniques related to water, freshwater and brackish water. He has specialization in designing artificial wetland systems for treatment of wastewaters. Besides this he is involved in teaching, research and extension experience of more than 15 years in the field of aquatic ecology. He has been instrumental in Instituting M.F.Sc & Ph.D Degree programmes in Aquatic Environment Management (AEM) in the Division of AEM, He has have more than 60 research publication in peer reviewed journals and have completed 02 externally funded and 03 university funded research projects. He has guided more than 20 students for their Masters and Ph.D degrees.

Dr. Monisa M. Malik, is presently working as Assistant Professor Cum Junior Scientist in the Division of Aquatic Environmental Management at Faculty of Fisheries, Rangil, Ganderbal, Sher-e-Kashmir University of Agricultural Sciences and Technology- Kashmir. An Undergraduate with Bachelors in Fisheries Science from University of Dr. B.S.K.K.V, Dapoli, she was awarded the Best Academic Performer and Best Student Award during her degree programme. She also completed her Masters and Ph.D in the same field from SKUAST-Kashmir. She is a Gold Medalist in Master's and recipient of National INSPIRE Fellowship during her Ph.D. Programme. Her areas of specialization include Limnology, Waste water treatment, Aquatic Ecology and Management. She is an active Editorial Member for reputed Journals like International Journal of Energy and Environmental Science, Journal of Zoo Biology and Nature Science e-Magazine. She is associated with Life Membership in Journal of Fisheries and Life Sciences and Journal of Environmental Biology. She has published more than 12 research publications in both national and international peer-reviewed research journals. An author of one book with International Publishing, she has also written a number of book chapters of significance. Dr. Malik has received one best poster presentation award from national organizations.

Dr. Adnan Amin, is presently working as Assistant Professor Cum Junior Scientist in the Division of Aquatic Environmental Management at Faculty of Fisheries, Rangil, Ganderbal, Sher-e-Kashmir University of Agricultural Sciences and Technology, Kashmir. He is an undergraduate from College of Fishery Science Muthukur, Nellore SVVU, Tirupati Andhra Pradesh. He did his Masters from College of Fisheries, Mangalore in the Department of Aquatic Environment Management. He completed Doctoral degree from College of Fisheries Ratnagiri in the Department of Aquatic Environment Management. His area of research interest is Emerging Contaminants, Aquatic Toxicology, Limnology. His major research contribution is heavy metal and pesticide toxicity in aquatic organisms, Impact of Climate change in aquatic biodiversity.

He is a recipient of university fellowship i.e. merit fellowship for perusing Masters. He is associated with Life Membership in Society of Fisheries and Life Science, College of Fisheries, Mangalore. He has published more than 25 research publications in both national and international peer-reviewed research journals. In addition, he has published five book chapter, Six review papers and Eighteen Magazine articles. Dr. Adnan Amin has received one best research paper awards from national organizations

Aquatic Ecosystem Management Strategies and Techniques

Adnan Abubakr
Professor and Head,
Division of Aquatic Environmental Management
Faculty of Fisheries, Rangil, Ganderbal
SKUAST-Kashmir, India

Monisa Mehboob Malik
Assistant Professor
Division of Aquatic Environmental Management
Faculty of Fisheries, Rangil Ganderbal
SKUAST-Kashmir, India

Adnan Amin
Assistant Professor
Division of Aquatic Environmental Management
Faculty of Fisheries, Rangil Ganderbal
SKUAST-Kashmir, India

NIPA® GENX ELECTRONIC RESOURCES & SOLUTIONS P. LTD.
New Delhi-110 034

NIPA® GENX ELECTRONIC RESOURCES & SOLUTIONS P. LTD.

101,103, Vikas Surya Plaza, CU Block
L.S.C. Market, Pitam Pura, New Delhi-110 034
Ph : +91-11-43860225, Mob.: +91 9717133558, 9540816132
E-mail: newindiapublishingagency@gmail.com
Website: www.nipaersources.com

Print ISBN: 978-93-58875-93-5
ebook ISBN: 978-93-58874-11-2

Composed and Designed by NIPA®.

Foreword

Aquatic ecosystem management involves the use of strategies and techniques to maintain, restore. And enhance the health and functionality of freshwater and marine ecosystems. These ecosystems include rivers, lakes, wetlands, estuaries and oceans, which support biodiversity, provide resources and offer ecosystems services essential for human and environmental wellbeing.

Integrated water resources management coordinates the management of water, land and related resources to maximize economic and social welfare without compromising ecosystems.

Ecosystem based management focuses on maintaining the ecological integrity of aquatic ecosystems. Considers cumulative impacts from multiple activities rather than isolated issues.

Watershed and catchment management ensures activities in the entire watershed do not degrade water bodies downstream.

Techniques for managing aquatic ecosystems utilizes organisms such as bacteria, algae and plants to clean up polluted water bodies.

Aquatic vegetation management controls excessive growth of aquatic plants (e.g., through harvesting or introducing herbivorous species).

Stakeholder engagement and education involves communities' governments and industries in decision making processes and conservation needs.

Effective aquatic management requires a multidisciplinary approach, integrating ecological science, engineering, policy and community involvement to ensure long term substantiality.

Looking to the multidimensional approach for managing the aquatic resources the present book serves as a valuable guide for researchers, practitioners, and policymakers, offering insights into the latest strategies and methodologies for the effective management of aquatic ecosystems.

The inclusion of case studies and region-specific strategies makes this book particularly relevant in a globalized yet diverse context. It highlights the importance of tailoring solutions to local environmental, social, and economic conditions. Moreover, the book's emphasis on emerging challenges, such as the impacts of microplastics and other contaminants, ensures that it remains at the forefront of contemporary environmental discourse.

I am confident that this book contributed by various authors will serve as a cornerstone for advancing knowledge and fostering collaboration among scientists, environmentalists, and decision-makers. It is not just a contribution to the field of aquatic sciences but a call to action for safeguarding one of our planet's most vital resources.

Prof. Ashwani Wanganeo
Former, Dean Faculty of Life Sciences
HOD, Department of Environmental Sciences and Limnology
Member SCAR (Scientific Community on Antarctic Research)

Preface

The management of aquatic ecosystems is a critical endeavour in an era of escalating environmental challenges. From the degradation of freshwater resources to the pollution of marine environments, the health of aquatic systems is intricately linked to human well-being, biodiversity conservation, and global sustainability. The book Aquatic Ecosystem Management: Strategies and Techniques seeks to address the complexities of managing these diverse ecosystems by presenting an integrated framework that merges scientific research with practical applications.

The book delves into various facets of aquatic ecosystems, encompassing freshwater bodies, estuarine environments, and marine habitats. Each chapter highlights the unique challenges faced by these ecosystems, such as water pollution, habitat destruction, invasive species, and climate change. Moreover, the book emphasizes the importance of multidisciplinary approaches, integrating ecological principles with cutting-edge technologies and sustainable management practices.

A significant portion of this work focuses on innovative strategies for mitigating anthropogenic impacts. Furthermore, the book sheds light on the role of community involvement and stakeholder engagement in fostering successful conservation initiatives, stressing the need for inclusive governance and participatory management.

This volume also addresses emerging concerns such as microplastic pollution, eutrophication, and the effects of toxic contaminants on aquatic life. By providing real-world case studies and evidence-based solutions, this book aims to equip policymakers, researchers, environmental managers, and students with the knowledge necessary to navigate the intricate dynamics of aquatic ecosystems.

The inspiration behind this book stems from a collective effort to preserve our planet's water resources for future generations. In compiling this comprehensive guide, we acknowledge the contributions of numerous scientists, practitioners, and organizations dedicated to aquatic conservation. Their insights and experiences have been instrumental in shaping the strategies and techniques presented within these pages.

It is our hope that this book serves as a valuable resource for those committed to safeguarding aquatic ecosystems. By fostering a deeper understanding of ecological processes and management strategies, we can collectively work towards a sustainable future where aquatic environments thrive in harmony with human development.

Editors

Contents

Forward .... *vii*
Preface .... *ix*

1. **Advancing Aquatic Sustainability: Innovations in Green Technology 1**
Ketan Makwana, H.L. Parmar, D.T. Vaghela, Mayur Bhadarka, Milan Ram and Gopal Paghadar

2. **Technological Innovations for Mitigating Water Pollution 19**
Nowsheen Mushtaq and Adnan Abubakr

3. **Use of Duckweed as a Potential Feed Additive in Poultry Nutrition for Sustainable Aquatic Management 31**
Subata Mehboob and Bushra Zaffar

4. **Climate Change Adaptation and Mitigation: Assessing Vulnerability and Promoting Resilient Aquaculture 39**
Hardik Sikotariya, T.H. Dave, D.T. Vaghela and Mayur Bhadarka

5. **Pesticides in Water: Unveiling the Hidden Threats to Aquatic Life 51**
Bhooleshwari and A.T. Ramachandra Naik

6. **Membrane Technologies for Heavy Metal Removal from Water 75**
Pinak Bamaniya, D.T. Vaghela and Mayur Bhadrka

7. **Microbial Bioprocesses in Aquatic Environments: Applications and Innovations 109**
Mayur Bhadarka, D.T. Vaghela, Alwinpeter M. and Hardik Sikotariya

8. **Microplastics in Water: Environmental Impacts, Detection and Mitigation 133**
D.T. Vaghela, Mayur Bhadarka and Milan Ram

9. Toxicants and Their Effects on Aquatic Systems Bioaccumulation, Biomagnification and Detoxification 153

Mayur Bhadarka, D.T. Vaghela, Rushikesh Valmik, Pinak Bamaniya

10. Emerging Contaminant in Aquatic System 179

Nazeefah Hassan, Monisa Malik, Ifra Tabassum, Adnan Abu bakr and Shabana Arjamand

11. Heavy Metal Pollution: A Hidden Threat to Aquatic Biodiversity 203

Gora Shiva Prasad, Gopikrishna, Shilpa Koleya, Suman Karma kar, Sulav Mandal and S.K. Rout

12. An Ambitious Carbon Sequestration by Phytoplankton in the Marine Coastal Systems 219

Narshivudu Daggula, Ramachandra Naik A.T. and Mukkeri Kran thirekha

13. Bioremediation: A Green Symphony for Environmental Healing 233

Kusumlata Goswami and Shubham Disodia

14. Sponge City: A Solution to Urban Flooding in Climate Change Prone Areas..247

Kurapati Nagendrasai, Kishore Kumar Krishnani and Guntapalli Sravani

15. Conservation of Aquatic Ecosystem: An Ecosystem-Based Approach..259

Rinkesh Nemichand Wanjari, Zaheer Abass, Hudisa Banoo and Dechen Chuskit

Index *..277*

1

Advancing Aquatic Sustainability Innovations in Green Technology

Ketan Makwana[1]*, H.L. Parmar[2], D.T. Vaghela[3], Mayur Bhadarka[4] Milan Ram[4] and Gopal Paghadar[5]

[1]PG Scholar, Department of Fisheries Resource Management*

[2]Assistant Professor and Head, Department of Fisheries Resource Management

[3]Associate Professor and Head, Department of Aquatic Environment Management

[4]PG Scholar, Department of Aquatic Environment Management

[5]PG Scholar, Department of Aquaculture, College of Fisheries Science, Kamdhenu University, Veraval (362265), Gujarat, India

Corresponding Author Email: ketankm2002@gmail.com

Abstract

Aquatic sustainability *is vital for maintaining the health and productivity of aquatic ecosystems. Human activities have significantly impacted these environments, leading to pollution, habitat destruction, and biodiversity loss. Green technologies offer innovative solutions to address these challenges. Key components of aquatic sustainability include biodiversity, water quality, ecosystem health, resource management, and social equity. Aquatic sustainability is crucial for food security, economic benefits, climate regulation, cultural and recreational values, and ecosystem services. Current challenges facing aquatic ecosystems include pollution, overfishing, habitat destruction, climate change, invasive species, plastic pollution, and habitat degradation. Green technologies play a crucial role in addressing these challenges by reducing pollution, promoting sustainable fisheries management, mitigating climate change, restoring habitats, conserving water, and monitoring aquatic ecosystems. Green technologies such as advanced water filtration systems, nanotechnology, bio-inspired filters, desalination technologies, UV-C light, ozone, eco-friendly wastewater management, integrated constructed wetlands, biological treatment systems, remediation technologies, natural infrastructure solutions, and biodegradable materials offer sustainable and effective solutions for aquatic sustainability.*

Keywords: *Aquatic sustainability, Pollution, Green energy, Nanotechnology*

1. Introduction

Aquatic sustainability refers to the ability of aquatic ecosystems to maintain their health, productivity, and biodiversity over time. It involves preserving the balance of aquatic environments while ensuring that their resources are used responsibly and equitably (Mansour *et al.*, 2022). Human activities have significantly impacted aquatic ecosystems, leading to pollution, habitat destruction, and biodiversity loss. To address these pressing challenges, innovative green technologies are emerging as crucial solutions. By harnessing the power of nature and cutting-edge technology, we can restore and protect our aquatic environments for future generations. These innovations range from advanced water filtration systems and sustainable aquaculture practices to renewable energy solutions and innovative habitat restoration techniques. As we continue to explore and develop these green technologies, we can create a more sustainable and resilient future for our aquatic ecosystems (Bradu *et al.*, 2023).

Key Components of Aquatic Sustainability: (Crute & Muir, 2011)

- Biodiversity: Maintaining a diverse range of aquatic species and habitats.
- Water Quality: Ensuring that water bodies are free from pollution and contaminants.
- Ecosystem Health: Protecting the structure and function of aquatic ecosystems.
- Resource Management: Wisely using aquatic resources like fish, shellfish, and water.
- Social Equity: Ensuring that the benefits of aquatic resources are distributed fairly.

Importance of Aquatic Sustainability: (Tacon & Metian, 2013)

- Food Security: Aquatic ecosystems provide a significant source of food, especially for coastal communities. Sustainable practices ensure that these resources are available for future generations.
- Economic Benefits: Healthy aquatic ecosystems support various industries, including fishing, tourism, and recreation.
- Climate Regulation: Oceans and other aquatic bodies play a crucial role in regulating the Earth's climate. Protecting these ecosystems helps mitigate the impacts of climate change.
- Cultural and Recreational Value: Aquatic environments offer cultural, spiritual, and recreational benefits to countless people.

- Ecosystem Services: Aquatic ecosystems provide essential services like water filtration, nutrient cycling, and flood control.

2. Current Challenges in Maintaining Healthy Aquatic Ecosystems

The preservation of healthy aquatic ecosystems is facing numerous challenges today. Here are some of the most significant ones: (Ramachandra *et al.*, 2005)

1. Pollution : (Ahmed *et al.*, 2021)
    - Industrial Pollution: Discharge of toxic chemicals and pollutants from factories and industries into rivers, lakes, and oceans.
    - Agricultural Runoff: Excess fertilizers, pesticides, and animal waste from farms that contaminate water bodies.
    - Municipal Wastewater: Untreated or inadequately treated sewage that can introduce harmful bacteria and pollutants into aquatic ecosystems.
2. Overfishing and Habitat Destruction: (Shao, 2009)
    - Unsustainable Fishing Practices: Overfishing, bycatch, and destructive fishing methods that deplete fish populations and damage habitats.
    - Coastal Development: Construction of infrastructure, such as ports, marinas, and resorts, that can destroy critical habitats like mangroves and coral reefs.
3. Climate Change: (Johnson & White, 2014)
    - Ocean Acidification: Increased carbon dioxide emissions leading to ocean acidification, which threatens marine organisms with calcified shells or skeletons.
    - Sea Level Rise: Rising sea levels inundating coastal habitats and displacing marine species.
    - Extreme Weather Events: More frequent and intense storms, hurricanes, and floods that can disrupt aquatic ecosystems.
4. Invasive Species: (Simberloff, 2010)
    - Introduction of Non-Native Species: Accidental or intentional introduction of species that can outcompete native species, disrupt food chains, and alter ecosystem dynamics.
5. Plastic Pollution: (Thushari & Senevirathna, 2020)
    - Microplastics: Tiny plastic particles that accumulate in aquatic ecosystems, harming marine life and disrupting food webs.
    - Macro Plastics: Large plastic items that can entangle and injure marine animals.

6. Habitat Degradation: (McCartney *et al.*, 2001)
    - Sedimentation: Excessive sedimentation from erosion and land use changes that can smother aquatic habitats and reduce water quality.
    - Dam Construction: Dams that alter water flow, disrupt ecosystems, and impact fish populations.

3. Innovative Green Technologies in Aquaculture

1. Sustainable Aquaculture Practices

Biofloc Technology (BFT) is a groundbreaking method in aquaculture that boosts sustainability and efficiency by harnessing microbial activity. It enhances water quality while lowering feed expenses by using microbial proteins as a natural food source for aquatic organisms.

Nutritional Benefits

BFT promotes a thriving microbial environment that enriches the water, providing a steady source of protein for aquatic species. This natural supplementation reduces the reliance on costly commercial feeds, allowing farmers to lower their expenses while enhancing the nutritional quality of their stock (Lal *et al.*, 2024). Furthermore, it optimizes feed efficiency, boosting sustainability in aquaculture.

Water Quality Enhancement

Biofloc systems transform harmful nitrogen compounds into microbial biomass, improving water quality and maintaining a more stable environment. This process reduces the likelihood of waterborne diseases and minimizes the ecological footprint of aquaculture by decreasing nutrient pollution in surrounding water bodies (Raza *et al.*, 2024). As a result, it supports more sustainable aquaculture practices.

Improved Growth and Health

Studies show that aquatic species such as red tilapia and shrimp cultivated in BFT systems experience enhanced growth, better feed conversion, and stronger immune systems. The microbial ecosystem helps strengthen disease resistance, leading to healthier, more resilient stocks (Sallam *et al.*, 2024; Joshna *et al.*, 2024). This ultimately leads to higher yields and lower mortality rates.

Economic Feasibility

Affordable Inputs

BFT relies on inexpensive agricultural by-products like molasses and cornmeal as carbon sources to regulate the carbon-nitrogen ratio. This not only reduces costs but also provides a sustainable solution by recycling waste from other industries (Melaku *et al.*, 2024). It contributes to a circular economy model in aquaculture.

Lowering Production Costs

BFT's ability to improve feed conversion efficiency results in lower feed consumption, which is typically the most significant cost in aquaculture. By cutting down on feed usage, overall production expenses are reduced, making the system more viable for both small-scale and large-scale operations (Raza *et al.*, 2024). This boost in efficiency is particularly beneficial in high-density farming setups.

Operational Challenges

Despite its many benefits, BFT requires meticulous management of key water parameters such as oxygen, pH, and the carbon-nitrogen ratio. If not carefully maintained, these variables can destabilize the system and diminish its efficiency (Lal *et al.*, 2024). As a result, BFT requires skilled operators and specialized equipment, which can increase operational complexity and costs.

2. Recirculating Aquaculture Systems (RAS)

RAS greatly reduces water usage, recycling up to 99% of the water, which is a significant improvement over conventional aquaculture systems (Lal *et al.*, 2024). This makes RAS particularly suited for regions facing water scarcity, where resource conservation is essential. By cutting down the need for continuous water input, RAS also alleviates the pressure on freshwater resources, promoting more sustainable farming practices.

Waste Management

The integration of microalgae into RAS systems helps capture excess nutrients and converts waste into useful products such as biofuels, animal feed, and even pharmaceuticals (Ende *et al.*, 2024). This approach to waste management not only reduces the environmental impact by lowering nutrient runoff, which can lead to issues like eutrophication, but also creates an additional revenue stream for farmers by turning waste into marketable goods.

Technological Advancements

Dynamic modeling in RAS provides precise control over key water quality factors such as temperature, pH, and oxygen levels, ensuring optimal conditions for fish growth and overall system performance (Montjouridès *et al.*, 2024). When combined with portable spectrophotometers, these technologies enable quick, on-site water quality assessments, allowing for timely interventions when needed (Amijar *et al.*, 2024). This real-time monitoring helps reduce risks, cut operational expenses, and increase productivity by maintaining the best possible environment for aquatic species.

3. Integrated Multi-trophic Aquaculture (IMTA)

IMTA systems drastically reduce waste by transforming the byproducts of fed species, such as fish and shrimp, into essential nutrients for extractive species like shellfish and seaweeds, which in turn improves water quality and minimizes pollution (Checa *et al.*, 2024). This recycling of nutrients strengthens the ecosystem's natural filtration abilities, reducing the accumulation of harmful substances. Additionally, IMTA contributes to the restoration of coastal ecosystems by enhancing biodiversity and lowering the environmental impact of aquaculture practices.

Research indicates that IMTA can significantly improve resource circularity in aquaculture, achieving up to a 90% increase in water reuse and 80-90% efficiency in bioremediation (Checa *et al.*, 2024). By closing nutrient cycles, IMTA promotes sustainability by reducing nutrient discharge into nearby waters, helping to prevent potential ecological harm.

Economic Viability

IMTA's approach of integrating different species offers farmers the advantage of multiple income streams by combining high-value fed species with extractive species like seaweed and shellfish, which can be marketed in various sectors (Naresh *et al.*, 2024). This diversification not only enhances profitability but also makes aquaculture more adaptable to market changes and price fluctuations.

Furthermore, the inclusion of multiple species reduces the risks commonly faced in monoculture systems, such as disease outbreaks, which can devastate single-species farms (Lal *et al.*, 2024). By distributing risk across different species and increasing system resilience, IMTA protects aquaculture investments and enhances long-term financial sustainability.

Challenges

Despite its environmental and economic benefits, IMTA faces several barriers to widespread adoption, particularly in the standardization of practices across different ecosystems and species (Checa *et al.*, 2024). Managing multiple species in one system requires advanced knowledge and monitoring, which adds complexity. Additionally, measuring circularity metrics—such as nutrient absorption and overall system performance—remains difficult, making it challenging for farmers and regulators to evaluate the system's efficiency.

4. Aquaponics

Aquaponics utilizes fish waste as a natural source of nutrients for plants, minimizing the reliance on artificial fertilizers and offering a more environmentally friendly cultivation method (Dongre, 2023). This method not only reduces the ecological impact of food production but also fosters a circular nutrient cycle, which enhances overall sustainability. The combined cultivation of fish and plants creates a mutually beneficial relationship that optimizes resource use and reduces waste.

The aquaponics system promotes biodiversity by establishing a complex ecosystem where fish, plants, and beneficial microorganisms interact and thrive together (Lal *et al.*, 2024). This variety helps maintain ecological balance and resilience, positioning aquaponics as a sustainable agricultural practice that supports environmental conservation.

Technological Integration

Recent advancements in aquaponics include the use of IoT-based monitoring systems that offer real-time data on water quality and nutrient levels, enabling precise adjustments to create ideal conditions for both fish and plants (Yau *et al.*, 2024). These technologies improve productivity by ensuring that the environment remains optimal for growth, thus enhancing the system's efficiency.

Additionally, machine learning technologies are being incorporated into aquaponics to facilitate data-driven decisions that boost system performance (Venkatraman and Surendran, 2023). By analyzing trends and patterns in data, machine learning can forecast potential problems, optimize the use of resources, and improve the adaptability of the system, leading to more effective and sustainable aquaponics practices.

System Components

Aquaponics systems consist of several essential components, including fish tanks, grow beds, and biofilters, which work together to sustain a balanced ecosystem (Ngo-Hoang, 2023). Fish tanks provide an environment for aquatic species, grow beds enable plant growth, and biofilters process fish waste into nutrients for the plants. This integrated setup creates a self-sufficient system that effectively recycles nutrients and conserves resources.

5. Smart Aquaculture: Data-Driven Technologies

Smart aquaculture utilizes advanced, data-driven technologies to enhance both productivity and sustainability in aquatic environments. By integrating IoT devices and advanced monitoring systems, these operations can precisely adjust environmental factors for aquatic organisms, thereby promoting better growth and overall well-being.

Real-Time Data Collection

IoT systems utilize a variety of sensors to continuously monitor critical factors such as pH, temperature, and dissolved oxygen, providing a steady flow of data necessary for maintaining ideal conditions (Arthi *et al.*, 2024; Megantoro *et al.*, 2024). This ongoing surveillance allows aquaculture operators to quickly identify and correct any deviations from optimal conditions, which supports the health and growth of aquatic organisms.

Automated Alerts

When these sensors identify unusual conditions, automated notifications are sent to operators without delay, enabling swift corrective measures (Arthi *et al.*, 2024). This proactive system not only enhances the well-being of the fish but also improves operational efficiency by addressing potential problems before they develop into significant issues, thus mitigating the risk of system failures and financial losses.

Enhanced Sustainability Practices

Precision Aquaculture

Precision aquaculture utilizes sophisticated data analytics and artificial intelligence (AI) to optimize feeding strategies and manage water quality (Lal *et al.*, 2024). By adjusting these parameters based on real-time information, this method reduces waste and lessens the environmental footprint of aquaculture, leading to a more sustainable and efficient production system.

Smart Biofloc Systems

Smart biofloc systems employ automated technology to continuously regulate water quality, creating ideal conditions for fish growth while reducing resource consumption (Podder *et al.*, 2023). This approach not only enhances fish production but also decreases the amount of water and feed required, supporting a more sustainable and resource-efficient aquaculture practice.

6. Solar-powered aquaculture systems

Solar-powered aquaculture systems represent a major advancement in sustainable fish farming, effectively addressing both energy efficiency and environmental impact. These systems harness renewable energy to boost productivity while reducing ecological footprints.

Efficiency and Performance

Studies show that solar-powered recirculating aquaculture systems (RAS) can achieve growth rates and survival rates similar to those of traditional petrol-powered systems. For example, solar-powered systems have demonstrated an 84% survival rate for African catfish fingerlings, compared to 64% in petrol-powered systems (Anyanwu-Akeredolu *et al.*, 2024).

Additionally, automated solar-powered water pumping systems use microcontrollers to regulate water levels efficiently, which reduces labor costs and simplifies operations (Sarwar and Iqbal, 2022).

Water Quality Management

Solar energy can also be used to operate water quality monitoring systems, providing ongoing supervision of essential parameters such as pH and dissolved oxygen, crucial for maintaining fish health (KP andHarigovindan, 2020). This use of solar technology further enhances the sustainability of aquaculture practices.

7. Innovations in Water Purification and Treatment

- **Advanced filtration systems** are designed to remove contaminants from water at a molecular level, providing a higher level of purification compared to traditional filtration methods. These systems often incorporate cutting-edge technologies to achieve superior results.

Nanotechnology in Water Filtration

- **Nanotechnology** involves the manipulation of matter at the nanoscale (one billionth of a meter). In water filtration, nanotechnology is used to create highly efficient filters with unique properties: (Gehrke *et al.*, 2015)
- **Nanomaterials:** Materials with unique properties at the nanoscale, such as carbon nanotubes, graphene, and nanofiltration membranes, can be used to remove contaminants more effectively.
- **Nanofiltration:** This process uses membranes with pores on the nanoscale to selectively remove contaminants based on their size and charge.
- **Nanoparticles:** Nanoparticles can be used to adsorb or degrade contaminants, such as heavy metals and organic pollutants.

Use of Bio-Inspired Filters

- **Bio-inspired filters** mimic the filtration mechanisms found in nature. These filters often incorporate natural materials or processes to purify water: (Bianciardi & Cascini, 2022)

- **Biomimicry:** Designing filters that mimic the structures and functions of biological organisms, such as the gills of fish or the filtration systems of plants.
- **Biofiltration:** Using biological processes, such as microbial activity, to degrade contaminants in water.
- **Natural Materials:** Incorporating natural materials like diatomaceous earth or activated carbon, which have strong adsorbent properties.

Desalination Technologies

- **Desalination** is the process of removing salt and other minerals from seawater or brackish water to produce freshwater. It has become an increasingly important solution for addressing water scarcity, particularly in arid regions and coastal areas(Darre & Toor, 2018).

Reverse Osmosis Advancements

- **Reverse osmosis** is one of the most widely used desalination technologies. It involves forcing seawater through a semipermeable membrane under high pressure, allowing freshwater to pass through while retaining salts on the other side. Recent advancements in reverse osmosis technology have improved its efficiency and reduced its energy consumption: (Habieeb *et al.*, 2023)
- **Energy-efficient membranes:** New membrane materials and designs have increased the water flux and reduced energy requirements.
- **Pressure optimization:** Advances in hydraulic design have improved the efficiency of pressure application.
- **Wastewater recovery:** Technologies have been developed to recover and reuse the concentrated brine produced during desalination, reducing environmental impact.

Solar-Powered Desalination Methods

- Solar energy offers a sustainable and renewable alternative for powering desalination plants. Several solar-powered desalination methods have been developed: (Ahmed et al, 2019)
- **Solar thermal desalination:** Using solar energy to heat seawater, causing it to evaporate. The vapor is then condensed to produce freshwater.
- **Solar-powered reverse osmosis:** Using solar energy to generate electricity for powering reverse osmosis pumps.

- **Solar-powered capacitive deionization:** A relatively new technology that uses solar energy to charge electrodes, which then remove salts from water.

Chemical-Free Water Treatment

- Chemical-free water treatment methods offer a sustainable and environmentally friendly approach to purifying water without relying on harsh chemicals. These methods often utilize physical or biological processes to eliminate contaminants (Westerhoff *et al.*, 2021).

UV-C Light and Its Applications

- **UV-C light** is a specific wavelength of ultraviolet light that can effectively inactivate microorganisms. It is commonly used in water treatment due to its ability to kill bacteria, viruses, and parasites without leaving behind any harmful residues (Urban *et al.*, 2016).

Applications of UV-C light : (Monteiro et al., 2021)

- **Drinking water disinfection:** UV-C light is used to disinfect municipal drinking water supplies, ensuring that it is safe for consumption.
- **Wastewater treatment:** UV-C light can be used to disinfect wastewater before it is discharged into the environment, reducing the risk of waterborne diseases.
- **Aquarium water treatment:** UV-C sterilizers are used in aquariums to control algae and pathogens.
- **Food processing:** UV-C light can be used to sanitize food surfaces and equipment.

Ozone and Other Chemical-Free Methods

- **Ozone** is a highly reactive gas that can oxidize contaminants in water. It is a powerful disinfectant that can effectively kill microorganisms (Staehelin *et al.*, 2001).

Other chemical-free methods : (Reeves, 2000)

- **Filtration:** Mechanical filtration processes, such as sand filtration and membrane filtration, can remove particulate matter and larger contaminants from water.
- **Boiling:** Boiling water can kill microorganisms, but it is not practical for large-scale water treatment.
- **Solar disinfection:** Exposing water to sunlight can help kill microorganisms, especially in developing countries with limited access to electricity.

8. Green Technology for Pollution Control

Eco-Friendly Wastewater Management

- Eco-friendly wastewater management practices aim to treat and reuse wastewater in a sustainable and environmentally responsible manner. These methods often utilize natural processes and biological systems to minimize the negative impact on aquatic ecosystems(Bharagava & Chowdhary, 2019).

Integrated Constructed Wetlands

- **Integrated constructed wetlands** are engineered ecosystems that mimic natural wetlands and utilize plants, microorganisms, and sediments to treat wastewater. They offer a sustainable and cost-effective solution for wastewater treatment, especially in rural and remote areas (Harrington & McInnes, 2009).

Key Components of Integrated Constructed Wetlands: (Scholz et al., 2007)

- **Submerged plants:** Plants that grow underwater, such as cattails and water lilies, help remove nutrients and organic matter.
- **Emergent plants:** Plants that grow in shallow water, such as reeds and rushes, provide additional filtration and habitat for aquatic organisms.
- **Sediment bed:** A layer of gravel, sand, or soil that provides a habitat for microorganisms that break down contaminants.
- **Zones:** The wetland is often divided into different zones, each with specific plant species and conditions to optimize treatment.

Biological Treatment Systems

- **Biological treatment systems** rely on microorganisms to break down organic matter and contaminants in wastewater. These systems are commonly used in wastewater treatment plants and can be designed to handle varying flow rates and wastewater compositions(Lawrence & McCarty, 1970).

Types of biological treatment systems: (Samer, 2015)

- **Activated sludge:** A process that involves aeration of wastewater to promote the growth of microorganisms that break down organic matter.
- **Trickling filters:** A system that uses a bed of media, such as rocks or plastic, to support the growth of microorganisms.
- **Sequencing batch reactors:** A system that operates in batches, allowing for better control of the treatment process.

Remediation Technologies

- **Remediation technologies** are used to clean up contaminated sites, including those affected by wastewater discharge. They aim to remove or reduce contaminants to safe levels (Khan *et al.*, 2004). **Phytoremediation** is the use of plants to remove or degrade contaminants from soil or water. Plants can absorb contaminants through their roots and accumulate them in their tissues(Ansari, 2016).

Types of Phytoremediation: (Pandey et al., 2021)

- **Phytoextraction:** Plants extract contaminants from the soil and concentrate them in their tissues.
- **Phytodegradation:** Plants break down contaminants into less harmful substances.
- **Rhizofiltration:** Plants remove contaminants from water as it passes through their root system.

Bioremediation utilizes microorganisms to break down contaminants into less harmful substances. These microorganisms can be naturally present in the environment or can be introduced to a contaminated site (Kensa, 2011).

Types of Bioremediation : (Wang et al., 2021)

- **In situ bioremediation:** Microorganisms are applied directly to the contaminated site.
- **Ex situ bioremediation:** Contaminated material is removed from the site and treated in a controlled environment.

9. Green Technologies for Coastal and Marine Conservation

Natural Infrastructure Solutions

Natural infrastructure solutions, or nature-based solutions, leverage natural processes and ecosystems to address environmental challenges. These solutions often provide more sustainable and resilient alternatives to traditional engineering approaches (Coates & Smith, 2012).

- **Green sea walls:** Instead of constructing concrete sea walls, green sea walls utilize living materials like vegetation, oysters, or coral reefs to protect coastlines from erosion and storm surge. These natural barriers offer enhanced biodiversity, wave attenuation, and nutrient cycling(Reddy *et al.*, 2024).
- **Mangrove restoration:** Mangroves are coastal ecosystems that provide invaluable ecosystem services, including shoreline protection, carbon sequestration, and habitat for diverse marine life. Restoring degraded

mangrove forests can help mitigate coastal erosion, improve water quality, and enhance coastal resilience (Lovelock, 2022).

10. Marine Pollution Management

Biodegradable Materials for Pollution Control

- Traditional pollution control methods often rely on synthetic materials that can persist in the environment for extended periods. Biodegradable materials offer a more sustainable alternative by breaking down naturally over time, reducing their environmental impact (Tian & Bilal, 2020).
- **Biodegradable oil sorbents:** These materials can absorb oil spills efficiently while decomposing harmlessly in marine environments. They are often made from natural fibers like coconut husks, seaweed, or agricultural waste (Idris *et al.*, 2014).
- **Biodegradable plastic alternatives:** Reducing the use of single-use plastics and replacing them with biodegradable alternatives can help prevent marine pollution and protect marine ecosystems (Mangal *et al.*, 2023).

Innovations in Oil Spill Response

Advancements in oil spill response technologies are crucial for minimizing the environmental damage caused by these incidents (Ivshina *et al.*, 2015).

- **Oil-eating bacteria:** Engineered bacteria can be deployed to degrade oil spills more efficiently and effectively. These microorganisms can accelerate the natural biodegradation process and reduce the long-term impact of oil contamination.
- **Remote sensing technologies:** Satellites and drones can be used to monitor oil spills in real-time, allowing for rapid response and containment efforts. These technologies can also help identify vulnerable areas and improve prevention strategies.

Marine Spatial Planning Tools

Geographic Information Systems (GIS) for Marine Management

- GIS is a powerful tool for analyzing and visualizing spatial data. It can be used to support marine management by: (Stanbury & Starr, 1999)
- **Mapping and monitoring marine resources:** GIS can help identify and track critical habitats, marine protected areas, and other valuable marine features.
- **Assessing environmental impacts:** GIS can be used to evaluate the potential impacts of human activities on marine ecosystems, such as coastal development, fishing, and shipping.

- **Developing sustainable management plans:** GIS can assist in the creation of comprehensive marine management plans that balance conservation and economic development.

Conclusion

Aquatic sustainability is crucial for preserving the health and well-being of our planet's aquatic ecosystems. By maintaining biodiversity, water quality, ecosystem health, and responsible resource management, we can ensure that these ecosystems continue to provide essential services for future generations. Human activities have significantly impacted aquatic ecosystems, leading to pollution, habitat destruction, and biodiversity loss. To address these challenges, innovative green technologies offer sustainable solutions. By harnessing the power of nature and cutting-edge technology, we can restore and protect our aquatic environments.Green technologies play a vital role in pollution reduction, sustainable fisheries management, climate change mitigation, habitat restoration, water conservation, and monitoring and data collection. Examples of these technologies include advanced water filtration systems, nanotechnology, bio-inspired filters, desalination technologies, UV-C light, ozone, eco-friendly wastewater management, integrated constructed wetlands, biological treatment systems, remediation technologies, natural infrastructure solutions, and biodegradable materials.By implementing and further developing these green technologies, we can create a more sustainable and resilient future for our aquatic ecosystems and ensure that they continue to provide the essential services upon which we rely.

References

Ahmed, F. E., Hashaikeh, R., & Hilal, N. (2019). Solar powered desalination–Technology, energy and future outlook. Desalination, 453, 54-76.

Ahmed, J., Thakur, A., & Goyal, A. (2021). Industrial wastewater and its toxic effects.

Amijar, M. I. S. M., Ramli, N. M., Nurulhuda, K., & Abd Aziz, S. (2024, June). Portable Spectrophotometer for Water Quality Monitoring in Recirculating Aquaculture Systems. In IOP Conference Series: Earth and Environmental Science (Vol. 1359, No. 1, p. 012028). IOP Publishing

Ansari, A. A. (2016). Phytoremediation (pp. 12-18). S. S. Gill, R. Gill, G. Lanza, & L. Newman (Eds.). Berlin, Germany:: Springer International Publishing.

Arthi, P., Mahesh, N., Saipurusoth, H., &Janardhanam, B. (2024). IoT-based smart aquaculture monitoring system for fish tank management. Journal of Ubiquitous Computing and Communication Technologies, 3(1). https://doi.org/10.36548/jucct.2024.3.001

Anyanwu-Akeredolu, B., Adeleke, M. L., Dada, A., &Oyedapo, F. (2024). Evaluating the Efficiency of Solar and Petrol-Powered Recirculating Aquaculture Systems for African Catfish (Clarias gariepinus) Fingerling Production. Agriculture, Forestry and Fisheries, 12(4), 70-80.

Bianciardi, A., & Cascini, G. (2022). A bio-inspired approach for boosting innovation in the separation technology sector. Proceedings of the Institution of Mechanical Engineers, Part C: Journal of Mechanical Engineering Science, 236(9), 4533-4550.

Bradu, P., Biswas, A., Nair, C., Sreevalsakumar, S., Patil, M., Kannampuzha, S., ... & Gopalakrishnan, A. V. (2023). Recent advances in green technology and Industrial Revolution 4.0 for a sustainable future. Environmental Science and Pollution Research, 30(60), 124488-124519.

Checa, D., Macey, B. M., Bolton, J. J., Brink-Hull, M., O'Donohoe, P., Cardozo, A., ... & Sánchez, I. (2024). Circularity Assessment in Aquaculture: The Case of Integrated Multi-Trophic Aquaculture (IMTA) Systems. Fishes, 9(5), 165.

Coates, D., & Smith, M. (2012). Natural infrastructure solutions for water security. Water and the green economy, 1, 167-181.

Crute, I. R., & Muir, J. F. (2011). Improving the productivity and sustainability of terrestrial and aquatic food production systems: future perspectives. The Journal of Agricultural Science, 149(S1), 1-7.

Darre, N. C., & Toor, G. S. (2018). Desalination of water: a review. Current Pollution Reports, 4, 104-111.

Dongre, S. (2023). Smart aquaponics system. International Journal for Research in Applied Science and Engineering Technology. https://doi.org/10.22214/ijraset.2023.57251

Ende, S., Henjes, J., Spiller, M., Elshobary, M., Hanelt, D., &Abomohra, A. (2024). Recent advances in recirculating aquaculture systems and role of microalgae to close system loop. Bioresource Technology, 131107.

Gehrke, I., Geiser, A., & Somborn-Schulz, A. (2015). Innovations in nanotechnology for water treatment. Nanotechnology, science and applications, 1-17.

Habieeb, A. R., Kabeel, A. E., Sultan, G. I., & Abdelsalam, M. M. (2023). Advancements in Water Desalination Through Artificial Intelligence: A Comprehensive Review of AI-Based Methods for Reverse Osmosis Membrane Processes. Water Conservation Science and Engineering, 8(1), 53.

Harrington, R., & McInnes, R. (2009). Integrated Constructed Wetlands (ICW) for livestock wastewater management. Bioresource Technology, 100(22), 5498-5505.

Idris, J., Eyu, G. D., Mansor, A. M., Ahmad, Z., & Chukwuekezie, C. S. (2014). A Preliminary Study of Biodegradable Waste as Sorbent Material for Oil-Spill Cleanup. The Scientific World Journal, 2014(1), 638687.

Ivshina, I. B., Kuyukina, M. S., Krivoruchko, A. V., Elkin, A. A., Makarov, S. O., Cunningham, C. J., ... & Philp, J. C. (2015). Oil spill problems and sustainable response strategies through new technologies. Environmental Science: Processes & Impacts, 17(7), 1201-1219.

Johnson, A., & White, N. D. (2014). Ocean acidification: The other climate change issue. American Scientist, 102(1), 60-63.

Joshna, M., Ahilan, B., Antony, C., &Ravaneswaran, K. Study on Biofloc Characteristics, Digestive Enzyme Activity and Physiological Responses in Polyculture Model Penaeus vannamei and GIF Tilapia (Oreochromis niloticus) Culture System-BFT Aquaculture System. Indian Journal of Animal Research, 1, 9.

Kensa, V. M. (2011). Bioremediation-an overview. Journal of Industrial Pollution Control, 27(2), 161-168

Khan, F. I., Husain, T., & Hejazi, R. (2004). An overview and analysis of site remediation technologies. Journal of environmental management, 71(2), 95-122.

KP, R. A. H., &Harigovindan, V. P. (2020, October). Water quality monitoring and reporting system for aquaculture with solar energy harvesting. In 2020 International conference on smart technologies in computing, electrical and electronics (ICSTCEE) (pp. 66-70). IEEE.

Lawrence, A. W., & McCarty, P. L. (1970). Unified basis for biological treatment design and operation. Journal of the Sanitary Engineering Division, 96(3), 757-778.

Lal, J., Vaishnav, A., Brar, K. S., Kumar, D., Jayaswal, R., Debbarma, S., & Kumar, S. (2024). Biofloc Technology: Optimizing Aquaculture through Microbial Innovation. Journal of Advances in Microbiology, 24(7), 11-24.

Lal, J., Vaishnav, A., Deb, S., Kashyap, S., Debbarma, P., Gautam, P., & Verma, D. K. (2024). Re-Circulatory Aquaculture Systems: A Pathway to Sustainable Fish Farming. Archives of Current Research International, 24(5), 799-810.

Lal, J., Vaishnav, A., Kumar, D., Jana, A., Jayaswal, R., Chakraborty, A., & Pavankalyan, M. (2024). Emerging Innovations in Aquaculture: Navigating towards Sustainable Solutions. International Journal of Environment and Climate Change, 14(7), 83-96.

Lovelock, C. E., Barbier, E., & Duarte, C. M. (2022). Tackling the mangrove restoration challenge. PLoS biology, 20(10), e3001836.

Mangal, M., Rao, C. V., & Banerjee, T. (2023). Bioplastic: an eco-friendly alternative to non-biodegradable plastic. Polymer International, 72(11), 984-996.

Mansour, A. T., Ashour, M., Alprol, A. E., & Alsaqufi, A. S. (2022). Aquatic plants and aquatic animals in the context of sustainability: Cultivation techniques, integration, and blue revolution. Sustainability, 14(6), 3257.

McCartney, M. P., Sullivan, C., Acreman, M. C., & McAllister, D. E. (2001). Ecosystem impacts of large dams. Background paper, 2.

Megantoro, P., Anugrah, A. W., Abdillah, M. H., Kustanto, B. J., Fadhilah, M., & Vigneshwaran, P. (2024). Smart measurement and monitoring system for aquaculture fisheries with IoT-based telemetry system. Bulletin of Electrical Engineering and Informatics, 13(3), 1555-1565.

Melaku, S., Getahun, A., Mengestou, S., Geremew, A., & Belay, A. (2023). Bioflocs Technology in Freshwater Aquaculture: Variations in Carbon Sources and Carbon-to-Nitrogen Ratios.

Montjourides, M. A., Roblitz, S., & Dahle, H. (2024). Optimizing alkalinity control in Recirculating Aquaculture Systems (RAS): a dynamic modelling approach. bioRxiv, 2024-06.

Monteiro, M. L. G., Rosario, D. K., de Carvalho, A. P. A., & Conte-Junior, C. A. (2021). Application of UV-C light to improve safety and overall quality of fish: A systematic review and meta-analysis. Trends in Food Science & Technology, 116, 279-289.

Naresh, R., Keer, S., Showkat, A., Dar, A., Abhiman, G., Chandra, G., Md. Idrish, R. K., Khan, C., Garg, R. K., & Manmohan, K. (2024). Integrated multi-trophic aquaculture (IMTA) system: Way forward towards sustainable aquaculture. DOI: 10.58532/nbennurch190

Ngo-Hoang, D. (2023). Aquaponics: System Layout and Components.

Pandey, N., Chandra, J., Xalxo, R., & Sahu, K. (2021). Concept and types of phytoremediation. Approaches to the remediation of inorganic pollutants, 281-302.

Podder, S., Anoy, M. F. S., Rafid, S. T. S., & Ajwad, A. J. (2023, December). Smart Biofloc System: Leveraging IoT for Enhanced Aquaculture Sustainability. In 2023 5th International Conference on Electrical, Control and Instrumentation Engineering (ICECIE) (pp. 1-7). IEEE.

Ramachandra, T. V., Ahalya, N., & Murthy, R. (2005). Aquatic ecosystems: conservation, restoration and management. Aquatic ecosystems-Conservation, restoration and management, 26-50.

Raza, B., Zheng, Z., & Yang, W. (2024). A Review on Biofloc System Technology, History, Types, and Future Economical Perceptions in Aquaculture. Animals, 14(10), 1489.

Reddy, M., Reddy, A., & Singh, S. (2024). Transitioning from "Grey to Green": Nature-Based Solution Seawall for Coastal Protection. European Journal of Engineering and Technology Research, 9(4), 18-24.

Reeves, G. G. (2000). Wastewater minimization with chemical-free water treating. Environmental progress, 19(4), 292-298.

Sallam, G. R., Shehata, A. I., Basuini, M. F. E., Habib, Y. J., Henish, S., Rahman, A. N. A., ... & Aly, H. A. (2024). Integrated biofloc technology in red tilapia aquaculture: Salinity-dependent effects on water quality, parental stock physiology, reproduction, and immune responses. Aquaculture International, 1-31.

Samer, M. (2015). Biological and chemical wastewater treatment processes. Wastewater treatment engineering, 150(10.5772), 61250

Sarwar, A., & Iqbal, M. T. (2022). Solar powered water pumping system automation and control using a microcontroller for aquaculture. European Journal of Electrical Engineering and Computer Science, 6(4), 17-22.

Scholz, M., Harrington, R., Carroll, P., & Mustafa, A. (2007). The integrated constructed wetlands (ICW) concept. Wetlands, 27(2), 337-354.

Shao, K. T. (2009). Marine biodiversity and fishery sustainability. Asia Pacific Journal of clinical nutrition, 18(4), 527-531.

Simberloff, D. (2010). Invasive species. Conservation biology for all, 1, 131-152.

Staehelin, J., Harris, N. R., Appenzeller, C., & Eberhard, J. (2001). Ozone trends: A review. Reviews of Geophysics, 39(2), 231-290.

Stanbury, K. B., & Starr, R. M. (1999). Applications of Geographic Information Systems (GIS) to habitat assessment and marine resource management. Oceanologica acta, 22(6), 699-703.

Tacon, A. G., & Metian, M. (2013). Fish matters: importance of aquatic foods in human nutrition and global food supply. Reviews in fisheries Science, 21(1), 22-38.

Thushari, G. G. N., & Senevirathna, J. D. M. (2020). Plastic pollution in the marine environment. Heliyon, 6(8).

Tian, K., & Bilal, M. (2020). Research progress of biodegradable materials in reducing environmental pollution. Abatement of environmental pollutants, 313-330.

Urban, L., Charles, F., de Miranda, M. R. A., & Aarrouf, J. (2016). Understanding the physiological effects of UV-C light and exploiting its agronomic potential before and after harvest. Plant Physiology and Biochemistry, 105, 1-11.

Venkatraman, M., & Surendran, R. (2023, November). Industrial 5.0 Aquaponics System Using Machine Learning Techniques. In 2023 7th International Conference on Electronics, Communication and Aerospace Technology (ICECA) (pp. 835-842). IEEE.

Wang, M., Chen, S., Jia, X., & Chen, L. (2021). Concept and types of bioremediation. In Handbook of bioremediation (pp. 3-8). Academic Press.

Westerhoff, P., Alvarez, P. J., Kim, J., Li, Q., Alabastri, A., Halas, N. J., ... & Wong, M. S. (2021). Utilizing the broad electromagnetic spectrum and unique nanoscale properties for chemical-free water treatment. Current opinion in chemical engineering, 33, 100709.

Yau, S. R., Jalani, J., Sadun, A. S., Rejab, S. M., & John, J. (2025). Development of an Aquaponics Farming Technology System Using Arduino Based on Internet of Things. Journal of Advanced Research in Applied Sciences and Engineering Technology, 45(2), 11-24.

2

Technological Innovations for Mitigating Water Pollution

Nowsheen Mushtaq and Adnan Abubakr*

Division of Aquatic Environmental Management, Faculty of Fisheries, Rangil SKUAST-Kashmir

Abstract

This chapter investigates the essential role of technological advancements in tackling the widespread challenge of water pollution. It includes the pressing necessity for innovative solutions in light of the increasing contamination of water sources globally, which threatens public health, ecosystems, and economic viability. The chapter outlines traditional strategies for mitigating water pollution, such as physical filtration, chemical treatments, and biological methods, while evaluating their effectiveness and shortcomings. Subsequently, it also provides thorough analysis of recent technological breakthroughs that present promising alternatives and improvements to conventional methods. These advancements encompass advanced oxidation processes, membrane technologies, nanotechnology, and smart sensing systems, etc, each offering distinct advantages and the potential to enhance water quality. By juxtaposing these emerging technologies with established practices, the chapter seeks to provide a holistic view of how state-of-the-art solutions can augment and, in certain instances, exceed traditional techniques.

Introduction

Water pollution is defined as the contamination of water sources, including rivers, lakes, and oceans, by harmful substances or pollutants. These pollutants originate from various sources, such as industrial discharges, agricultural runoff, and improper waste management. The impacts of water pollution are extensive, influencing both aquatic ecosystems and human health. It represents a significant global environmental issue that threatens both ecosystems and public health (WHO, 2020). As industrial and urban development continues to grow, the levels of pollutants entering water bodies have become increasingly concerning (United Nations, 2021). The introduction of these contaminants has resulted in numerous adverse effects, including the degradation of aquatic habitats, the endangerment of marine species, and the pollution of drinking water supplies (EPA, 2020). The ongoing progress in civilization, industrialization, and urbanization underscores the necessity for promoting sustainability to ensure that humans can lead longer, healthier, and more fulfilling lives. Various human activities contribute to the release of hazardous substances, known as pollutants, into our environment, jeopardizing the potential for long, healthy, and happy lives. Achieving a better and more sustainable existence is closely tied to adhering to environmental management principles, where pollution control and advanced treatment technologies are essential for tackling sustainability challenges. Recent technological advancements have equipped us with effective tools and strategies to combat water pollution, providing optimism for a cleaner and more sustainable future (Scientific American, 2020).

What is the Need For Mitigating Water Pollution?

Aquatic ecosystems, especially freshwater environments, face greater levels of contamination compared to other ecosystems due to their involvement in various industrial activities and the discharge of pollutants from both industrial and urban developments (Demirak *et al.,* 2006; Fernandesa *et al.,* 2007). Globally, approximately 80% of municipal wastewater is released into water bodies without treatment, while industries contribute millions of tons of heavy metals, solvents, toxic sludge, and other forms of waste annually (WWAP 2017; Mateo-Sagasta *et al.*, 2017). Agriculture accounts for 70% of global water usage and significantly contributes to water pollution Although most aquatic ecosystems can naturally dilute some pollutants, severe contamination can lead to significant changes in the community's flora and fauna (Mateo-Sagasta *et al.,* 2017). Pollutants can inflict immediate damage with clear environmental impacts or cause secondary damage through subtle disruptions in the biological food web, which may only become apparent over extended periods (Sharma 2012; Al Naggar *et al.,* 2014; Ghani 2015). Therefore, preserving the quality of aquatic ecosystems is one of the most significant challenges confronting global society in the twenty-first century (Hampel *et al.*, 2015).

What are the Challenges in Mitigating Water Pollution?

The challenges associated with addressing water pollution can be outlined as follows:

Varied Pollution Sources: Water pollution originates from multiple sources, including industrial effluents, agricultural runoff, improper waste management, and urban expansion, complicating the identification and effective management of specific pollutants (Pedersen *et al.,* 2020).

Insufficient Monitoring and Data: The lack of real-time monitoring systems and inadequate water quality data impede the precise evaluation of pollution levels and the execution of targeted remediation efforts (Scavia *et al.,* 2021).

Developing Affordable Solutions: The implementation of effective pollution control measures can be costly, particularly for developing nations or areas with limited financial resources. It is vital to create affordable technologies and strategies to promote widespread adoption and ensure long-term sustainability (Ambe *et al.,* 2020).

Regulatory and Policy Challenges: Weak enforcement of current regulations, disjointed governance frameworks, and insufficient coordination among stakeholders can obstruct effective water pollution management. Enhancing policies and regulatory structures is crucial for ensuring compliance and fostering collaborative action (Verlicchi *et al.,* 2020).

Climate Change and New Contaminants: The effects of climate change, including changes in precipitation patterns and increased temperatures, can intensify water pollution issues. Furthermore, the emergence of new contaminants, such as pharmaceuticals and microplastics, complicates efforts to mitigate pollution (Stalter *et al.,* 2020).

Conventional Methods of Mitigating Water Pollution

Conventional approaches to addressing water pollution typically encompass established techniques that have been in practice for many years. These strategies are commonly employed to treat water, ensuring it is safe for human consumption. Key methods include:

Conventional sewage treatment: This process integrates coagulation, sedimentation, filtration, and disinfection to deliver clean drinking water to communities. Coagulation entails the addition of iron or aluminum salts to the water, which neutralizes the negative charges of dissolved and suspended particles. This reaction causes the particles to clump together, forming larger aggregates known as floc, which are dense and settle quickly to the bottom of the water source. This settling phase is referred to as sedimentation. The resulting coagulated water can be filtered through various mediums, such as sand and gravel, or through microfiltration or ultrafiltration membranes, or it may be directed to a settling tank. In the settling tank, heavier particles settle out and are removed, allowing the water to proceed to the filtration stage. Filtration eliminates any remaining particles and impurities, followed by disinfection to eradicate any lingering bacteria, viruses, or pathogens.

Chemical coagulation and flocculation: This technique involves the introduction of chemicals to the water to facilitate the coagulation and flocculation of suspended particles, which are subsequently removed through sedimentation or filtration.

Disinfection process: This method focuses on eliminating or inactivating harmful microorganisms present in the water. The most prevalent disinfection technique is chlorination, which involves the addition of chlorine to the water to effectively kill bacteria and viruses.

Sedimentation tank: This technique entails allowing water to remain still for a designated duration, enabling suspended particles to settle at the bottom. Once the particles have settled, the water is drained from the top to remove the accumulated solids.

Sand filtration: This process consists of directing water through a sand bed, effectively eliminating any residual particles and contaminants present in the water.

Lime softening: This approach involves the addition of lime to the water, which serves to soften it and eliminate dissolved minerals. The reaction between lime and the minerals results in the formation of solid particles that can subsequently be removed through sedimentation or filtration.

Limitations of Conventional Methods

Conventional methods of reducing water pollution face several challenges, including:

Incomplete Elimination: Conventional techniques may struggle to eliminate all pollutants, particularly newer contaminants such as pharmaceuticals and personal care items.

High Infrastructure Expenses: The construction and upkeep of sewage treatment facilities and related infrastructure can incur substantial costs.

Energy Demands: Many established methods necessitate considerable energy for their operation, leading to higher operational expenses and a greater environmental footprint.

Chemical Dependency: Processes like coagulation and flocculation often involve chemicals, which can pose environmental risks.

Scalability Issues: These methods may not be easily adaptable to accommodate rising pollution levels and growing populations.

Maintenance Difficulties: Components such as sedimentation tanks require consistent maintenance, and their efficiency may decline without proper oversight.

Emerging Technologies for Mitigating Water Pollution

The advent of emerging technologies has significantly transformed efforts to combat water pollution. The increasing scarcity of clean, fresh water is a pressing global issue, prompting businesses to leverage technological advancements to tackle this problem. These innovative technologies present effective alternatives for various wastewater treatment methods and distribution systems, enabling the reduction of contaminants to acceptable levels. While technology has been utilized in water treatment for many years, recent advancements have notably reduced costs, making treatment more accessible (UNEP, 2020). The following outlines some of the emerging technologies employed in water pollution mitigation:

Nanotechnology

Nanotechnology has surfaced as a valuable approach for tackling water pollution by utilizing nanoscale materials and techniques to effectively eliminate contaminants from water sources (Goh *et al.*, 2020). The nanoparticles have the ability to attract and bond with pollutants, simplifying their removal process. Various methods within nanotechnology are employed to extract contaminants from water. These include:

- **Contaminant Adsorption:** Nanomaterials, including activated carbon nanoparticles and graphene derivatives, possess extensive surface areas and exceptional adsorption capabilities, allowing them to efficiently capture and eliminate pollutants from water (Wang *et al.,* 2020).
- **Catalytic Breakdown:** Nanoscale catalysts, such as zero-valent iron nanoparticles, can enhance the breakdown of organic pollutants in water through advanced oxidation techniques (Zhang *et al.,* 2019).
- **Nanofiltration Technology:** Membranes developed using nanotechnology, featuring accurately sized pores at the nanoscale, can effectively filter out particles, bacteria, and contaminants from water, thereby enhancing water quality (Chen *et al.,* 2017).

Smart Water Management

Smart water technology encompasses a range of advanced technologies and components designed to monitor, manage, and conserve water resources effectively. Key technologies in this field include sensors, the Internet of Things (IoT), and artificial intelligence (AI).

- **Sensors:** These devices are essential for measuring various physical and chemical properties of water, including temperature, pH levels, turbidity, dissolved oxygen, and conductivity. Sensors can be strategically placed at various locations and depths within water bodies, allowing for real-time data collection and wireless transmission to a central server or cloud platform (Celine Bernard, 2022).

- **Internet of Things (IoT):** IoT refers to a network of interconnected devices that communicate and share data with one another. This technology facilitates the integration of sensors and other devices to monitor and manage water quality and quantity across different scenarios. It can effectively identify leaks, bursts, or contamination in water pipelines, promptly alerting operators or users (Ajith *et al.,* 2020).
- **Artificial Intelligence (AI):** AI systems are capable of performing tasks that typically require human intelligence. They analyze data gathered from sensors and IoT devices, providing valuable insights for enhancing water management practices. AI can detect sources and patterns of water pollution, recommending effective strategies for prevention or reduction. Additionally, it aids in forecasting future water demand and supply, enabling optimal allocation of water resources (Krishnan *et al.,* 2022).

Floating Solar Installations

Floating solar systems, or photovoltaic (PV) panels, are installed on bodies of water. Their primary function is to produce renewable energy; however, they also play a role in reducing water pollution (Al-Widyan *et al.,* 2021). Recent research conducted in the UK indicates that these floating solar systems may lessen the effects of climate change on aquatic environments and aid in maintaining water quality, although further investigation is necessary (Edith Hancock, 2021). Floating solar systems can address water pollution through various mechanisms: 1. Decreased Evaporation 2. Temperature Regulation 3. Prevention of Algal Blooms 4. Enhanced Water Quality Monitoring.

Membrane Technology (Electrified Membranes)

Electrified membranes (EMs) provide solutions for the removal of emerging toxic contaminants that possess low molecular weight and neutral charge by incorporating electroactivity as an additional function of the membrane. These membranes can effectively target a variety of pollutants, including heavy metals, new organic contaminants, pharmaceuticals, personal care products, pathogenic microorganisms, and uncharged molecules. EMs have been utilized as porous flow-through electrodes in a two-electrode configuration (comprising working and counter electrodes) by applying a potential difference across them. The improved efficacy of EMs is linked to the simultaneous influence of various field effects (such as thermal, fluid, and electric fields) within the confined nanoscale intraporous structures, which facilitates rapid contaminant removal (Meng Sun *et al.,* 2021).

Bioremediation

Bioremediation refers to a biological method aimed at reducing the impact of pollutants through biodegradation, wherein complex toxic substances are broken down or transformed into simpler, less harmful compounds with the aid of microorganisms (Ajona and Vasanthi, 2021). Microorganisms utilize enzymes or metabolic pathways to decompose the organic components found in pollutants, depending on their growth and metabolic processes. Various microbial groups, including bacteria, fungi, yeasts, and algae, are frequently employed for bioremediation efforts (Das, 2021). These microbes are capable of degrading a wide range of pollutants, such as heavy metals, dyes, hydrocarbons, and polychlorinated biphenyls. Additionally, genetically modified microorganisms exhibit enhanced degradation capabilities and can be effectively applied to mitigate or eliminate different types of pollutants under controlled environments (Tran *et al.,* 2021).

Advanced Oxidation Processes

Advanced oxidation processes (AOPs) have garnered significant attention for their effectiveness in addressing a variety of pollutants, including emerging contaminants found in wastewater (Wang *et al.,* 2016). These processes are characterized by the generation of highly reactive species, particularly hydroxyl radicals (HO°), which play a crucial role in the degradation of organic compounds. The unique advantage of hydroxyl radicals lies in their remarkable reactivity and ability to interact with a wide range of both organic and inorganic materials (Bracamontes-Ruelas *et al.,* 2022). Hydroxyl radicals are often referred to as "universal oxidants" due to their non-selective nature, allowing them to effectively break down complex organic molecules that are typically resistant to conventional treatment methods. This includes a variety of pollutants such as pharmaceuticals, personal care products, and industrial chemicals, which are increasingly being recognized as emerging contaminants in wastewater streams. The presence of these substances poses significant environmental and health risks, making their removal a priority for wastewater treatment facilities. AOPs can be implemented through various methods, including ozonation, photocatalysis, and electrochemical oxidation, each of which has its own operational parameters and effectiveness depending on the specific contaminants present.

- **Photocatalysis:** In recent years, photocatalytic technology has emerged as a sustainable method for the degradation of pollutants (Subhiksha *et al.,* 2022). This innovative approach utilizes photocatalysts to produce highly reactive oxygen species, including hydroxyl radicals (•OH) and superoxide radicals ($*O_2^-$), when exposed to light (Acharya & Parida 2020). These reactive species are capable of oxidizing and breaking down organic contaminants in water, converting them into benign byproducts such as carbon dioxide and water (Jiang *et al.,* 2021).

- **Ozonation:** One highly effective method for treating wastewater is Ozonation. This process involves the generation of ozone, a molecule composed of three oxygen atoms, using an ozone generator. The produced ozone gas is then dissolved in water, where it serves to eliminate microorganisms and remove both organic and inorganic contaminants. Acting as a potent oxidizing agent, ozone breaks down organic materials and neutralizes harmful bacteria and viruses through a safe chemical reaction. This technique is commonly employed in both industrial and municipal wastewater treatment facilities. Once dissolved, ozone molecules bond with pollutants or microorganisms, leading to their effective destruction. The application of ozone can occur at various stages of the treatment process, offering significant flexibility for wastewater treatment plants. Additionally, ozone can be combined with other oxidants, such as hydrogen peroxide, to form Advanced Oxidation Processes (AOPs) in water purification efforts.
- **Electrochemical Oxidation**: The primary factors contributing to the reliability and widespread application of this technique are its efficiency and environmentally friendly characteristics (Radjenovic *et al.,* 2020). Electrochemical oxidation (EO) processes can be categorized into two main types: direct electro-oxidation, also known as anodic oxidation, and indirect electro-oxidation. In the direct EO method, the degradation of pollutants occurs through a direct electron transfer (DET) between the anode surface and the contaminants (Salvestrini *et al.,* 2020). Conversely, the indirect EO process involves the reaction of organic pollutants with potent oxidants, such as Cl_2, H2O2, and O3, which are generated during the electrolysis process.

Floating Treatment Wetland

Floating Treatment Wetland (FTW) systems are engineered solutions that mimic the natural water purification processes found in floating vegetated islands. These systems enhance the water quality of storage ponds by facilitating various treatment mechanisms, including adhesion, filtration, nutrient absorption by plants, and sequestration. An FTW typically comprises a floating platform that supports emergent hydrophytes. The above-water shoots of these hydroponic plants thrive while their roots extend into the water through a growth medium. The macrophytes present in FTWs effectively eliminate both organic and inorganic pollutants, such as nutrients, potentially harmful metals, and suspended solids (Colares *et al.,* 2020). The roots of these plants spread throughout the floating structure and into the water, forming dense root columns with significant surface area. In addition to directly absorbing nutrients and contaminants, the roots and the floating material create ample surface area for microbial growth, resulting in a biofilm layer. This biofilm slows down water flow, promoting sedimentation and the entrapment of suspended particles (sediment trapping) (DEC, 2020). Most nutrient uptake and degradation processes in an FTW system occur within this biofilm.

Conclusion

The development of technological innovations aimed at reducing water pollution is not merely beneficial but essential for the protection of our vital water resources. As the lifeblood of our ecosystems, water plays a crucial role in sustaining both human life and the myriad forms of flora and fauna that depend on it. The necessity for such measures arises from the alarming and detrimental effects that contaminated water has on aquatic organisms, vegetation, human health, and the broader environment. Polluted water can lead to the decline of biodiversity, disrupt food chains, and contribute to the degradation of habitats, ultimately resulting in significant repercussions for ecosystems and the services they provide. Despite the critical importance of addressing water pollution, the process of mitigating its effects presents a myriad of challenges. One of the primary difficulties lies in the diverse range of pollutants and their various origins, which complicates the issue. Pollutants can stem from industrial discharges, agricultural runoff, urban wastewater, and even atmospheric deposition, each requiring tailored strategies for effective management. Furthermore, the emergence of new contaminants, such as microplastics and pharmaceuticals, poses additional hurdles for traditional water treatment methods, which often struggle to keep pace with these evolving threats.

References

Acharya R. & Parida K. 2020. A review on TiO2/g-C3N4 visible-light-responsive photocatalysts for sustainable energy generation and environmental remediation. Journal of Environmental Chemical Engineering 8 (4), 103896.

Al-Widyan, M., Khasawneh, M., & Abu-Dalo, M. (2021). Potential of Floating Photovoltaic Technology and Their Effects on Energy Output, Water Quality and Supply in Jordan. Energies, 14(24), 8417.

Ambe, I. M., Nnaji, P. C., & Odigure, J. O. (2020). Challenges and strategies for sustainable water pollution control in developing countries. Sustainability, 12(9), 3593.

Bracamontes-Ruelas, A.R.; Ordaz-Díaz, L.A.; Bailón-Salas, A.M.; Ríos-Saucedo, J.C.; Reyes-Vidal, Y.; Reynoso-Cuevas, L. Emerging Pollutants in Wastewater, Advanced Oxidation Processes as an Alternative Treatment and Perspectives. Processes 2022, 10, 1041. https://doi.org/10.3390/pr10051041

Chen, L., Zhu, L. D., & Zhang, H. (2017). A review on the fabrication of membranes with nanocellulose for nanofiltration and reverse osmosis. Green Energy & Environment, 2(1), 18-34.

Colares, G.S., Dell'Osbel, N., Wiesel, P.G., Oliveira, G.A., Lemos, P.H.Z., P. da Silva, F.,Lutterbeck, C.A., Kist, L.T., & Machado, E.L. (2020). Floating treatment wetlands: A review and bibliometric analysis. Science of The Total Environment, 714, 136776.

Das, S. (2021). Biosorption and removal of toxic heavy metals by metal tolerating bacteria for bioremediation of metal contamination: A comprehensive review. Journal of Environmental Chemical Engineering, 9(1), 104686.

Department of Environment Conservation (DEC). (2020). Wetland Functions And Values: Surface And Ground Water Protection. Vermont Department of Environmental 14 Conservation.

Giovanella, P., Gabriela A.L., Igor, V., Pellizzer, E., Fontes, B., & Lara D. (2020). Metal and organic pollutants bioremediation by extremophile microorganisms. Journal of Hazardous Materials, Volume 382, 121024.

Goh, K. H., Lim, T. T., & Dong, Z. (2020). Nanostructured materials in water purification: Opportunities and challenges. Nano Today, 35, 100926.

Guo, Y.; Wang, J.; Shinde, S.; Wang, X.; Li, Y.; Dai, Y.; Ren, J.; Zhang, P.; Liu, X. Simultaneous wastewater treatment and energy harvesting in microbial fuel cells: An update on the biocatalysts. RSC Adv. 2020, 10, 25874–25887.

Hancock, E. (2021, March 25). New floating solar study demonstrates water quality improvements.

International Institution for Sustainability Development (IIISD) (2017). Floating Treatment Wetland. IISD.org. https://www.iisd.org/story/floating-treatment-wetlands/ (March 11th,2020).

Jain, H.; Kumar, A.; Verma, A.K.; Wadhwa, S.; Rajput, V.D.; Minkina, T.; Garg, M.C. Treatment of textile industry wastewater by using high-performance forward osmosis membrane tailored with alpha-manganese dioxide nanoparticles for fertigation. Environ. Sci. Pollut. Res. 2022, 29, 80032–80043.

Jiang D., Otitoju T. A., Ouyang Y., Shoparwe N. F., Wang S., Zhang A. & Li S. 2021 A review on metal ions modified TiO2 for photocatalytic degradation of organic pollutants. Catalysts 11 (9), 1039.

M. Ajona, P. Vasanthi. (2021). Bioremediation of petroleum contaminated soils – A review,Materials Today: Proceedings,45(7), pp.7117-7122. https://doi.org/10.1016/j.matpr.2021.01.949.

Malik, S., Kishore, S., Dhasmana, A., Kumari, P., Mitra, T., Chaudhary, V., Kumari, R., Bora, J., Ranjan, A., Minkina, T., & Rajput, V. (2023). A Perspective Review on Microbial Fuel Cells in Treatment and Product Recovery from Wastewater. Water 15(2) 316.

Pedersen, E. K., & Binning, P. J. (2020). The challenges of addressing multiple sources of water pollution. Water Resources Research, 56(3), e2019WR025795.

Radjenovic, J.; Duinslaeger, N.; Avval, S.S.; Chaplin, B.P. Facing the Challenge of Poly- and Perfluoroalkyl Substances in Water: Is Electrochemical Oxidation the Answer? Environ. Sci. Technol. 2020, 54, 14815–14829.

Salvestrini, S.; Fenti, A.; Chianese, S.; Iovino, P.; Musmarra, D. Electro-Oxidation of Humic Acids Using Platinum Electrodes: An Experimental Approach and Kinetic Modelling. Water 2020, 12, 2250

Scavia, D., & Field, J. C. (2021). Advances in water quality monitoring and assessment for enhancing ecosystem management. Frontiers in Ecology and the Environment, 19(4), 197-205.

Stalter, D., Magdeburg, A., Wagner, M., & Oehlmann, J. (2020). What do we know about emerging organic contaminants in the water cycle? In Chemical Water and Wastewater Treatment (pp. 3-42).

Subhiksha V., Kokilavani S. & Khan S. S. 2022 Recent advances in degradation of organic pollutant in aqueous solutions using bismuth based photocatalysts: a review. Chemosphere 290, 133228.

Sun, M., Wang, X., Winter, L.R., Zhao, Y., Ma, W., Hedtke, T., Kim, J., & Elimelech, M. (2021). Electrified Membranes for Water Treatment Applications. ACS EST Eng, 725–752.

Tran, B.M., Prabha, H., Iyer, A., Byrne, C., Abee, T., & Poolman, B. (2021).Measurement of Protein Mobility inListeria monocytogenesReveals aUnique Tolerance to Osmotic Stress and Temperature Dependence of Diffusion. Frontiers in Microbiology,12:640149. doi: 10.3389/fmicb.2021.640149

Wang, N.; Zheng, T.; Zhang, G.; Wang, P. A review on Fenton-like processes for organic wastewater treatment. J. Environ. Chem. 2016, 4, 762–787.

Wang, Q., Huang, Y., Xu, S., Liu, H., Li, J., & Sun, D. D. (2020). Nanoscale photocatalysts for water purification: State of the art and future prospects. Critical Reviews in Environmental Science and Technology, 50(16), 1659-1726.

Wu, Feng Li, Shengwei Yi, & Fei Ge. (2021). Genetically engineered microbial remediation of soils co-contaminated by heavy metals and polycyclic aromatic hydrocarbons: Advances and ecological risk assessment. Journal of Environmental Management, 15, 113185.

Yaroshenko, I., Kirsanov, D., Marjanovic, M., Lieberzeit, P.A., Korostynska, O., Mason, A.; Frau, I., Legin, A. Real-Time Water Quality Monitoring with Chemical Sensors. Sensors 2020, 20, 3432.

Zhang, L., Li, Q., Chen, J., Zhang, J., Xu, D., Ding, Y., ... & Wang, Z. (2019). Progress in the application of zero-valent iron nanoparticles for remediation of pollutants in aqueous systems. Chemical Engineering Journal, 373, 306-327.

Zhao B., Johnston F. H., Salimi F., Kurabayashi M., Negishi K. (2020). Short-term exposure to ambient fine particulate matter and out-of-hospital cardiac arrest: a nationwide case-crossover study in Japan. Lancet Planet. Health 4 (1), e15–e23.

3

Use of Duckweed as a Potential Feed Additive in Poultry Nutrition for Sustainable Aquatic Management

***Subata Mehboob*[1*] *and Bushra Zaffar*[2]**

[1]*Assistant Professor (Animal Nutrition) MRCVS& Research Centre, Jhajjar, Haryana*

[2]*PhD Scholar, Poultry Science, SKUAST-K, Shuhama*

[*]*Corresponding Author Email: subatamalik@gmail.com*

Abstract

Aquatic weeds, such as duckweed, present a dual challenge and opportunity within aquatic ecosystems and the poultry feed industry. These fast-growing plants, often deemed nuisances due to their potential to disrupt ecosystems, offer significant potential as alternative feed sources. Duckweed, a floating aquatic plant, is particularly noteworthy due to its high protein content, rapid growth rate, and nutrient recycling capabilities. In freshwater systems, excessive growth of aquatic weeds can lead to decreased ecosystem productivity, hypoxia, and anoxia, making effective management crucial. However, duckweed's ability to thrive in nutrient-rich conditions and its high nutritional value, including substantial protein and essential vitamins, position it as a valuable feed ingredient for poultry. The Indian poultry sector, growing at an annual rate of 8-10%, faces challenges such as rising feed costs and limited availability of conventional feed ingredients. Integrating duckweed into poultry diets can provide a cost-effective, sustainable alternative that reduces reliance on traditional feed sources and supports ecosystem management. This approach not only addresses feed shortages but also contributes to environmental sustainability by recycling nutrients and mitigating pollution. Despite its potential, challenges such as lack of standardized regulations and market acceptance remain. Continued research, policy support, and market promotion are essential to harness the full benefits of aquatic weeds.

Keywords: Aquatic, Duckweed, Poultry, Sustainable and Feed Additives

Introduction

The Indian poultry sector has evolved into a vibrant agribusiness spurred by domestic economic growth and consumption dynamics. The sector has been growing at around 8-10% annually over the last decade with annual growth rates of 5.57 percent and 11.44 percent in egg and broiler production, respectively driven by increased domestic consumption (Wakcaure *et al.,* 2016). It accounts for about one percent of India's GDP and 11.70 per cent of the GDP from the livestock sector (Sasidhar and Murari, 2015). The revolution of Indian poultry sector is contributing to improved nutrition and poverty reduction. The poultry sector in India is providing direct or indirect employment to 6.5 million people (Rajendran *et al*., 2014). Poultry has helped in filling the animal protein gap in most of the developed and developing countries. Poultry meat is the fastest growing animal protein in India. The cost of poultry meat is low for consumers, relative to other meat products, and has comparatively wider acceptability across regions and religions (Manning and Baines, 2004). The market demand for poultry meat and eggs in the state of J&K, particularly in Kashmir is very high because of the food habits of people and their likeness for non-vegetarian food and tourist influx. But the present set up in J&K State is too scanty to meet the challenges thrown open by the changing lifestyle of the common Kashmiri and the expected rumble in the tourism industry. Despite substantial qualitative and quantitative advancements in different sectors of the poultry industry, the industry faces numerous challenges like rise in the cost of inputs such as feed cost, disease problems etc. In poultry rearing, feed cost accounts for 60-70% of the total cost of either egg or poultry meat production and protein accounts 15% of feed cost (Singh, 1990). The world is now facing a severe scarcity of poultry feed and very limited number of feed ingredients are available to choose for the formulation of balanced diet. Non availability and higher price of feed ingredients are main constrains of poultry production. The profit of poultry farming mainly depends on economic feeding of balanced ration. From the economic point of view, the poultry should be supplied with cheaper feed to get maximum return with minimum cost. It is therefore, a recent trend among the poultry nutritionists to explore the unconventional cheaper ingredients towards reducing feed cost to maximize profit from poultry farming. So, locally produced available unconventional feed resources may reduce feed cost and in turn increase profitability of poultry enterprise.

Aquatic biodiversity is comprehensive term that comprises freshwater ecosystems with lakes, ponds, reservoirs, rivers, streams, groundwater, and wetlands. The other part of aquatic biodiversity has marine ecosystems, which makes up an ocean, estuaries, salt marshes, coral reefs, mangroves and algal colonies. Different kinds of phytoplanktons, zooplanktons, aquatic plants, insects, fishes, birds and mammals are also an important part of aquatic biodiversity. Amongst these are aquatic weeds which are undesirable plants that cause harm to the aquatic organisms and the environment (Ganguly *et al.*, 2012). These weeds hold a significant position in aquaculture systems and their control represents a global challenge. Because of their fast growth rate, ease of adaptation, and nonexistence of natural adversaries make the plants aggressive, disturbing, and problematic to control (Dissanayaka *et al.*, 2023). Failure to effectively control of excessive aquatic weed growth can result in decreased ecosystem productivity, as the weeds consume nutrients and potentially obstruct sunlight, leading to cause hypoxia or anoxia (Giri, 2020). Resultantly, adequate management or conversion of them into valuable products becomes crucial. Near about 50 species of aquatic weeds are known to be used as either direct or indirect feed for both fishes and animals (Naseem *et al.*, 2020). The aquaculture industry may profit from the sustainable use of these plant ingredients as feed additives in animal industries. The utilization of aquatic weeds as feed additives for animals especially poultry is a viable and eco-friendly strategy that lessens the environmental hazards, maintains the balance of the ecosystem, and offers cheap protein source in the industry. Therefore, the identification and development of sustainable protein source from aquatic plants can play a significant role in enhancing global poultry production.

Duckweed as Feed Additive in Poultry

Duckweed is a monocotyledon species of the family *Lemnaceae* adapted to grow in water at temperatures between 6 and 33°C (Leng *et al.,* 1995). It is a small floating aquatic plant that grows very well on stagnant ponds and is commonly found throughout tropical countries in natural ponds, lakes and flooded rice fields (Leng *et al.,* 1995). It can tolerate high nutrient stress and appears to be more resistant to pests and diseases than other aquatic plants. Duckweed (*Lemna L.)* has for years been the subject of interest in many regions of the world. Its use in folk medicine has been scientifically substantiated. It exhibits cholagogic, carminative, expectorant, diaphoretic, diuretic, and anticarcinogenic properties. For more than a decade, duckweed has been successfully used in biotechnology to produce biologically active compounds of importance in dietetics, phytotherapy, and phyto cosmetics (Czerpak and Piotrowska, 2005). Duckweed is also used in ecotoxicology, and in biogas and ethanol production (Sikora and Malinowski, 2015). Moreover, it has high protein and carotene contents (Xuan *et al.,* 1995). Crude protein yields between 6 and 10 tonnes/ha/year have been recorded when the N content in the water is in the range of 10 to 30 mg/litre (Anh and Preston, 1997b; Rodriguez and Preston, 1996). It can be grown to recycle nutrients from waste water and it provides a good source of proteins and can be utilized for the production of some products such as animal feed and fuel ethanol (Cheng and Stomp, 2009). Despite these numerous benefits, the fast paced growth of these weeds results in a thick surface coverage over the water bodies often called duckweed mats. These mats can lead to ill effects in the aquatic ecosystem like poor penetration of sunlight to lower depths of the water body, inadequate exchange of gases between the hydrosphere and atmosphere, including a few others, finally affecting the life and existence of several submerged aquatic organisms. Traditionally spoken, these duckweed overgrowths are seen as a menace and several physical and chemical methods are applied to get rid of the floating duckweeds, which is usually a tough job—with damaging effects on the ecological system. And, it is also noticed that an increase of the density and coverage of the duckweed on the water body can keep recurring with seasonal changes and with the changes in the physico-chemical properties of the water body. Duckweeds also take up nutrients for their growth, harvesting biomass which finally results in removing the excessive nutrients from the surface water.

There is an increasing research effort to evaluate the nutritive value of different non-conventional feed resources, including aquatic weeds as feed additives in poultry industries (Mondal and Ray, 1999). Inclusion of these feed resources in poultry ration could be one of alternatives which may reduce the cost of poultry production along with maintaining the ecosystem of water bodies. Many trials have been carried out using duckweed as the major feed to raise fish, pig, chicken and also ducks. Duckweed has high crude protein content and a well-balanced amino acid profile and is also a good source of vitamins and minerals for livestock and poultry (Men *et al.*, 2001). The use of duckweed as poultry feed has been recognized by many authors (Islam *et al.,* 1997; Leng, 1999; Samnang, 1999 Haustein *et al.*, 1994). Even though the moisture content of duckweed can be the first limiting factor for chickens, it can play important role in poultry feeding. It contains 28% Crude protein, 3.7% Crude fat, 33.8% ash,11.5% fiber and 42.6% carbohydrates (Tania *et al.*, 2009). The plant has been successfully used to replace protein sources such as sesame oil cake (Ahammad *et al.*, 2003) and fishmeal (Effiong *et al.*, 2009) at graded levels in poultry industry. The use of duckweed as a feed additive could offer a sustainable solution to the feeding of broiler chickens (Ahammed *et al.*, 2003; Khandaker*et al.*, 2007). These weeds can be sample tested for their composition and can accordingly be utilized as feed additives for integrated poultry farming. This can be adopted by individuals or by communities and villages depending on the size of the water body as. The conversion of these aquatic weeds into valuable products could play a crucial role in sustainability of the aquaculture and poultry industry and mitigating environmental hazards as well as pollution.

Summary

Aquatic weeds present a promising alternative to traditional poultry feed ingredients due to their abundance, sustainability, nutritional value, and cost-effectiveness. These plants grow prolifically in aquatic environments, offering a renewable resource that can support the poultry and aquaculture industries. Their use could also help mitigate environmental issues associated with nutrient pollution in water bodies. However, the integration of aquatic weeds into poultry feed faces several challenges. These include the risk of ecosystem disruption if not managed properly, potential health concerns for poultry, and the lack of standardized regulations and guidelines, which hinder large-scale adoption and market acceptance.

Conclusion

The potential benefits of using aquatic weeds in poultry feed are significant, offering a sustainable and economically viable alternative to traditional feed sources. However, realizing these benefits requires overcoming key challenges, including proper management practices, ensuring poultry health, and establishing clear regulatory frameworks. Continued research and development are crucial to address these challenges and optimize the use of aquatic weeds in feed formulations. By fostering collaboration among researchers, industry stakeholders, and regulatory bodies, and by implementing sustainable harvesting practices, the poultry industry can effectively integrate aquatic weeds into feed systems, contributing to a more sustainable and resilient agricultural sector.

References

Ahammad, M.U., Swapon, M.S.R., Yeasmin, T., Rahman, M.S. and Ali, M.S. 2003.Replacement of Sesame oil cake by Duckweed (Lemna minor) in Broiler diet. Pakistan Journal of Biological Sciences 6(16):1450-1453.

Anh, N.D. and Preston, T. R. 1997b.Evaluation of protein quality in duckweed (Lemna spp.) using a duckling growth assay. Livestock Research for Rural Development 9(2): 28-31.

Cheng, J.J. and Stomp, A.M. 2009.Growing Duckweed to recover nutrients from wastewaters and for production of fuel ethanol and animal feed. Clean37(1):17-26.

Cui.,W., C h e n g J.J. (2015). Growing duckweed for biofuel production: A review. Plant Biol., 17: 16–2

CzerpakR., Piotrowska. (2005). Wolfa arrhiza – the smallest plant with the highest adaptation ability and applications (in Polish). Kosmos. Probl. Nauk Biol., 54: 241–250.

Dissanayaka DMNS, Udumann SS, Dissanayake DKRPL, Nuwarapaksha TD, Atapattu AJ. 2023. Review

Dissanayaka DMNS, Udumann SS, Dissanayake DKRPL, Nuwarapaksha TD, Atapattu AJ. 2023. Review on Aquatic Weeds as Potential Source for Compost Production to Meet Sustainable Plant Nutrient Management Needs. In Waste. 1(1): 264-280. MDPI.

Effiong, B. N., Sanni, A. and Fakunle, J. O. 2009.Effect of partial replacement of fish meal with duckweed (Lemna pauciscostata) meal on the growth performance of Heterobranchuslongifilis fingerlings.Report and Opinion1(3):76-81.

Ganguly, A., Chatterjee, P.K., Dey, A., 2012. Studies on ethanol production from water hyacinth—a review. Renew. Sustain. Energy Rev. 16, 966–972.

Giri A. 2020. Various types of Aquatic Weeds in A Village Fish Pond and Their Control. International Journal of Environmental Sciences & Natural Resources. 25(3): 142-146.

Haustein, A.T., Gillman, R.H., Skillicorn, P.W., Hannan, H., Dias, F., Guevana, V., Vergara, V., Gastanaduy, A. and Gillman, J.B. 1994. Performance of broiler chickens fed diets containing duckweed (Lemnagibba). Journal of Agricultural Science122(2): 288-289.

Islam, K.M. S., Shahjalal, M., Tareque, A. M. M. and Howlider, M. A. R. 1997. Complete replacement of dietary fish meal by duckweed and soybean meal on the performance of broilers. Asian-Australasian Journal of Animal Sciences10(6):629-634.

Khandaker, T., Khan, M.J., Shahjalal, M. and Rahman, M.M. 2007. Use of Duckweed (Lemnaperpusilla) as a protein source feed item in the diet of semi-scavenging Jinding layer ducks. The Journal of Poultry Science 44:314- 321.

Leng, R.A. 1999.Duckweed - A tiny aquatic plant with enormous potential for agriculture and environment. FAO, Rome (Italy). Animal Production and Health Div. University of Tropical Agriculture Foundation, Phnom Penh (Cambodia).

Leng, R.A., Stambolie, J.H. and Bell, R. 1995. Duckweed - a potential high- protein feed resource for domestic animals and fish. Livestock Research for Rural Development 7(5): 20-23.

Manning, L. and Baines, R.N. 2004. Globalisation: a study of poultry meat supply chain. British Food Journal106(10/11):819-836.

Men, B.X., Ogle, R.B. and Lindberg, J.E. 2001.Studies on intergrated duck-rice systems in the Mekong Delta of Vietnam. Journal of Sustainable Agriculture20(3): 21-32.

Mondal, T. K. and Ray, A. K. 1999. The nutritive value of Acaciaauri culiformis leaf meal in compounded diets for Labeorohita fingerlings. In: The Fourth Indian Fisheries Forum Proceedings. Kochi. pp. 295–298.

Naseem, Shahida & Bhat, Sami & Gani, Adil & Bhat, Farooz A. (2020). Perspectives on utilization of macrophytes as feed ingredient for fish in future aquaculture. Reviews in Aquaculture. 13. 10.1111/raq.12475.

Rajendran, K., Mani, K., Shamsudeen, P. and Vasanthakumar, T. 2014. Broiler Industry – understanding the integration and role of private industries. In: Proceedings of the 31st Annual Conference and National Symposium of Indian Poultry Science Association (IPSACON 2014), Dec. 18-20, Namakkal, India.

Rodriguez, L. and Preston, T. R. 1996.Use of effluent from low cost plastic biodigesters as fertilizer for duckweed ponds. Livestock Research for Rural Development8(2):60-69.

Sikora, J., Malinowsk i M. (2015). Effect of duckweed in the substrate for the energy use of digest- ate (in Polish). Proc. ECOpole, 9: 2, DOI: 10.2429/proc.2015.9(2)085.

Samnang, H. 1999. Duckweed versus ground soya beans assupplement for scavenging native chickens in an integrated farming system. Livestock Research for Rural Development 11: Article # 1, Retrieved July 23,2012.

Sasidhar, P.V.K and Murari, S. 2015. Intrgrated contract broiler farming: An evaluation case study in India. Modernizing extension and advisory services.

Singh, R.A., 1990. Poultry production.3 rd Ed.,Kalyani Publishers, 1/1 RajinderNagar,Ludhiana-141008,New Delhi, India.

Tania, M., Khan, M.A., Khatun, R., Rahman, M. And Khalil, I. 2009. An investigation on the nutritional composition of unconventional poultry feed resources. Bangladesh Journal of Life Science21(2):21-28.

Wakchaure, R., Babu, G. P., Ganguly, S., Para, P. A., Praveen, P. K., Kumar, A. and Sharma S. 2016. Dairy, poultry and fisheries for overcoming food security: A review. World Journal of Biology and Medical Sciences 3(1): 36-42.

Xuan, M.B., Ogle, B. and Preston, T.R. 1995. Use of duckweed (lemnaspp) as replacement for soyabean meal in a basal diet of broken rice for fattening ducks. Livestock Research for Rural Development7(3):10-15.

4

Climate Change Adaptation and Mitigation: Assessing Vulnerability and Promoting Resilient Aquaculture

Hardik Sikotariya[1], T.H. Dave[2], D.T. Vaghela[3] and Mayur Bhadarka[4]

[1]PG Scholar, Department of Fisheries Resource Management*

[2]Assistant professor, Department of Fisheries Resource Management

[3]Associate Professor and Head, Department of Aquatic Environment Management

[4]PG Scholar, Department of Aquatic Environment Management

College of Fisheries Science, Kamdhenu University, Veraval (362265), Gujarat, India

Abstract

The aquaculture industry is increasingly vulnerable to the impacts of climate change, particularly in regions like the UK, USA, and Nigeria. The study highlights the pressing need for innovative adaptation strategies, such as Nanosolar technology and integrated farming systems, to safeguard this vital industry against fluctuating temperatures, extreme weather events, and other environmental stressors. Through an in-depth analysis of vulnerability, the research underscores the importance of understanding the complex interplay between environmental, economic, and social factors in developing effective adaptation and mitigation strategies. Case studies from around the world illustrate both the impacts of climate change on aquaculture and the diverse strategies employed to enhance resilience. By examining the role of technological innovations, adaptive infrastructure, and policy-oriented strategies, the study offers comprehensive insights into how aquaculture can be made more sustainable and resilient in the face of ongoing climate challenges.

Keywords: *Climate, Aquaculture Sustainability, Risk Management, Sustainable Practices.*

1. Introduction

Climate change has a significant impact on marine ecosystems, disrupting oceanic processes and posing a global threat to biodiversity (Makwana *et al.*, 2024a). In India, where fisheries and aquaculture play a crucial role in the economy, the country's large coastal regions and heavy reliance on marine resources heighten its vulnerability to climate change (Makwana *et al.*, 2024b). Climate change is having a profound impact on aquaculture, putting the productivity and long-term sustainability of this vital industry at risk in various regions around the world. Vulnerability assessments highlight that countries like the UK, USA, and Nigeria are particularly at risk, as they heavily depend on aquaculture systems that are increasingly susceptible to climate-related challenges such as fluctuating temperatures and extreme weather conditions (Awotunde, 2024). To counter these threats, adopting innovative adaptation strategies is essential. Techniques like Nanosolar technology and integrated farming systems play a crucial role in boosting the resilience of aquaculture, helping to safeguard food security in the face of these growing challenges (Paul *et al.*, 2024). Moreover, a deep understanding of the socio-ecological factors, including the responsiveness and adaptability of local communities, is vital for crafting effective strategies to mitigate the impacts of climate change on aquaculture (Vögler *et al.*, 2023).

2. Understanding Vulnerability in Aquaculture

Understanding vulnerability in aquaculture requires defining the concept and identifying the key factors that contribute to it. In this context, vulnerability refers to the susceptibility of aquaculture systems to various stressors, including environmental, social, and economic challenges.

2.1 Defining Vulnerability in Aquaculture

Vulnerability in aquaculture is determined by the interaction of multiple stressors, such as climate change, natural disasters, and socio-economic pressures. It involves the sensitivity of aquaculture systems to these stressors and their ability to adapt to them (Islam & Chuenpagdee, 2022; Macusi *et al.*, 2021).

2.2 Key Factors Contributing to Vulnerability

2.2.1 Environmental Factors

Changes in water quality, including pollution and the introduction of invasive species, can severely affect fish health and populations, leading to economic instability for those dependent on fisheries (Nair & Nayak, 2023).

2.2.2 Economic and Social Factors

Economic challenges, such as market volatility and the impacts of the COVID-19 pandemic, have heightened vulnerabilities, especially in small-scale fisheries (Islam & Chuenpagdee, 2022; Macusi *et al.*, 2021).

2.2.3 Technological and Governance Issues

Limited access to technology and weak governance structures impede effective management and adaptation efforts, further increasing the vulnerability of aquaculture systems (Berio & Salugsugan, 2022).

3. Impacts of Climate Change on Aquaculture

Climate change has a profound impact on aquaculture, affecting it across physical, environmental, economic, and social dimensions. These effects are complex, influencing the viability of species, the efficiency of production, and the livelihoods of communities dependent on aquaculture.

3.1 Physical and Environmental Impacts

3.1.1 Temperature and Water Quality

Rising temperatures are particularly detrimental to species like Rohu (*Labeo rohita*), which thrive best in temperatures between 28°C and 32°C. Any deviation from this range can cause thermal stress, leading to lower survival rates (Mridul *et al.*, 2024).

3.1.2 Temperature and Salinity Changes

Rising global temperatures can push aquaculture species to their upper tolerance limits, leading to higher mortality rates and increased susceptibility to disease outbreaks (Silva *et al.*, 2013). In deltaic regions, rising sea levels cause saline water intrusion, further complicating aquaculture operations by making the environment less suitable for certain species (Silva *et al.*, 2013).

3.1.3 Ocean Acidification and Algal Blooms

Increased levels of CO2 contribute to ocean acidification, which negatively affects shellfish such as oysters, as they are highly sensitive to these environmental changes (Neokye *et al.*, 2024).

3.1.4 Extreme Weather Events

The increasing frequency of storms and flooding poses significant risks to aquaculture infrastructure, resulting in considerable production losses (Vergara-Solana *et al.*, 2024).

3.2 Economic and Social Impacts

3.2.1 Production Declines

In regions like Latin America, even a 1% increase in temperature can result in a 0.08% decrease in aquaculture production, illustrating the economic vulnerabilities tied to climate change (Muniz & Del Rio, 2023).

3.2.2 Community Livelihoods

Coastal communities that depend on aquaculture for food and income are particularly at risk. The threats posed by climate change to their livelihoods emphasize the urgent need for adaptation strategies (Neokye *et al.*, 2024; Vergara-Solana *et al.*, 2024).

3.2.3 Feed Supply Challenges

Aquaculture's heavy reliance on fishmeal and fish oil, which are becoming increasingly scarce due to climate-induced declines in wild fish populations, poses a serious threat to its sustainability (Hall, 2015; Silva *et al.*, 2013). This situation underscores the urgent need to develop alternative feed sources, such as underutilized crops that are more resilient to climate stressors (Hall, 2015).

4. Assessing Vulnerability in Aquaculture

Assessing vulnerability in aquaculture is essential for grasping the impacts of climate change and other stressors on this critical industry. The research underscores the complex nature of vulnerability, highlighting the importance of developing targeted adaptation strategies to address the diverse challenges faced by aquaculture systems.

4.1 Frameworks for Vulnerability Assessment

4.1.1 Vulnerability Scoping Diagram

This framework was applied in China to assess fishery systems, uncovering how spatiotemporal variations in vulnerability are influenced by factors such as funding and technological advancements (Wang *et al.*, 2023).

4.1.2 Semi-Quantitative Approaches

In Uruguay, a risk-based assessment approach highlighted the interaction between environmental and social factors, revealing that fishers face high vulnerability due to their low adaptive capacity (Vögler *et al.*, 2023).

4.1.3 Water Quality Analysis

Research conducted in Chilika Lagoon, India, demonstrated that changes in water quality have increased the vulnerabilities of small-scale fisheries. This study stressed the importance of integrated assessments that take both environmental and social factors into account (Nair & Nayak, 2023).

4.2 Data Collection and Analysis

4.2.1 Data Collection Methods

4.2.1.1 Productivity-Susceptibility Analysis (PSA)

This method assesses species vulnerability by examining life history traits and susceptibility factors, making it particularly useful in data-limited situations like the marine aquarium trade (Baillargeon *et al.*, 2024).

4.2.1.2 Multi-Indicator Framework

Used in China's coastal provinces, this framework combines a vulnerability scoping diagram with the entropy weight method to systematically evaluate vulnerabilities across different times and locations (Wang *et al.*, 2023).

4.3 Analysis Techniques

4.3.1 Expert Knowledge Integration

In extremely data-poor contexts, such as Peruvian coastal fisheries, integrating expert insights is crucial for completing accurate vulnerability assessments (Castagnino *et al.*, 2024).

4.3.2 Literature Review and Meta-Analysis

A systematic review of the impacts of climate change on fisheries in Asia underscores the importance of adaptive management strategies to address both social and economic vulnerabilities (Habib *et al.*, 2023).

4.4 Case Studies

4.4.1 Climate Change Impacts

The UK aquaculture sector has been identified as highly vulnerable due to its large-scale operations and significant exposure to climate-related risks, with the USA and Nigeria also facing substantial challenges (Awotunde, 2024). In Zambia, more than 80% of aquaculture producers are aware of the effects of climate change, with rising temperatures and extreme weather events notably driving up production costs (Maulu *et al.*, 2024).

4.4.2 Regional Vulnerability Assessments

In China, coastal fishery systems show varying levels of vulnerability, with provinces like Liaoning and Hainan facing particularly high risks due to financial and technological limitations (Wang *et al.*, 2023). Similarly, the pink shrimp fishery in Uruguay is highly vulnerable, influenced by environmental factors and the low adaptive capacity of local fishers (Vögler *et al.*, 2023).

4.4.3 Data-Driven Insights

In Peru, a productivity susceptibility analysis has highlighted the high vulnerability of several coastal groundfish species to overfishing, emphasizing the urgent need for better data collection and management practices (Castagnino *et al.*, 2024).

5. Adaptation Strategies for Resilient Aquaculture

5.1 Technological Innovations in Aquaculture

5.1.1 Water Quality Monitoring

The use of wireless sensor networks (WSNs) and IoT-based solutions enables real-time monitoring of critical water parameters, improving fish health and overall productivity (Rashmi and sangeeta, 2024).

5.1.2 Recirculating Aquaculture Systems (RAS)

RAS creates a controlled environment that mitigates the effects of climate variability, promoting sustainable fish production while minimizing environmental impact (Ahmed & Turchini, 2021).

5.1.3 Species Adaptation

Research on species like the Indian pompano highlights the potential for selecting and breeding resilient species that can adapt to changing conditions (Divu *et al.*, 2024).

5.1.4 Precision Aquaculture

Technologies like ICT tools enable real-time data collection, enhancing decision-making and resource management. This approach helps reduce vulnerability to climate impacts by allowing more precise and efficient aquaculture practices (Rajesh *et al.*, 2024).

5.2 Adaptive Infrastructure

5.2.1 Submergible Cages

In response to marine heatwaves, adapting infrastructure such as deeper net pens can help maintain optimal conditions for fish farming (Atalah *et al.*, 2024).

5.3 Policy-Oriented Strategies

5.3.1 Research and Monitoring

Policies should focus on advancing research into species vulnerability to climate change and developing accessible tools, such as WebGIS, to provide valuable climate data (Portilla-Cabrera *et al.*, 2024).

5.3.2 Government Incentives

In Kenya, government support packages have been crucial in aiding aquaculture stakeholders to recover from disruptions, such as those caused by the COVID-19 pandemic (Munguti *et al.*, 2022).

5.3.3 India's Adaptation Policies

India's agricultural and fisheries sectors are heavily impacted by climate change, requiring robust institutional frameworks. Current policies focus on crop diversification, water management, and community-based adaptation strategies. However, many adaptation needs remain unmet due to the country's diverse socio-cultural contexts (Gao *et al.*, 2022).

5.3.4 Artisanal Fisheries in Colombia

In San Andrés de Tumaco, Colombia, policy-oriented strategies emphasize educational and behavioral adaptations, along with research into fish species' vulnerabilities. These strategies support sustainable fisheries policies, helping to enhance local resilience against climate change (Portilla-Cabrera *et al.*, 2024).

5.3.5 Critical Appraisal of Governance

Effective governance is crucial for successful climate adaptation, yet challenges like insufficient funding and fragmented policies continue to hinder progress. A cross-sectoral approach is needed to effectively address and mitigate the impacts of climate change (Pandey *et al.*, 2024).

6. Mitigation Strategies in Resilient Aquaculture

6.1 Emission Sources and Mitigation Strategies

6.1.1 Scope of Emissions

A large share of greenhouse gas (GHG) emissions in aquaculture comes from Scope 3 activities, particularly feed production and transportation. These often surpass the direct emissions generated by aquaculture operations themselves (Hammer *et al.*, 2022).

6.1.2 Technological Innovations

Adopting integrated multi-trophic aquaculture and recirculating systems can significantly improve resource efficiency and lower emissions (Abisha *et al.*, 2022).

6.1.3 Cost-Effectiveness Analysis

Applying cost-effectiveness analysis is crucial for identifying the most effective mitigation strategies, and ensuring that resources are utilized in the most impactful way (MacLeod *et al.*, 2019).

6.2 Best Practices in Manure Management

6.2.1 Manure Management Techniques

Effective manure management strategies, such as frequent removal and anaerobic digestion, can greatly reduce methane and ammonia emissions (Yan *et al.*, 2024).

6.2.2 Nutritional Adjustments

Optimizing feed formulations not only decreases emissions linked to feed production but also minimizes the environmental impact of animal waste (Liu & Liu, 2018).

6.3 Carbon Sequestration Techniques

6.3.1 Biochar Utilization

Biochar, created from organic materials, offers a long-term solution for carbon sequestration in aquaculture by enhancing soil carbon storage, with the potential to last for centuries (Kumar *et al.*, 2022).

6.3.2 Microalgae Cultivation

Cultivating microalgae is an effective way to sequester organic carbon from wastewater, simultaneously reducing greenhouse gas emissions and improving water quality (Kumar *et al.*, 2022).

6.4 Integrated Aquaculture Systems

6.4.1 Integrated Multi-Trophic Aquaculture (IMTA)

IMTA fosters biodiversity and enhances carbon sequestration by integrating multiple species that utilize each other’s waste products, creating a more sustainable and productive system (Nazar & Rifqi, 2024).

6.4.2 Recirculating Aquaculture Systems (RAS)

RAS is designed to minimize water usage while optimizing carbon capture through efficient waste management practices, making it an eco-friendly option for sustainable aquaculture (Nazar & Rifqi, 2024).

6.5 Policy and Technological Innovations

6.5.1 Carbon Pricing Mechanisms

Introducing strong carbon pricing mechanisms can drive aquaculture operations to adopt carbon sequestration methods, encouraging more sustainable practices (Guddaraddi *et al.*, 2023).

6.5.2 Research and Development

Ongoing innovation in carbon capture technologies and the development of advanced monitoring systems are essential for the successful implementation of these practices (Guddaraddi *et al.*, 2023).

7. Case Studies in Resilient Aquaculture

7.1 Financing Risk Resilience

A case study on the Homey Group underscores the critical role of market diversification and sustainable practices in mitigating financing risks, beyond merely having financial resources. The study identified weaknesses in the risk management system, pointing to the need for specialized oversight to bolster the group's resilience capabilities (Zhang *et al.*, 2024).

7.2 Supply Chain Integration

The experience of the Mino Ngremboko Cultivator during the COVID-19 pandemic highlights the importance of supply chain resilience. Despite suffering a 50-70% drop in revenue, the group successfully adapted by leveraging digital marketing and enhancing stakeholder alignment, demonstrating effective strategies for recovery (Cahyacipta *et al.*, 2023).

7.3 Climate Adaptation

In Kenya, aquaculture is severely impacted by climate challenges such as drought, which has been shown to reduce growth by 40%. The study emphasizes the need for better infrastructure and economic incentives to help farmers implement effective adaptation measures (Adekola *et al.*, 2022).

7.4 COVID-19 Impacts

South Africa's mussel farming sector faced significant challenges due to market closures during the COVID-19 pandemic. The study highlights the necessity of financial support and long-term planning to enhance resilience, with a focus on diversified revenue streams and maintaining reserve funds as key factors for recovery (Brown-Webb *et al.*, 2022).

Conclusion

In light of the growing challenges posed by climate change, it is clear that the aquaculture industry must adapt to secure its future sustainability. Understanding the complex vulnerabilities that affect aquaculture from environmental and socio-economic pressures to technological and governance issues is essential for developing targeted strategies that enhance resilience. The integration of technological innovations, adaptive infrastructure, and policy-oriented approaches offers a pathway to mitigate these impacts and promote a more sustainable industry. Moreover, a strong emphasis on mitigation strategies, such as reducing emissions and improving manure management, combined with carbon sequestration techniques, is crucial for reducing the industry's overall environmental footprint. The case studies highlighted in this analysis underscore the importance of adaptability, innovation, and proactive governance in building a resilient aquaculture sector that can withstand both current and future challenges.

References

Abisha, R., Krishnani, K. K., Sukhdhane, K., Verma, A. K., Brahmane, M., & Chadha, N. K. (2022). Sustainable development of climate-resilient aquaculture and culture-based fisheries through adaptation of abiotic stresses: a review. Journal of Water and Climate Change, 13(7), 2671-2689.

Adekola, O., Gatonye, M., & Orina, P. (2022). Status and outlook for climate resilient aquaculture in Kenya: Stakeholders perspectives. Case Studies in the Environment, 6(1), 1544759.

Ahmed, N., & Turchini, G. M. (2021). Recirculating aquaculture systems (RAS): Environmental solution and climate change adaptation. Journal of Cleaner production, 297, 126604.

Atalah, J., Ibañez, S., Aixalà, L., Barber, X., & Sánchez-Jerez, P. (2024). Marine heatwaves in the western Mediterranean: Considerations for coastal aquaculture adaptation. Aquaculture, 588, 740917.

Awotunde, M. (2024). Climate Change Impacts on Aquaculture Systems in Key Producing Nations: Assessing Vulnerability and Proposing a Novel Adaptation Measure. Asian Journal of Fisheries and Aquatic Research, 26(2), 20-36.

Baillargeon, G. A., Wynn, A. A., Baldisimo, J. G. P., Tlusty, M. F., & Rhyne, A. L. (2024). Evaluating Species at Risk in Data-Limited Fisheries: A Comprehensive Productivity-Susceptibility Analysis of the Most Traded Marine Aquarium Fish. bioRxiv, 2024-03.

Berio, M. J. M., & Salugsugan, J. C. P. (2022). Analysis of the Sustainability, Vulnerability, and Challenges of Aquaculture Livelihood Projects. Technium Business and Management, 2(3), 38-62.

Brown-Webb, B., Nesamvuni, A. E., de Bruyn, M., Van Niekerk, J. A., & Pillay, P. (2022). Assessing the impacts of COVID-19 on the aquaculture industry in South Africa, using the marine mussel farming sector as a case study. Technium Soc. Sci. J., 29, 363.

Cahyacipta, S. A., Mulyo, J. H., & Suryantini, A. (2023). Pokdakan Mino Ngremboko's Business Recovery Strategy: Resilience Analysis and Supply Chain Integration. Jurnal Perikanan Universitas Gadjah Mada, 25(1), 57-66.

Castagnino, F., Caillaux, M., & Ramos, E. (2024). Productivity susceptibility analysis in extremely data-poor scenarios: The case of Peruvian coastal groundfish. Fisheries Research, 276, 107044.

Divu, D. N., Mojjada, S. K., Anil, M. K., Gopidas, A. P., Sundaram, S. L. P., Mahalingam, A., ... & Gopalakrishnan, A. (2024). Thermal tolerance and environment adaptability of Indian pompano: Discovery of a resilient candidate species for sustainable mariculture production in a climate change scenario. Aquaculture, 584, 740665.

Gao, M., & Liu, J. (2022). Plant genetic transformation promotes modern agriculture and the food safety of genetically modified plants. Advances in Modern Agriculture, 2(2).

Guddaraddi, A., Goutam, P. K., Sonaniya, P., Saikanth, D., Abirami, S., & Dheer, V. (2023). Role of biological carbon mitigation in carbon sequestration: a review. Int J Environ Clim Chang, 13(11), 749-757.

Habib, A., Borazon, E. Q., Macusi, E. D., Islam, M. M., Nor, S. A. M., Kamaruzzaman, Y. N., ... & Rahman, M. M. (2023). Climate change vulnerability and impacts on tropical fisheries and aquaculture in Asia: A bibliometric and meta-analysis.

Hall, G. M. (2015). Impact of climate change on aquaculture: the need for alternative feed components. Turkish Journal of Fisheries and Aquatic Sciences, 15(3), 569-574.

Hammer, A. J., Millar, C., & Hennige, S. J. (2022). Reducing carbon emissions in aquaculture: Using Carbon Disclosures to identify unbalanced mitigation strategies. Environmental Impact Assessment Review, 96, 106816.

Islam, M. M., & Chuenpagdee, R. (2022). Towards a classification of vulnerability of small-scale fisheries. Environmental Science & Policy, 134, 1-12.

Kumar, A. V., Bharti, V. S., Mukhopadhyay, R., & Krishna, V. H. (2022). Carbon Sequestration in Aquaculture System. In Impact of Climate Change on Hydrological Cycle, Ecosystem, Fisheries and Food Security (pp. 145-150). CRC Press.

Liu, Z., & Liu, Y. (2018). Mitigation of greenhouse gas emissions from animal production. Greenhouse Gases: Science and Technology, 8(4), 627-638.

MacLeod, M., Hasan, M. R., Robb, D. H., & Mamun-Ur-Rashid, M. (2019). Quantifying and mitigating greenhouse gas emissions from global aquaculture. Food and Agriculture Organization of the United Nations.

Macusi, E. D., Geronimo, R. C., & Santos, M. D. (2021). Vulnerability drivers for small pelagics and milkfish aquaculture value chain determined through online participatory approach. Marine Policy, 133, 104710.

Makwana, K., Parmar, H. L., Sikotariya, H., Biswas, S., & Bhadarka, M. (2024a). A review on climate change impacts on marine ecosystems: From oceanic dynamics to biodiversity. AgriGate: An International Multidisciplinary e-Magazine, 4(2), 160-166.

Makwana, K., Parmar, H. L., Sikotariya, H., & Bhadarka, M. (2024b). Climate change's catch: Navigating the impact on fisheries sector sustainability & food security. EIACP: Geodiversity and Impact on Environment, 29(4), 11-15.

Maulu, S., Hasimuna, O. J., Chibesa, M., Bbole, I., Mphande, J., Mwanachingwala, M., ... & Mweetwa, H. M. (2024). Perceived effects of climate change on aquaculture production in Zambia: status, vulnerability factors, and adaptation strategies. Frontiers in Sustainable Food Systems, 8, 1348984.

Mridul, M. M. I., Zeehad, M. S. K., Aziz, D., Salin, K. R., Hurwood, D. A., & Rahi, M. L. (2024). Temperature induced biological alterations in the major carp, Rohu (Labeo rohita): Assessing potential effects of climate change on aquaculture production. Aquaculture Reports, 35, 101954.

Munguti, J., Iteba, J. O., Outa, N., Kirimi, J., Mungai, D., Kyule, D., ... & Ogello, E. O. (2022). Impacts of Covid-19 Pandemic and Adaptive Strategies to Build Resilience in the Kenyan Aquaculture Sector. AfricArXiv Preprints.

Muniz, R. J., & Del Rio, J. A. J. (2023). Impact of the Climate Change on the Production of the Fishing and Aquaculture Sectors of Latin America. International Journal of Energy Economics and Policy, 13(3), 80-91.

Nair, N. V., & Nayak, P. K. (2023). Uncovering water quality and evaluating vulnerabilities of small-scale fisheries in Chilika Lagoon, India. Frontiers in Marine Science, 10, 1087296.

Nazar, F., & Rifqi, M. (2024). Realizing Low-carbon and Climate-resilient Development on Aquaculture. Advances in Research, 25(2), 39-49.

Neokye, E. O., Wang, X., Thakur, K. K., Quijón, P. A., & Nawaz, R. A. (2024). Climate change impacts on oyster aquaculture-Part II: Impact assessment and adaptation measures. Environmental Research, 119535.

Pandey, V. K., Anshu, & Sharma, K. K. (2024). Climate Change Adaptation Policies and Governance: A Critical Appraisal. In Climate Change and Human Adaptation in India: Sustainability and Public Policy (pp. 199-206). Cham: Springer Nature Switzerland.

Paul, P., Rahi, M. F., Saiquit, I. H., Sarower, M. G., Rouf, M. A., Parvez, M. S., ... & Rahman, S. M. (2024). Climate Change Effects on the Small-Scale Fisheries in the Northern Part of Bangladesh and Associated Adaptation Measures. Khulna University Studies, 33-46.

Portilla-Cabrera, C. V., Cifuentes-Ossa, M. A., & Selvaraj, J. J. (2024). Policy-oriented adaptation strategies for artisanal marine fisheries under climate and non-climate stressors in San Andres de Tumaco, Nariño, Colombia. Marine Policy, 161, 105993.

Rajesh, C. M., Jadhav, A., Manohar, K. N., Bhat, P. P., Prasad, R., Anil, K., & Pavan, V. (2024). A Review on Adaptive Strategies for Climate Resilience in Agricultural Extension Services in India. Archives of Current Research International, 24(6), 140-150.

Rashmi, Yadav., Sangeeta, Mookherji. (2024). Advancing sustainable aquaculture through water quality monitoring and technological innovations. 166-177. doi: 10.58532/v3bcag15p2ch3

Silva, S. D. (2013). Climate change impacts: challenges for aquaculture.

Vergara-Solana, F. J., Garza, F. A., & Martinell, D. P. (2024). Aquaculture and climate change. In An Introduction to Sustainable Aquaculture (pp. 71-90). Routledge.

Vögler, R., Soto, D., Quinones, R. A., Martínez, G., D'Ambrosio, L., Alaggia, S., ... & Pérez, F. (2023). Vulnerability assessment of the pink shrimp small-scale fishery to climate change in southwestern Atlantic brackish coastal lagoons (Uruguay). Ocean & Coastal Management, 245, 106864.

Wang, Y., Du, P., Wang, J., Liu, Y., Xu, Z., & Chen, C. (2023). Vulnerability assessment of the fishery system in China's coastal provinces since 2000. Frontiers in Marine Science, 10, 1323033.

Yan, X., Ying, Y., Li, K., Zhang, Q., & Wang, K. (2024). A review of mitigation technologies and management strategies for greenhouse gas and air pollutant emissions in livestock production. Journal of Environmental Management, 352, 120028.

Zhang, S. T., Jiang, M. M., & Chu, H. R. (2024). PLTS/ARAS-based financing risk resilience capability evaluation for fisheries enterprise: A case study of green transformation and upgrading. Journal of Environmental Management, 366, 121785.

5

Pesticides in Water Unveiling the Hidden Threats to Aquatic Life

Bhooleshwari and A.T. Ramachandra Naik

Department of Aquatic Environment Management, Karnataka Veterinary, Animal and Fisheries Sciences University, College of Fisheries, Mangaluru–575 002, Karnataka

Abstract

Pesticides, which are extensively employed in agriculture and pest management, pose serious dangers to the environment and public health because they frequently enter aquatic ecosystems through runoff, leaching, and inappropriate disposal. Acute and chronic toxicity in aquatic creatures caused by pesticide poisoning in water bodies can disturb ecosystem services and biodiversity. As sentinel species, fish have immune-toxic consequences, such as impaired immune function and increased vulnerability to illness. Aquatic animals and human health are at risk due to the bioaccumulation of pesticides in fish tissues and their subsequent biomagnification in the food chain. In order to control pesticide exposure and guarantee safety, important parameters such as Maximum Residual Limits (MRLs) and Acceptable Daily Intake (ADI) are essential. In order to control pesticide, use and lessen its negative effects, government regulations and international standards are essential. In order to protect aquatic ecosystems and human health, future research should give priority to environmentally suitable pesticide substitutes and strong regulatory frameworks.

Keywords: *Pesticides, Toxicity, Aquatic Environment, Bioaccumulation and Biomagnification.*

Introduction

A misguided attempt to boost productivity in response to rising population levels and food demands, the over use of pesticides in agriculture is causing dangerously widespread devastation of aquatic habitats. The frequency and intensity of pesticide treatments have increased significantly to support agricultural production and profitability, with usage rising by 15–25% globally. Crop losses have persistently stayed the same over the previous

40 years despite this concerning trend. In order to preserve our priceless ecosystems and guarantee adequate supply of food in the face of growing demands for food, feed, fiber, biofuels and other bio-based goods, we must embrace environmentally conscious farming techniques. Every stage of a pesticide's lifespan, which includes production, authorization, transportation, retention, implementation and elimination, has the potential to endanger the environment and public health. While agricultural pesticides are essential for protecting crops from dangerous pests, pesticides intended for health-conscious uses are essential for vector control, household use and commercial pest management (WHO, 2019). However, one of the biggest worries about the use and spread of pesticides is the harmful consequences they have on people who are exposed to them. It is crucial to comprehend maximum residue limits (MRL), which specify the permissible concentrations of residues in food and feed, because these compounds frequently leave behind hazardous residues on treated agricultural goods. The most widely used pesticides, such as pyrethroids, carbamates and organophosphates, illustrate the scope of this problem, claims Gilbert (2012). With around four million tonnes of pesticides used annually worldwide, mostly herbicides (56%) followed by fungicides (25%) and insecticides (19%), there has never been a greater need for ethical application and more stringent controls. We can shield our ecosystem and our wellness from the unexpected effects of pesticide use by emphasizing ethical procedures.

The manufacturing and Use of Pesticides Throughout Chronology

Three periods can be distinguished in the history of pesticides: (1) the history of the use of natural pesticides, such as sulfur in ancient Greece, prior to the 1870s; (2) the time of inorganic synthetic pesticides, which lasted from the 1870s until 1945. The utilization of natural materials and inorganic compounds was predominant during this time; (3) the third phase, which began in 1945, is known as the organic-synthetic pesticide era. Man-made organic pesticides, such as Dichlorodiphenyltrichloroethane (DDT), 2,4-Dichlorophenoxyacetic acid (2,4-D) and later Hexachlorocyclohexane (HCH) and dieldrin, have been produced by humans since 1945 and are referred to as chemical pesticides. Over time, the use of pesticides has steadily risen. Between 2000 and 2021, it increased significantly, from 434 tonnes to 62193 tonnes (GOI, 2021). In comparison to Japan (12 kg), China (12 kg), the United Kingdom (7 kg) and Germany (3 kg), India consumes a very small amount of pesticides per acre on a worldwide scale (600g) (Parajuli *et al.*, 2021). The largest user of pesticides in India is Maharashtra, which consumed 13,243 tons of pesticides in 2020–2021 or 21.2% of the total amount consumed. Jammu & Kashmir,

West Bengal, Telangana, Haryana, Uttar Pradesh and Punjab are some of the other top consumer states (GOI, 2021).

Pesticides Poisoning in India

A few particularly notable cases of accidental pesticide poisoning have occurred. Pesticides are accountable even though they account for a small but significant percentage of acute human poisonings. Particular attention should be focused on an assortment of instances associated with accidental pesticide poisoning.

Table 1. Chart illustrating India's history of widespread poisoning of pesticides

Place	Year	Affected number of people	Deaths	Causative agents	Sources
Kerala	1958	360	100	Parathion	Karunakaran, 1958
Madhya Pradesh	1967	35	5	Malathion	Sethuraman, 1977
Madhya Pradesh	1967-68	12	12	Aldrin and HCH	Gupta, 1975
Uttar Pradesh	1977	8	-	HCH	Nag *et al.*, 1977
Uttar Pradesh	1978	250	4	HCH	Nag *et al.*, 1977
Madhya Pradesh	1978	-	6	Phosgene	Gupta, 1986
Madhya Pradesh	1984	250000	8000	Methyl isocyanate	Pandey, 1986; Jain and Dave, 1986
Uttar Pradesh	1992	8	6	Aluminum phosphide	No documented report available

More than 100 people died after consuming wheat flour contaminated with parathion in Kerala in 1958, marking the first pesticide poisoning in India (Karunakaran, 1958). Ethyl parathion, marketed by Bayer under the name Folidol E 605, was the substance utilized. In Kerala, poisoning killed 102 individuals in the same year. The primary cause of this was the negligent handling and storage of wheat. Since then, there have been several reports of poisonings in humans and animals in addition to fish and bird deaths (Sethuraman, 1977, Banerjee, 1979). In a different outbreak in 1977, eight major mass strokes were documented from an Uttar Pradesh hamlet after the inadvertent consumption of wheat tainted with HCH (Nag *et al.*, 1977). There have occasionally been reports of multiple toxtin instances in humans, cows, buffaloes and heifer calves. According to Gupta (1986), exposure to phosgene gas caused the deaths of six people in Bhopal in 1978. There has never been a disaster like the Bhopal gas tragedy in the history of the industry. Early on December 3, 1984, a deadly gray mist was pushed past the Union Carbide plant wall in Bhopal, Madhya Pradesh, India, by a rolling wind. The Union Carbide Bhopal plant emitted forty tonnes of hazardous vapors, which dispersed over the city. The lethal cloud was gradually dispersed over 40 km^2 by a mild wind, resulting in a massive loss of life. Nearly 15,000 animals sustained the harmful gas effect, at least 8,000 people died and more than 4,000 animals died within minutes of being exposed to the gas. Pulmonary edema was one of the side effects that led to the majority of deaths, with subsequent respiratory infections such bronchitis and bronchial pneumonia accounting for a large number of deaths (Pandey, 1986, Jain and Dave, 1986).

Pesticide Contamination in Aquatic Environment

Pesticides in riverine systems have grown to be a significant issue in India. A significant amount of India's fertile agricultural land is located next to rivers. As a result, the likelihood of agricultural runoff contaminating the waterways is extremely high. Significant riverine systems, such the Ganges and its tributaries, like the Yamuna, have been found to be seriously contaminated (Asim *et al.*, 2021). In another Ganga tributary, drinking water contains pesticides such as DDT, endosulfan, endrin, aldrin, dieldrin and heptachlor. Lindane, endrin, heptachlor epoxides and DDT are examples of the accumulated sediments in the riverbed. (Malik *et al.*, 2009). Studies on a number of Indian rivers, such as the Hindon River in Uttar Pradesh, the Cauvery River in Karnataka, and the Ghaggar River in Haryana, have consistently shown a similar trend of contamination with pesticides. (Kaushik *et al.*, 2010; Agrawal *et al.*, 2015). DDT, HCH and 15 other organochlorine pesticide residue levels were found to be over allowed limits in the southern Indian perennial Tamiraparani river basin, according to an investigation into contamination (Durairaj *et al.*, 2021; Raymond *et al.*, 2021). Surface water and sediment samples from the Tapi River contained endosulfan, chlorpyrifos and methyl parathion. In fish samples, the amounts of endosulfan, chlorpyrifos and methyl parathion were determined to be 101.28, 0.392 and 3.49 ng/g, respectively. It indicates that the region continues to use exceedingly harmful pesticides and stringent enforcement of laws limiting their production and use is necessary (Hashmi *et al.*, 2020). An investigation into the Hooghly River basin in West Bengal, India, revealed that insecticides were more prevalent in the river water samples than fungicides and herbicides. While organochlorine pesticides were prevalent in the nearby agricultural sectors, most river samples exhibited pesticide levels above the European drinking water standards (Mondal *et al.*, 2018).

What are Pesticides?

An important agricultural tool for preventing or managing pests, such as undesirable plants and animals, is a pesticide. It aids in safeguarding our food and agricultural goods as they are being produced, stored, and processed. Furthermore, pesticides are essential for maintaining healthy ecosystems and food security since they efficiently control external parasites in animals. According to Dhaliwal *et al.* (2015), the pesticides are compounds, either natural or artificial, employed in a variety of agronomic activities to control plant diseases, weeds and pests. Pesticides include insecticides, fungicides, herbicides, rodenticides, nematicides and other substances. Pest control products are described using a variety of characteristics, such as their toxicity, the pest species they kill and how they work as pesticides, their chemical makeup and entry point, their mode of action, how or when they work, compositions and their origins (Tudi *et al*., 2021).

Classification of Pesticides (Source: Nayak and Solanki, 2021)

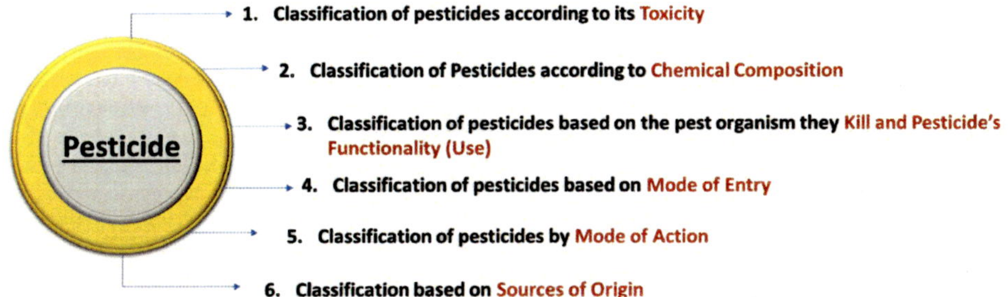

Pesticides are categorized based on their toxicity

The two main determinants of pesticide toxicity are dose and time. Acute and chronic toxicity are the two types of toxicity that result from the quantity of this chemical (dose) and the frequency (time) of exposure.

Table 1. Pesticides classification on the basis of its toxicity.

WHO Class		**LD50 for rats (mg/kg of body weight)**	
		Oral	**Dermal**
Class-la	Extremely Hazardous	Less than 5	Less than 5
Class-1	Highly Hazardous	5 to 50	5 to 200
Class-II	Moderately Hazardous	50 to 200	200 to 2000
Class-III	Slightly Hazardous	Over 2000	Over 2000
Class-V	Unlikely to present acute hazard	5000 or higher	

(*Source*: WHO 2019)

Pesticide Categorization Based on Chemical Structure

This approach is particularly popular and useful way to group pesticides based on their chemical structure. Insecticides, fungicides, herbicides and rodenticides are among the pesticides that are categorized according to their chemical nature.

Table 2. Pesticides classification on the basis of its chemical composition.

Pesticides	Chemical Composition	Pesticides	Chemical Composition
Insecticides	1. Carbamates (Carbaryl), 2. Organochlorine (Endosulfan), 3. Organophosphorus (Monocrotophos), 4. Pyrethroids (Permethrin), 5. Neonicotinoids (Imidacloprid), 6. Spinosyns (Spinosad), 7. Benzolureas (Dilubenzuron), 8. Antibiotics (Abamectin)	**Fungicides**	1 Aliphatic nitrogen fungicides (dodine), 2 Amide fungicides (carpropamid), 3 Aromatic fungicides (chlorothalonil), Dicarboximide fungicides (famoxadone), 4 Dinitrophenol fungicides (dinocap), and others.
Herbicides	1. Anilide herbicides (lufenacet), 2. Phenoxyacetic herbicides (2, 4-D), 3. Quaternary ammonium herbicides (Paraquat), 4. Chlorotriazine herbicides (atrazine), 5. Sulfonylurea herbicides (chlorimuron), and others.	**Rodenticides**	1. Inorganic rodenticides (Zinc phosphide, Aluminium Phosphide) 2. Organic coumarin rodenticides (bromadiolone, coumatetralyl).

Pesticides are Categorized According to the Pest Organisms they Eliminate and the Effectiveness with which they Work (Use)

This is how pesticides are classified according to the pest organisms they destroy and how they work. Below is a list of many pesticide types:

Table 3. Categorization of pesticides according to the pest organisms they destroy and how they function.

Pesticides	Function
Insecticides	Chemicals that are used to kill insects and other arthropods.
Fungicides	Chemicals that kill fungi.
Acaricides	Pesticides that kill mites and ticks.
Algicides	Chemicals that kill or suppress algae.
Herbicides	Chemicals that are used to kill undesired plants.
Antifeedants	Chemicals that stop insects and other pests from eating.
Avicides	Poisonous chemicals used to kill birds.
Larvicides	Stop larvae from growing.
Bactericides	Substances that kill or inhibit bacteria.
Repellents	Substances that repel bugs based on their taste or odour.
Virucides	Antiviral agents.

Ovicides	Inhibits the growth of insect and mite eggs.
Nematicides	Chemicals that kill nematodes, which are plant parasites.
Termiticides	Chemicals that kill termites.
Chemosterillants	Chemicals that make an insect sterile and hence prevent it from reproducing.

Sorting pesticides according to their route of entry

Pesticide routes of administration are the various ways that pesticides come into contact with or penetrate the target.

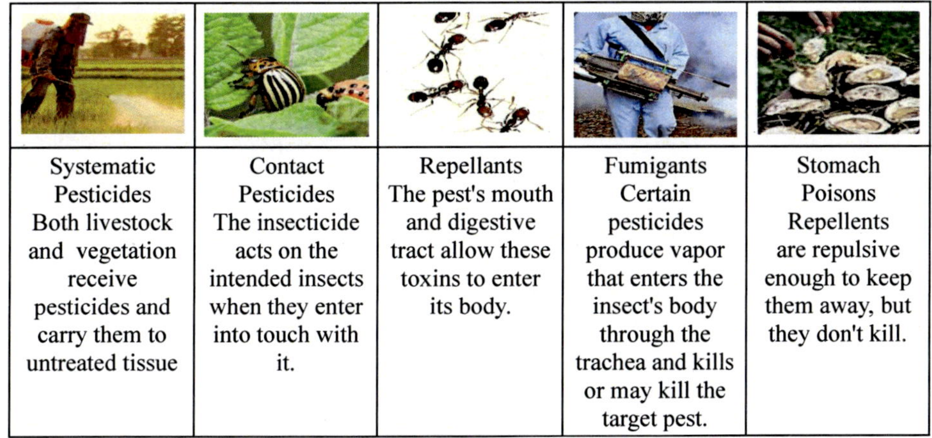

Systematic Pesticides	Contact Pesticides	Repellants	Fumigants	Stomach Poisons
Both livestock and vegetation receive pesticides and carry them to untreated tissue	The insecticide acts on the intended insects when they enter into touch with it.	The pest's mouth and digestive tract allow these toxins to enter its body.	Certain pesticides produce vapor that enters the insect's body through the trachea and kills or may kill the target pest.	Repellents are repulsive enough to keep them away, but they don't kill.

Fig. 1. Pesticides classification on the basis of its mode of entry.

Categorization According to Origin

Depending on where they come from, pesticides are classified as either conventional or environmentally friendly pesticides. Bio-pesticides (bacteria, viruses, fungi and nematodes) are pesticides that come from natural sources like plants, animals and microorganisms and chemical pesticides (organophosphate, organochlorine, carbamate and pyrethroid). Inorganic pesticides include calcium arsenate, copper sulfate and others (Hext, 1973; Wani, 2018).

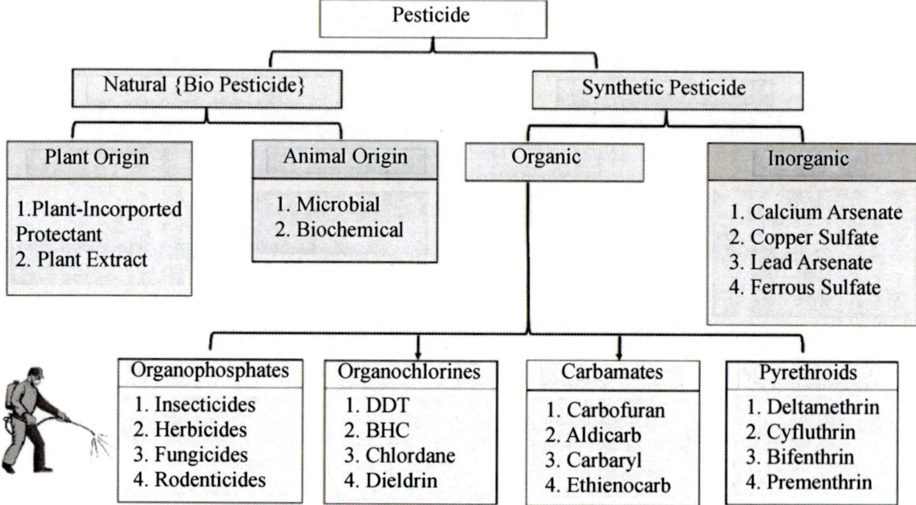

Fig. 2. Pesticides classification on the basis of its sources of origin.

Pathway of Contamination

Sources: The types of pesticides used vary greatly, and there are many ways for them to circulate into lakes, rivers, streams, and oceans. Runoff from farmland, leaching, soil erosion, aerosol drift, atmosphere fallout and domestic and commercial sewage, careless empty container dumping, equipment washing, and ground infiltration are some of the ways that pesticides may diffuse into aquatic environments. Pesticide pollution threatens over 64% of agricultural land worldwide, with 31% at significant risk (Kadiru *et al.*, 2022).

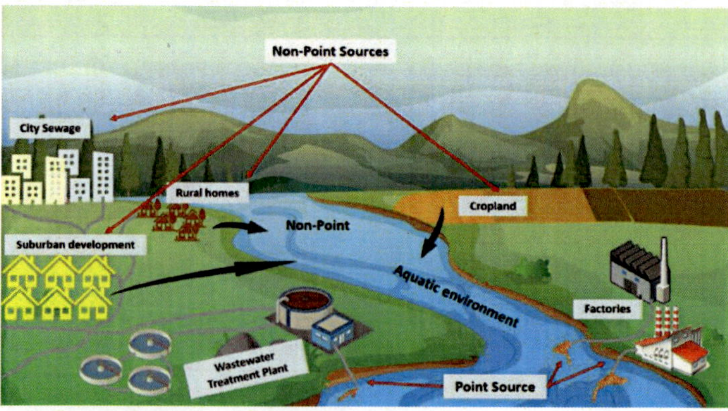

Fig. 3. Sources of the pesticide.

Numerous environmental compartments have become contaminated as a result of their extensive use, including:

Soil Contamination

- **Leaching:** Pesticides can leach into the soil, contaminating groundwater and affecting plant health.
- **Persistence:** Some pesticides are persistent in the soil, remaining active for extended periods.
- **Bioaccumulation:** Pesticides can accumulate in the soil, posing risks to organisms that ingest it.

Water Contamination

- **Runoff:** Pesticides can be carried into water bodies through runoff from agricultural fields.
- **Groundwater contamination:** Sources of drinking water may be impacted by pesticides that seep into groundwater.
- **Aquatic ecosystems:** Pesticide contamination can harm aquatic organisms, including fish, invertebrates, and amphibians.

Air Contamination

- **Volatilization:** Some pesticides can volatilize into the air, contributing to atmospheric pollution.
- **Long-range transport:** The atmosphere has the capacity to carry pesticides over great distances, affecting ecosystems far from their application sites.
- **Bioaccumulation:** Pesticides can accumulate in the atmosphere and eventually return to the earth's surface, contaminating soil and water.

Factors Affecting Pesticide

- **Pesticide properties:** The physical and chemical properties of pesticides, such as solubility, volatility and persistence, influence their movement and distribution in the environment.
- **Climate:** Rainfall, temperature and wind patterns can affect pesticide transport and degradation.
- **Soil characteristics:** Pesticide leaching and persistence can be influenced by soil quality, pH, and organic matter content.
- **Land use:** Agricultural practices, such as tillage and irrigation, can affect pesticide runoff and erosion.

Monitoring and Assessment

- **Sampling:** Collecting samples of soil, water and air to assess pesticide levels.

- **Analytical techniques:** Using laboratory methods to detect and quantify pesticide residues. (Add detection techniques)
- **Risk assessment:** Evaluating the potential risks associated with pesticide contamination, including human health and ecological impacts.

Mitigation Strategies

- **Integrated Pest Management (IPM):** Implementing sustainable practices that minimize pesticide use while maintaining crop yields.
- **Best management practices:** Adopting practices that reduce pesticide runoff and erosion, such as cover crops and no-till agriculture.
- **Buffer strips:** Establishing buffer zones between agricultural fields and water bodies to prevent pesticide contamination.
- **Pesticide regulations:** Enforcing strict regulations to control the use and disposal of pesticides.

Pesticides Residue

The chemical residues in the food might provide either immediate or prolonged health risks, depending on the dosage. The potential for health risks is determined by the amount of toxin consumed per kilogram of body weight. "Acceptable Daily Intake" (ADI) refers to the highest level of pesticide residues that can be consumed every day for the rest of one's life without posing a significant risk. The ADI is based on a scientific assessment of all available information at the exact moment of evaluation to establish a threshold below which no adverse effects are anticipated. Milligrams of substance per kilogram of body weight is the ADI's unit.

"Any chemical or combination of chemicals found in meals for individual or organisms caused by the use of a pesticide and includes any specified derivatives, such as degradation and conversion products, metabolites, reaction products, and impurities that are considered to be of toxicological significance," is how the World Health Organization (WHO) defines a pesticide residue (Sharma & Choudhury, 2018).

Acceptable Daily Intakes (ADI)

The Food and Agriculture Organization (FAO) and the World Health Organization (WHO) have established tolerable daily intakes of pesticide residues, which are the highest amounts of a chemical that can be consumed in a lifetime and seem to pose no significant risks. Toxicological, pharmacological and biochemical data on experimental animals as well as any available information on individuals form the basis of the decision. ADIs are typically estimated to be 1% of the most sensitive animal species' no-effect level. The

WHO Core Assessment Group on Pesticide Residues evaluates toxicological data to determine acute reference doses (ARfDs) and acceptable daily intakes (ADIs), which is essential for protecting the general population's health. As part of the Codex Committee on Pesticide Residues (CCPR) periodic review program, the team recently analyzed 15 pesticides, including two that had been re-evaluated and five new compounds (FAO, 2021). Notably, maternal toxicity results from a rat developmental study led the European Union to lower its ADI and ARfD to 0.005 mg/kg body weight (bw) per day in 2018. A dedication to safety is evident in this modification, as the previous levels were 0.125 mg/kg bw and 0.006 mg/kg bw daily. Based on a NOAEL (no-observed-adverse-effect threshold) of 1.1 mg/kg bw from a one-year dog study and an ARfD of 0.1 mg/kg bw from an acute neurotoxicity investigation, JMPR established an ADI of 0–0.01 mg/kg bw in 2005. These assessments demonstrate the dedication to pesticide safety (FAO, 2023). Despite their frequent misuse and detrimental effects on the environment, pesticides remain essential instruments for raising agricultural output. However, their extensive environmental distribution can jeopardize human health and constitute a major hazard to non-target creatures. The overwhelming majority of these chemicals contribute to environmental contamination and it is believed that just around 0.1% of them achieve their intended goals (Gill and Garg, 2014). Some pesticides and their acceptable daily intake and maximum residual limits are given below in the table 5.

Table 5. A few pesticides, along with the maximum residual limits and tolerable daily consumption.

AMOXICILLIN (antimicrobial agent)			
Microbiological acceptable daily intake	0–0.002 mg/kg body weight (bw)/day, depending on how amoxicillin affects the gut bacteria.		
Acute reference dose	0.005 mg/kg bw/day according to gastrointestinal microbial impacts		
Estimated chronic dietary exposure	0.14 μg/kg bw daily (for the general population), which is equivalent to 7% of the mADI's upper limit.		
Estimated acute dietary exposure	This represents 1.4 μg/kg bw/day, or 28% of the microbiological ARfD, for the overall population. For children, 1.6 μg/kg bw/day represents 31% of the microbiological ARfD.		
Species	**Tissue**	**MRL (μg/kg)**	**Note**
Finfish	Fillet	50	All fish species are referred to be "finfish" because of their natural proportion of muscle to skin.
	Muscle	50	All fish species are included in the word "finfish."
AMPICILLIN (antimicrobial agent)			

Microbiological acceptable daily intake	0–0.003 mg/kg bw/day		
Acute reference dose	According to the microbiological end-point, 0.012 mg/kg bw/day		
Estimated chronic dietary exposure	0.29 µg/kg bw per day (for the general population), which represents 10% of the upper bound of the ADI.		
Estimated acute dietary exposure	16% of the ARfD, or 1.9 µg/kg bw daily (for the general population). Children's daily dose of 1.7 µg/kg bw, or 14% of the ARfD		
Species	**Tissue**	**MRL (µg/kg)**	**Note**
Finfish	Fillet	50	All fish species are referred to be "finfish" because of their natural proportion of muscle to skin.
	Muscle	50	All fish species are included in the word "finfish."
CHLORTETRACYCLINE/OXYTETRACYCLINE/TETRACYCLINE (antimicrobial agent)			
Acceptable daily intake	16% of the ARfD, or 1.9 µg/kg bw daily (for the general population). Children's daily dose of 1.7 µg/kg bw, or 14% of the ARfD		
Species	**Tissue**	**MRL (µg/kg)**	**Note**
Finfish	Muscle	200	Applies only to oxytetracycline
Giant prawn (*Paeneus monodon*)	Muscle	200	Applies only to oxytetracycline
DELTAMETHRIN (insecticide)			
Acceptable daily intake	0–10 µg/kg bw/day		
Species	**Tissue**	**MRL (µg/kg)**	**Note**
Salmon	Muscle	30	
DIFLUBENZURON (insecticide)			
Microbiological acceptable daily intake	0–0.02 mg/kg bw and 2 mg/kg bw per day		
Acute reference dose	The immediate oral toxicity is negligible, and a single dose is unlikely to result in brain damage or any other neurotoxic effects.		
Estimated chronic dietary exposure	As an average person, it is 0.84 µg/kg bw daily, or 4% of the upper bound of the ADI; for children, it is 2.85 µg/kg bw daily, or 14% of the highest bound of the ADI.		
Species	**Tissue**	**MRL (µg/kg)**	**Note**
Salmon	Muscle plus skin in natural proportions	10	

EMAMECTIN BENZOATE (antiparasitic agent)			
Microbiological acceptable daily intake	ADI of 0–0.5 μg/kg bw/day		
Estimated acute dietary exposure	11 micrograms per person per day, or roughly 37% of the ADI's upper limit		
Species	**Tissue**	**MRL (μg/kg)**	**Note**
Salmon	Fillet	100	Skin and muscle in their normal ratio
Salmon	Muscle	100	Skin and muscle in their normal ratio
Trout	Fillet	100	Skin and muscle in their normal ratio
Trout	Muscle	100	Skin and muscle in their normal ratio
FLUMEQUINE (antimicrobial agent)			
Acceptable daily intake	0–30 μg/kg bw/day		
Species	**Tissue**	**MRL (μg/kg)**	**Note**
Trout	Muscle	500	Muscle, with the appropriate amount of skin
LUFENURON (insecticide)			
Microbiological acceptable daily intake	0–0.02 mg/kg bw/day		
Acute reference dose	Superfluous, considering that lufenuron has minimal immediate oral toxicities and no developmental toxicity or other hazardous consequences that could be triggered by a single dose.		
Estimated chronic dietary exposure	5.5% of the upper bound of the ADI, or 1.1 μg/kg bw daily (for the overall public).		
Species	**Tissue**	**MRL (μg/kg)**	**Note**
Salmon	Fillet	1350	Skin and muscle in their normal ratio
Trout	Fillet	1350	Skin and muscle in their normal ratio
LUFENURON (insecticide)			
Acceptable daily intake	0–5 μg/kg bw/day		
Estimated chronic dietary exposure	Based on an individual weighing 60 kg, the EDI is 42.9 μg/person per day, or almost 14% of the ADI's upper limit.		
Species	**Tissue**	**MRL (μg/kg)**	**Note**
Salmon	Fillet	400	Skin and muscle in their normal ratio
Salmon	Muscle	400	Skin and muscle in their normal ratio

Transport and Fate

Analyzing the movement and fate of pesticides in aquatic settings requires consideration of the various aquatic environment types as well as the diverse spectrum of pesticide forms. Coastal and deep ocean waters, rivers, lakes, streams, estuaries and other unique environmental features are all considered aquatic ecosystems. Organophosphates, organochlorine, carbamates, juvenile hormones, and viruses are only a few of the chemical and biological types of pesticides. In the many aquatic ecosystems, these varied chemicals and natural materials may interact in different ways. The physical and chemical contrasts between freshwater and saltwater systems make this particularly true. As illustrated in fig., concentration, dilution and degradation are the three primary parameters that can impact the transport and destiny of pesticides in the aquatic environment (Ketchum, 1967).

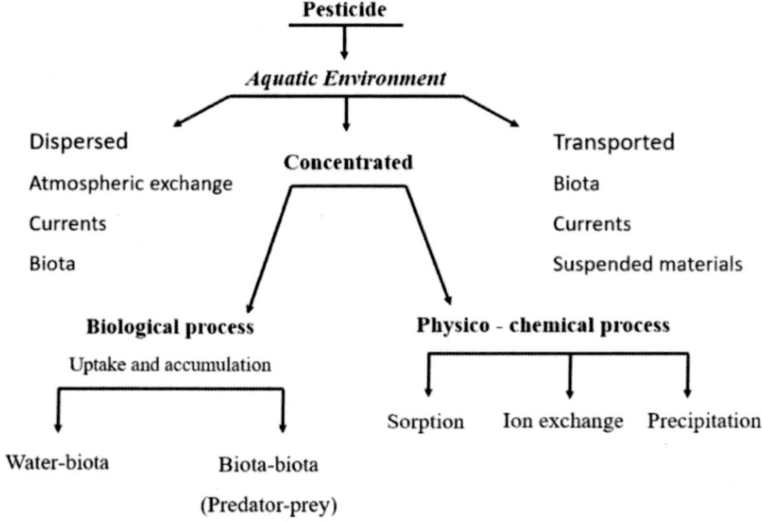

Fig. 4. Movement of Pesticides in the aquatic environment.

A specific kind of bioaccumulation in a particular environmental component is concentration. The chemicals may be transported from organism to higher trophic levels via the food chain or they may be concentrated directly by the organism from the water. Certain pesticides, like DDT levels in fish-eating birds, are acquired through predator-prey relationships. Biogeochemical processes can potentially concentrate pesticides in the environment. It is common for chemicals to be extracted from water via sorption onto suspended particles, which are subsequently placed on the bottom to form the bottom substrate. Chemical processes in marine waters have the potential to precipitate pesticides from the water. At the mouth of an estuary, when freshwater stream water meets saltwater from the sea, this can lead to the buildup of pesticides. Dilution can happen when a pesticide spreads from a point source through biogeochemical processes in an environment. Through the migration of creatures that have concentrated the chemical from their inshore environment, pesticide concentrations can be transferred from estuaries to the open ocean. Additionally, pesticides accumulated by suspended materials may be carried to different locations by erosion and air exchange from water to the atmosphere (Khan, 2013).

Pesticide Bioaccumulation and Biomagnification in Fish

Bioaccumulation is linked to both the environmental persistence of substances and the logarithmic octanol-water partition coefficient (log K_{ow}), which indirectly assesses lipid solubility. Pesticides are highly likely to concentrate on aquatic creatures if their log K_{ow} is greater than three and their half-life in soil is longer than thirty days (Ernst *et al.*, 2018; Rossi *et al.*, 2020). As they go through the food chain, pesticides often associate with lipids and show biomagnification. Higher trophic level organisms, including humans, may experience higher concentrations of some chemicals, with levels in their bodies reaching magnitudes that are roughly 10 million times greater than those in the surrounding water (Osuala *et al.*, 2020; Xie *et al.*, 2022). According to Rossi *et al.* (2020), Glyphosate, bifenthrin, azoxystrobin and cyproconazole are among the pesticides used in the cultivation of rice that are in fact building up in fish and affecting their physiology, biochemistry and ultimately their health. It was also mentioned that because of the way these chemicals bioaccumulate in fish, the combination of pesticides may be harmful to human health through the food chain (Ghosh & Sarower, 2024).

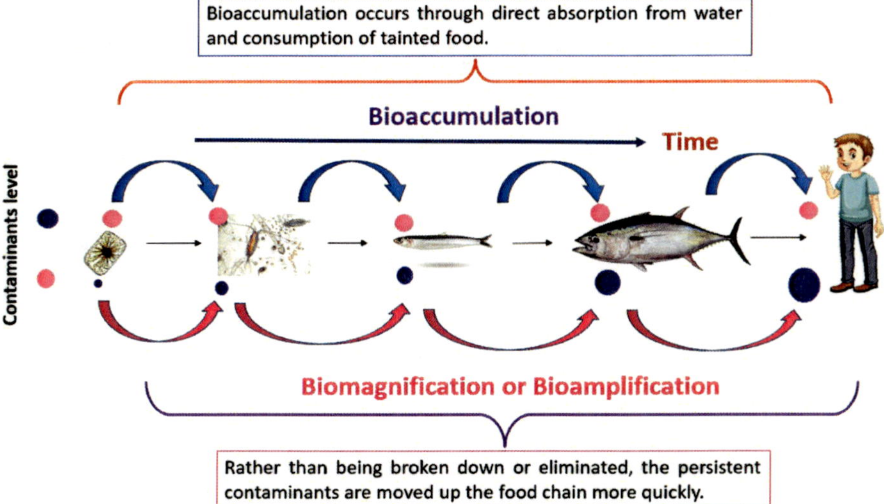

Fig. 5. A diagram showing the bioaccumulation and biomagnification of pesticides in fish. Within an organism, bioconcentration and bioaccumulation take place, but biomagnification happens at different food chain levels.

Pesticide-Induced Immunotoxin Consequences in Fish

Environmental exposure to pesticides from the agriculture industry affects the physiology of many aquatic species, including fish, algae, shellfish and mammals. Chemicals like pesticide are subsequently passed to higher trophic levels in the food chain when they absorb in aquatic environments. Due to their location at the top of the aquatic food chain, fish are frequently found to exhibit biomagnification of pesticides. Fish immune responses can be weakened by pesticides discharged into water, making them more vulnerable to diseases like bacteria, viruses and metazoans. This can lead to environmental destruction. Fish are commonly employed as biological models to investigate how external stimuli impact the immune response, since they are the first evolutionary group to possess both an innate and an adaptive immune system. Analyzing the immunotoxicity of pesticides involves looking at a variety of factors, including immunoglobulin level s, immune cell changes and cytokine production (Yang *et al*., 2021). Variables including photoperiod, temperature and oxygen concentration affect fish immunological parameters like haematology, lysozyme levels and immunoglobulin M (IgM) concentration (de Souza *et al*., 2016; Yang *et al*., 2021). The immune-toxicological effects of pesticides on aquatic species, especially fish, have served as the subject of numerous research (Ghosh & Sarower, 2024).

Toxicity of Pesticides in Fish

Both mammalian and aquatic organisms have been shown to have pesticide traces. Regulatory agencies and organizations have set maximum residue levels (MRLs) for each pesticide to guarantee the safety of food and feed consumption (Crépet *et al*., 2021). Pesticide poisoning adversely affects the normal functions of vital organs in fish, including the gills, liver, kidneys and digestive tract, leading to physiological disruptions (Khatib *et al*., 2022). According to Pourgholum *et al*. (2013), fish oxygenation and osmoregulation are hampered by pesticide toxicity, which also results in histological alterations in the gill tissues, including hyperemia, hyperplasia, necrosis, membrane separation and lamellar fusion. Pesticides can enter the bodies of aquatic species, like fish, primarily through the skin, gills and mouth.

Pesticides can enter fish bodies through three primary pathways. Firstly, they are absorbed directly through the skin when fish swim in contaminated waters. Secondly, the gills facilitate pesticide uptake during the respiration process. Lastly, pesticides are ingested through the mouth, either by consuming contaminated water or prey, a process referred to as “secondary poisoning” (Poopak *et al*., 2023).

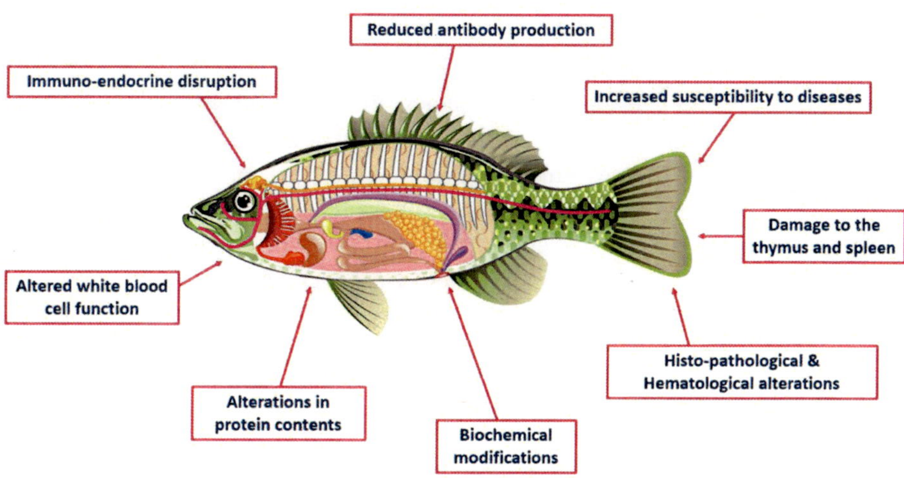

Fig. 6. A diagram showing an Illustration depicting the impact of Pesticides on Fish Immunity.

Toxicity of Pesticides in Human Health

In addition to contaminating food, excessive and uncontrolled pesticide use also polluted the aquaculture, farming and the surroundings (Cech *et al*., 2023). traces of pesticides can be found in fruits, vegetables, processed foods, water, air, and soil. Particularly in developing nations, food exposure and the acute and long-term health impacts of agricultural pesticides pose major public health risks. Chemical pesticides have the potential to be mutagenic, cytotoxic, and carcinogenic to humans. Pesticides usually kill or damage organisms other than pests, including people, due to their non-species-specific method of action (Fang *et al*., 2020).

A report by WHO and the United Nations Environment Programme (UNEP) highlights that pesticide exposure poisons three million people annually, with 200,000 fatalities, primarily in developing countries (Boedeker *et al*., 2020). Pesticides generate reactive oxygen species (ROS), which reduce antioxidant levels, reducing their capacity to defend cells against oxidative damage. This imbalance affects cellular signaling pathways by damaging proteins, lipids and nucleic acids, leading to oxidative stress and ROS's for years negative health effects. The widespread, imprecise use of pesticides results in various adverse health outcomes, ranging from acute poisoning to chronic illnesses (Ahmad *et al*., 2024). Additionally, food products, particularly vegetables, frequently have elevated amounts of pesticide residues from organophosphates. Considering dietary supplements are usually only evaluated on a limited sample, the risk of pesticide contamination continues to pose a direct and indirect threat to human health (Kadiru *et al*., 2022).

Toxicity of Pesticides on Environment

The degree to which various pesticides bioaccumulate in aquatic organisms is determined by their polarity and water solubility. The solubility of pesticide compounds in water and their bioaccumulation in fish are inversely correlated. As the pesticide becomes more soluble in water, the amount of concentration diminishes. Water solubility is therefore a crucial factor in reducing the dynamics of pesticides in aquatic settings. It seems that the pace at which pesticides are eliminated or desorbable varies by species. The rate of absorption and rate of elimination reactions define the amount of pesticides present in a species. The spread of pesticide residues in the aquatic ecosystem has been the subject of extensive research and documentation. On the other hand, higher and more humid temperatures may be anticipated in temperate zones, which could lead to additional prospects for aquaculture cultivation. With proper pollution control methods for nutritional recycle, wastewater purification and infiltration loss management, up to 60% of the water extracted for inland aquaculture might be recycled. Contamination and wastewater discharge must be managed sustainably; sub-regions with few or no freshwater operational control mechanisms need particular consideration (Stentiford *et al.*, 2020). The wellness of humans, wildlife and the natural world may all suffer as a result. The usage of pesticides can have a direct or indirect impact on organisms:

1. By upsetting food chains and upsetting an ecosystem's natural equilibrium, pesticides have an indirect impact on the environment and non-targeted organisms. Herbicides, for instance, will limit the amount of weeds and their seeds that bird chicks may access, which frequently results in their starvation. Herbicides may also make it harder for predators to find suitable habitat, which could lead to a decrease in predators and an outbreak of pests.
2. Direct contact with pesticides can also have an impact on non-target organisms and predators. Direct impacts on people include cancer, infertility, poisonings, and even neurological issues in organophosphate users. The effects of pesticides include:

Effect on Microorganisms

The microorganisms in aquatic environments, such as rivers and estuaries, can be impacted by pesticides through shifts accidents and runoff from farms. The form and function of microbial communities can be affected by pesticide toxicity. Pesticides can persist in the environment or be broken down by microbes. Various pesticide poisoning processes exist in microorganisms depending on the type of toxin and the microbial species exposed to it. The quantity of nitrogen-fixing bacteria is impacted by excessive pesticide use, which has a negative effect on soil quality. Soil microbial respiration was reduced when conventional concentrations of the pyrethroid insecticide λ-cyhalothrin and the organophosphate pesticides chlorpyrifos, phosalone and dimethoate were applied of them, chlorpyrifos had the highest levels of toxicity and persistence (Kapun *et al.*, 2021)

Effect on Aquatic Invertebrates

In both municipal and farming environments, pesticides are used to manage a range of pests, including plants, fungus, algae, mammals like rats and invertebrates like insects. They can influence the target organism in a number of ways, but they frequently have a detrimental effect on non-target organisms through direct toxicity or secondary or indirect effects on food sources or surroundings. A pesticide may be carried far from its application site by drainage, shifts and spillage.

Effect on Aquatic Vertebrates

Numerous studies in India have examined the effects of commonly used pesticides on aquatic vertebrates. Chlorpyrifos, an organophosphate pesticide, is widely applied in agriculture (Yen *et al.*, 2011), while bifenthrin is used in both cultivation and public health campaigns, such as mosquito control. Research investigating the genotoxic effects of a pesticide combination—bifenthrin and chlorpyrifos—on the major Indian carp *Labeo rohita* revealed that DNA damage increased in a time-dependent manner for up to 56 days following exposure to a sub-lethal concentration (33% LC50). This was followed by a slight reduction over the subsequent 14 days. Additionally, chlorfenapyr, a halogenated pyrrole-based pesticide, is employed in agriculture to control insects and mites (Bano *et al.*, 2021). In several nations, acetamiprid is widely used in agriculture to control pests. A broad-ranging an organophosphate pesticide, dimethoate is applied to combat a variety of pests and mites (Tarbah *et al.*, 2007). The outcomes of a study comparing the effects of acetamiprid, dimethoate and chlorfenapyr on *Cirrhinus mrigala* during chronic exposure revealed a disrupted thyroid profile and a substantial difference in blood profile. Additionally, it was noted that the fish treated with pesticides had higher serum levels of liver biomarker enzymes. Most of the groups given treatment showed notable changes in the histopathological study of their liver and gills. These three pesticides were graded according to their toxicity trend: dimethoate > acetamiprid > chlorfenapyr. According to the study's findings, the careless application of these pesticides endangers ecosystems, non-target creatures and human health (Ghayyur *et al.*, 2021). A different approach an organophosphate insecticide that is widely used to treat a variety of agricultural pests is malathion. This insecticide is extremely poisonous in addition to being resistant to biodegradation. It could stay in the human body for two generations after entering (Kalantary *et al.*, 2016). Subsequently prolonged periods, the effects of a mild dose of malathion on *Channa punctatus* fish (Bloch) were assessed. The results demonstrated that longer exposure times to malathion result in a progressive decline in morphometric indices, different biochemical parameters and enzyme levels, particularly of antioxidant systems (Bharti *et al.*, 2021).

Fourth-generation halogenated synthetic pyrethroids, including cypermethrin, are used extensively in both domestic and agricultural pest control. In addition to causing problems with growth in embryos, the variety of contaminants has an immediate effect on adult aquatic bird breeding. Mortality, decreased hatchability, chick failure and other teratological consequences, including as skeletal deformities and defective differentiation of the reproductive and neurological systems through methods of endocrine imitation of estrogens, are among the impacts on oocytes. Acute death, sublethal stress, decreased fertility, inhibition of egg formation, eggshell thinning and compromised incubation and chick nurturing habits are among the several chemical impacts on adult birds (Kadiru *et al.*, 2022).

India's Laws to Combat Pesticide Contamination

India has implemented various laws to protect its environmental resources and address pesticide pollution. The Wildlife Protection Act, 1972, introduced by the National Committee on Environmental Planning and Coordination, was the first such legislation. Subsequently, three central laws targeting water pollution were enacted:

Legislation	**Year**	**Key feature**	**Ammendments/ Updates**
Wildlife Protection Act	1972	Introduced to safeguard wildlife and biodiversity by the National Committee on Environmental Planning.	N/A
Water (Prevention and Control of Pollution) Act	1974	Established Central and State Pollution Control Boards (PCBs).	N/A
Water (Prevention and Control of Pollution) Cess Act	1977	Allows PCBs to levy cess on water usage for financial support.	Last amended in 2003
Environment (Protection) Act	1986	Comprehensive legislation integrating all environmental safeguards. Prohibits pollution of water bodies. Requires SPCB approval for potentially polluting activities.	Last amended in 1991

Insecticides Act	1968	Regulates import, manufacture, registration, sale and use of pesticides to mitigate risks. Mandatory pesticide approval by Central Insecticides Board & Registration Committee (CIB & RC). Prohibits use of agricultural pesticides in households. Allows periodic reviews for banning harmful pesticides.	N/A
Pesticide Management Bill (PMB)	2020	Seeks to replace the Insecticides Act, 1968. Aims to regulate manufacture, import, sale, use, and disposal of pesticides. Ensures safe pesticide availability while minimizing environmental and health risks. Introduced stricter controls for modernized pesticide management. Under review by the Standing Committee on Agriculture.	N/A

(Kadiru *et al.*, 2022)

Conclusion

It illustrates how widespread it is in the aquatic habitats of India. Since aquatic animals have a tendency to collect these pesticides along the food chain, it is imperative that present restrictions be strictly enforced and mitigation solutions be developed in India. A few issues need immediate action in order to promote the domestic pesticide industry and guarantee the safe use of pesticides. Priority one should be given to regulating and encouraging the use of reasonably priced, environmentally acceptable insecticides. Old, hazardous pesticides must be deregistered and testing procedures (labs, parameters, etc.) must be consistent to avoid the adverse impacts. Additionally, raising farmer knowledge and promoting safe application techniques are major concerns.

References

Agarwal, A., Prajapati, R., Singh, O.P., Raza, S.K. and Thakur, L.K. 2015. Pesticide residue in water-a challenging task in India. Environ. Monit. Assess., 187: 1-21.

Ahmad, M.F., Ahmad, F.A., Alsayegh, A.A., Zeyaullah, M., AlShahrani, A.M., Muzammil, K., Saati, A.A., Wahab, S., Elbendary, E.Y., Kambal, N. and Abdelrahman, M.H. 2024. Pesticides impacts on human health and the environment with their mechanisms of action and possible countermeasures. Heliyon.

Asim, M. and Nageswara Rao, K. 2021. Assessment of heavy metal pollution in Yamuna River, Delhi-NCR, using heavy metal pollution index and GIS. Environ. Monit. Assess., 193(2): 103.

Banerjee, S.N., 1979. Agricultural production through increased pesticide usage. Pestic Inf, 4(4): 17-19.

Bano, N., Nadeem, A., Maalik, S., Mushtaq, S., Iqbal, N. and Alfarraj, S. 2021. Effect of pesticides on erythrocytes of indigenous fish Labeo rohita. J King Saud Univ Sci., 33(7): 101586.

Boedeker, W., Watts, M., Clausing, P. and Marquez, E. 2020. The global distribution of acute unintentional pesticide poisoning: estimations based on a systematic review. BMC public health, 20: 1-19.

Cech, R., Zaller, J.G., Lyssimachou, A., Clausing, P., Hertoge, K. and Linhart, C. 2023. Pesticide drift mitigation measures appear to reduce contamination of non-agricultural areas, but hazards to humans and the environment remain. Sci. Total Environ., 854: 158814.

Crépet, A., Luong, T.M., Baines, J., Boon, P.E., Ennis, J., Kennedy, M., Massarelli, I., Miller, D., Nako, S., Reuss, R. and Yoon, H.J. 2021. An international probabilistic risk assessment of acute dietary exposure to pesticide residues in relation to codex maximum residue limits for pesticides in food. Food Control., 121: 107-563.

de Souza, K. B., Asker, N., Jönsson, E., Förlin, L., & Sturve, J. 2016. Increased activity of lysozyme and complement system in Atlantic halibut exposed to elevated CO_2 at six different temperatures. Mar. Environ. Res., 122: 143-147.

Dhaliwal, G. S., Jindal, V. and Mohindru, B. 2015. Crop Losses due to insect pests: Global and Indian Scenario. Ind. J. Entom., 77(2): 165–165.

Durairaj M, Arasan S, Pandurengan P, Paulraj J, George R. 2021. Occurrence of Pesticide Residues in Freshwater and Estuarine Fishes of Thamirabarani River. Indian J. Anim. Res., 55(10): 1233-1239.

Ernst, F., Alonso, B., Colazzo, M., Pareja, L., Cesio, V., Pereira, A. and Pérez-Parada, A. 2018. Occurrence of pesticide residues in fish from south American rainfed agroecosystems. Sci total env., 631: 169-179.

Fang, L., Liao, X., Jia, B., Shi, L., Kang, L., Zhou, L. and Kong, W. 2020. Recent progress in immunosensors for pesticides. Biosensors and Bioelectronics, 164: 112255.

FAO 2023. Maximum Residue Limits (Mrls) and Risk Management Recommendations (Rmrs) For Residues of Veterinary Drugs in Foods, CXM 2-2023 fao.org/fao-who-codexalimentarius/sh-proxy/en/?lnk=1&url=https%253A%252F%252Fworkspace.fao.org%252Fsites%252Fcodex%252FStandards%252FCXM%2B2%252FMRL2e.pdf.

FAO. 2018. Pesticide Use Data-FAOSTAT. Retrieved from http://www.fao.org/faostat/en/#data/RP.

Ghosh, A. K., & Sarower, M. G. 2024. Potential toxicity of pesticides and its transformation products to aquatic organisms. Khulna University Studies., 21-32.

Gill, H. K., & Garg, H. 2014. Pesticides: Environmental Impacts and Management Strategies. In and others (Ed.), Pesticides- Toxic Aspects. Intech Open. https://doi.org/10.5772/57399.

GOI 2021. Directorate of Plant Protection, Quarantine and Storage. Available online: http://ppqs.gov.in/statistical-database.

Gupta, P. C. 1975. Neurotoxicity of chronic chlorinated hydrocarbon insecticide poisoning-a clinical and electroencephalographic study in man.

Gupta, P. K. 1986. Pesticides in Indian environment interprint. New Delhi, 4.

Gupta, P.K. 1985. Pesticides. In: Gupta, P.K., Salunkhe, D.K. (Eds.), Modern Toxicology. Metropolitan Book Company, New Delhi., 1–60.

Hashmi, T.A., Qureshi, R., Tipre, D. and Menon, S. 2020. Investigation of pesticide residues in water, sediments and fish samples from Tapi River, India as a case study and its forensic significance. Environ Forensics, 21(1): 1-10.

Hext, H. D. 1973. Pesticides in the aquatic environment (Doctoral dissertation, University of British Columbia).

Kadiru, S., Patil, S., & D'Souza, R. 2022. Effect of pesticide toxicity in aquatic environments: A recent review. Int J Fish Aquat Stud., 10: 113-118.

Karpun, N. N., Yanushevskaya, E. B., Mikhailova, Y. V., Díaz-Torrijo, J., Krutyakov, Y. A., Gusev, A. A., & Neaman, A. 2021. Side effects of traditional pesticides on soil microbial respiration in orchards on the Russian Black Sea coast. Chemosphere, 275: 130040.

Karunakaran, C. O. 1958. The Kerala food poisoning. Indian Med. Assoc., 31: 204-205.

Khan, M. (Ed.). 2013. Pesticides in aquatic environments. Springer Science & Business Media, 10: 2-4).

Khatib I., Rychter P., Falfushynska H. 2022. Pesticide Pollution: Detrimental Outcomes and Possible Mechanisms of Fish Exposure to Common Organophosphates and Triazines. J. Xenobiot., 12(3): 236- 265. https://doi.org/10.3390/jox12030018.

Mondal, R., Mukherjee, A., Biswas, S. and Kole, R. K. 2018. GC-MS/MS determination and ecological risk assessment of pesticides in aquatic system: a case study in Hooghly River basin in West Bengal, India. Chemosphere, 206: 217-230.

Nag, D., Singh, G. C. and Senon, S. 1977. Epilepsy epidemic due to benzahexachlorine. Tropical and geographical medicine, 29(3): 229-232.

Nations,U. 2019.World population prospects Highlights, 2019 revision Highlights, 2019 revision.

Nayak, P., & Solanki, H. 2021. Pesticides and Indian agriculture—a review. Int J Res Granthaalayah, 9(5): 250-263.

Osuala, F. I., Abiodun, O. A. and Alebiosu, E. A. 2020. Contamination levels of organochlorine pesticides in Tympanotonus fuscatus and sediment of Lagos Lagoon, Nigeria. FUW Trends Science and Technology, 5(2): 552-556.

Pandey, N. N. 1986. Report on 'Immediate and residual effects of MIC gas exposure on animals of Bhopal Gas Tragedy'. Indian Veterinary Research Institute, Izatnagar.

Poopak H., Raeeszadeh m. and Salimi B. 2023. Accumulation of heavy metals in meat and their relationship with water and food intake of aquatic animals in Kermanshah, western Iran. Int. J. Environ. Health Res., 7: 1-13.

Popp, J., Pető, K. and Nagy, J. 2013. Pesticide productivity and food security. A review. Agronomy for Sustainable Development, 33(1): 243–255. Retrieved from https://dx.doi.org/ 10.1007/s13593-012-0105-x 10.1007/s13593-012-0105-x.

Pourgholam, R., Ghiyasi M., Rezai, M., Nasrollahzadeh H., Saiedi, A.A. and Safari, R. 2013. Histopathological studies of some organs of Grass carp (Ctenopharyngodon idella) exposed to sublethal concentrations of diazinon. J. Mazand. Univ. Med. Sci., 22(98): 282-287.

Reymond, D.J. and Sudalaimuthu, K. 2021. Assessment of Heavy Metal Pollution in Sediment for Thamiraparani River, India. In IOP Conference Series: Materials Science and Engineering, 1130(1): 012045).

Rossi, A. S., Fantón, N., Michlig, M. P., Repetti, M. R. and Cazenave, J. 2020. Fish inhabiting rice fields: Bioaccumulation, oxidative stress and neurotoxic effects after pesticides application. Ecol. Indic., 113: 106186.

Sethuraman, V. 1972. case of arsenic poisoning in a heifer calf. Indian Veterinary Journal, 49: 216-217.

Shah, Z.U. and Parveen, S. 2021. Pesticides pollution and risk assessment of river Ganga: A review. Heliyon, 7(8).

Sharma, D. and Choudhury, P. P. 2018. Pesticide use and residue management in vegetables. Technical Bulletin, (74): 40.

Stentiford, G. D., Bateman, I. J., Hinchliffe, S. J., Bass, D. 1., Hartnell, R., Santos, E. M. and Tyler, C. R. 2020. Sustainable aquaculture through the One Health lens. Nature Food, 1(8): 468-474.

Tarbah, F. A., Shaheen, A. M., Benomran, F. A., Hassan, A. I. and Daldrup, T. 2007. Distribution of dimethoate in the body after a fatal organphosphate intoxication. Forensic Sci Int., 170(2-3): 129-132.

Tudi, M., Ruan, H. D., Wang, L., Lyu, J., Sadler, R., Connell, D., Chu, C. and Phung, D. T. 2021. Agriculture Development, Pesticide Application and Its Impact on the Environment.

Int J Environ Res Public Health., 18(3): 1112–1112. Retrieved from https://dx.doi.org/10.3390/ijerph18031112 10.3390/ ijerph18031112.

Wani, K. A. (Ed.). 2018. Handbook of Research on the Adverse Effects of Pesticide Pollution in Aquatic Ecosystems. IGI Global., 5-12.

WHO. 2019. Global situation of pesticide management in agriculture and public health: report of a 2018 WHO-FAO survey. World Health Organization.

Xie, J., Tao, L., Wu, Q., Bian, Z., Wang, M., Li, Y. and Lin, T. 2022. Bioaccumulation of organochlorine pesticides in Antarctic krill (Euphausia superba): Profile, influencing factors, and mechanisms. J. Hazard., 426: 128115.

Yang, C., Lim, W. and Song, G. 2021. Immunotoxicological effects of insecticides in exposed fishes. Comparative Biochemistry and Physiology Part C. Toxicol. Pharmacol., 247: 109064.

Yen, J., Donerly, S., Levin, E. D. and Linney, E. A. 2011. Differential acetylcholinesterase inhibition of chlorpyrifos, diazinon and parathion in larval zebrafish. Neurotoxicol. Teratol., 33(6): 735-741.

6

Membrane Technologies for Heavy Metal Removal from Water

Pinak Bamaniya, D.T. Vaghela and Mayur Bhadrka

[1]*PG Scholar, Department of Aquatic Environment Management*

[2]*Associated Professor and Head, Department of Aquatic Environment Management*

College of Fisheries Science, Kamdhenu University, Veraval (362265), Gujarat, India

Introduction

Ensuring environmental sustainability through wastewater treatment is a crucial part of the United Nations' Sustainable Development Goals (SDGs). The global shortage of clean water and the potential for future water scarcity, particularly in developing countries, represent a major crisis for humanity in this century and beyond. The UN reports that 1.2 billion people currently lack access to clean drinking water, and half of the global population does not have sufficient water purification. By 2025, it is expected that 1.8 billion people will live in regions facing severe water shortages (WHO, 2018). Currently, one-third of the world's population lives in water-stressed countries, and this figure is expected to increase to two-thirds by 2025. By 2050, between 2 and 7 billion people may face water shortages. The worsening water situation is largely due to inadequate water supplies and rising demand in developing countries driven by rapid population growth and industrialization. Contaminated water and poor sanitation contribute to nearly 60% of global illnesses. The World Health Organization (WHO) estimates that about 2.4 billion people lack basic sanitation facilities, and over 1 billion lack access to safe drinking water. Unclean water can lead to diseases such as diarrhea, cholera, dysentery, guinea worm infections, typhoid, intestinal worms, and trachoma (WHO/UNICEF, 2010). The World Health Organization (WHO) reports that four billion people suffer from diarrhea each year, leading to nearly 1.8 million deaths, with 90% of these deaths occurring in children under five years old. The quality of available water is declining because of rising pollution from industrial, agricultural, and other human activities. This issue is becoming a significant environmental problem.

Water pollution increases with the number of pollutants present, so it's essential to regularly monitor both subsurface and surface water quality. Heavy metals pose significant risks due to their stability, accumulation in the environment, and potential to become more concentrated in biological systems. Despite stricter environmental regulations in recent decades, heavy metal contamination remains a serious environmental issue in many areas, including affluent regions (Dökmeci., 2020).

A recent study by the European Environment Agency (EEA) estimates that heavy metals are found at high concentrations in 75% to 96% of European waters. Excessive levels of heavy metals in drinking water can lead to poisoning. Therefore, it is crucial to investigate heavy metal toxicity in water and implement various solutions to address contamination and reduce health risks (Kapoor and Singh., 2021). There are about 50 elements classified as heavy metals, with 17 being particularly hazardous. Arsenic, which is a known carcinogen at levels above 10 parts per billion (ppb), naturally occurs in groundwater in various countries. This issue is severe in Bangladesh, where 60 million people are significantly affected by arsenic in drinking water. Additionally, endocrine-disrupting compounds from wastewater generated by pharmaceutical, cosmetic, and food processing industries, even in trace amounts (nanograms per liter), are carcinogenic and often not fully removed by conventional primary and secondary treatment processes. Notably, steroid hormones, which are continuously excreted by humans and animals, are among these endocrine-disrupting contaminants.

Reverse osmosis (RO) membranes have long been used for desalinating seawater and brackish water, as well as producing high-purity water for food and pharmaceutical processing and industrial waste treatment. The advent of nanofiltration (NF) membranes has expanded their use to include water softening, removing disinfection by-product precursors (such as trihalomethanes), reducing total organic carbon (TOC), and applications in food processing and industrial water treatment.

The reasons membrane filtration is becoming the preferred technology for water treatment, alongside RO or NF membrane systems, are outlined below:

- Development of high-frequency backwash ultrafiltration (UF) and microfiltration (MF) processes to mitigate membrane fouling using high-integrity hollow fiber membranes.
- Application of dead-end MF/UF systems, particularly suitable for treating municipal raw water due to their efficiency in handling low particle loads.
- Consistent production of high-quality water with effective removal of parasites and pathogens.

- Minimal chemical usage, ensuring compliance with stringent water pollution regulations.
- Reliable and well-established membrane system operations.
- Automated system controls, leading to reduced labor and operational costs.
- Compact and modular designs that allow for easy future expansion.
- Increasing cost-efficiency, with the price of reverse osmosis (RO) membranes decreasing significantly over the past two decades, while permeate production has tripled (Fane, 2008, pp. 239–270).
- Proven membrane integrity, though polymeric membranes have a lifespan of 3–5 years. Ceramic membranes, which offer greater longevity, are also being implemented but come at a higher cost.

Water Pollution

Types of contaminants

Inorganic Contaminants: Includes substances such as carbonate/non-carbonate hardness, fluoride, arsenic, lead, copper, chromium, mercury, antimony, and cyanide.

Organic Contaminants: Originates from sources like pesticides, domestic waste, and industrial effluents.

Biological Contaminants: Comprises living organisms, including algae, bacteria, protozoa, and viruses.

Radiological Contaminants: Consists of radioactive elements, such as Uranium-226 (U-226), Radium-226 (Ra-226), Radium-228 (Ra-228), and Radon-228 (Rn-228).

Among the various pollutants of concern, heavy metals are particularly significant due to their inability to be easily degraded through biological processes, unlike other pollutants. Heavy metals are generally defined as elements with atomic weights ranging from 63.5 to 200.6 and densities greater than 5 g/cm^3 (Abdullah *et al.*, 2019). While certain heavy metals, such as copper (Cu), iron (Fe), zinc (Zn), manganese (Mn), and cobalt (Co), are essential for vital biochemical functions in the human body, excessive exposure to these metals can have harmful, even lethal, effects. Moreover, heavy metals such as cadmium (Cd), arsenic (As), mercury (Hg), lead (Pb), and chromium (Cr) are highly toxic even in trace amounts, such as parts per billion (ppb). These metals are non-biodegradable and have the potential to bioaccumulate in human tissues, leading to serious health risks (Landaburu-Aguirre *et al.*, 2006).

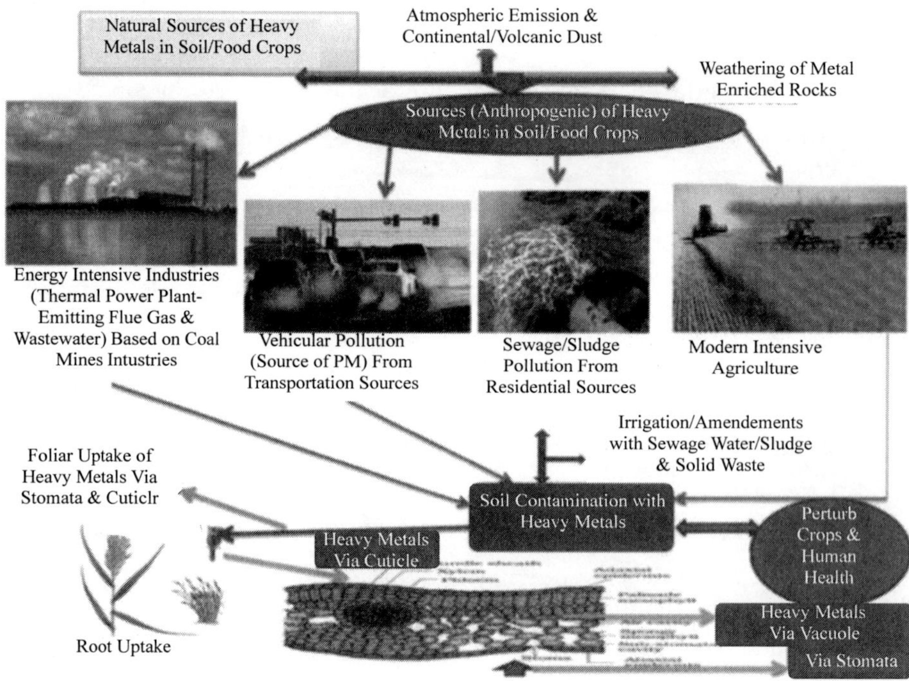

Figure 1. Sources of Water Contamination by Heavy Metals

Natural Sources

Natural processes such as rock weathering, mineral degradation, forest fires, volcanic eruptions, biogenic activity, sea-salt sprays, evaporation from soil and water surfaces, and wind-driven soil particles contribute significantly to the release of metals into the environment. Annually, approximately 1.72 × 10- mg/kg of arsenic (As) is emitted as volcanic ash, while the Earth's crust contains around 4.01×10^{16} mg/kg of As. Additionally, undersea volcanic activity releases another 4.87×10^{6} mg/kg of As. Industrial activities extract metals from natural sources, subsequently releasing them into the atmosphere. Metals can also be mobilized from their natural habitats through weathering processes and transported to various environments, including aquatic systems like seas and rivers, where they contaminate water (Kapoor & Singh, 2021). Heavy metals often exist in complex forms such as oxides, sulfides, silicates, sulfates, hydroxides, organic compounds, and phosphates. The most common heavy metals of concern include lead (Pb), arsenic (As), copper (Cu), zinc (Zn), cadmium (Cd), nickel (Ni), mercury (Hg), and chromium (Cr) (Masindi & Muedi, 2018).

Anthropogenic Sources

Pollutants are introduced into various environments through sources such as wastewater, industrial processes, agriculture, mining, and runoff. Activities like metal ore smelting, extraction, electroplating, mining practices, coal-fired power generation, leather production, and the burning of fossil fuels contribute significantly to heavy metal pollution. Metallurgical processes, automobile exhaust (which releases lead), the use of insecticides, waste incineration, and landfills are also major contributors. Heavy metals can enter ecosystems through agricultural, residential, and industrial waste, posing a significant environmental and health risk.

Diverse Sources

Sources of Waste Materials

Electronic Waste/ (E-Waste), Industrial Effluents, Biomedical Waste, Agriculture Wastes

Sources from Industrial Processes

Mining, Mineral Extraction, Electroplating, Smelting, Automobile Exhaust Fumes, Geothermal Sources, Power Plants, Urban Surfaces, Plumbing and Fittings, Irrigation Water, Tannery Activities. According to studies, zinc smelting activities in Hezhang released over 50 tonnes of mercury and 450 tonnes of cadmium into the environment between 1989 and 2001. Heavy metals come from several sources, including natural sources, human activities (anthropogenic), and a range of industrial processes that generate waste. The different sources of heavy metals are outlined in Table 1.

Table 1. Natural and anthropogenic sources of heavy metal contamination

S. No.	Heavy Metals	Sources
1	Arsenic (As)	Atmospheric deposition, mining, pesticides, smelting of gold, industrial waste
2	Lead (Pb)	Batteries, coal combustion, paint industry, ceramics, PVC, pesticides
3	Mercury (Hg)	Coal combustion, fish, mining, paint industry, paper industry
4	Cadmium (Cd)	Plastic, fertilizers, electronics, pigments and paints, wastewaters
5	Chromium (Cr)	Steel fabrication, electroplating, textile, tanning, chemical industry
6	Aluminium (Al)	Rock and soil leaching, minerals, solid wastes associated with industrial processes
7	Iron (Fe)	High intake of iron, supplements & oral consumption

Analytical Methods for Heavy Metals Detection

A metal ion detector is a device or instrument used to identify and quantify metal ions in environmental samples. Before removing these harmful metal ions from water, it is crucial to employ technologies that can detect their presence and provide quantitative measurements of pollution levels to guide the selection of the most effective removal strategy. Therefore, it is essential to develop detection methods that are not only time and cost-efficient but also environmentally sustainable. Additionally, the detection technique must be highly sensitive to accurately measure even trace amounts of metal ions. According to the literature, numerous analytical methods have been developed and applied for detecting heavy metals in both environmental and biological samples (Jagaba, A *et al.*,2021)

Heavy metals can be detected using various methods, including chromatographic analysis, electrical analysis, instrumental neutron activation analysis, spectrometry, and combined techniques. Hybrid methods, which integrate multiple techniques, are commonly used for identifying different forms of metals. Inductively Coupled Plasma Mass Spectrometry (ICP-MS) and Inductively Coupled Plasma Atomic Emission Spectrometry (ICP-AES) are becoming popular due to their ability to detect multiple elements simultaneously, shorter processing times, higher throughput, and lower sample requirements. ICP-MS, in particular, offers high sensitivity, a wide dynamic range, and strong resistance to interference (Jagaba *et al.*, 2020).

Highly sensitive methods for detecting heavy metal ions include Graphite Furnace Atomic Absorption Spectrometry, Atomic Fluorescence Spectrometry (AFS), X-ray Fluorescence Spectrometry, Atomic Absorption Spectroscopy (AAS), Neutron Activation Analysis, and both ICP-MS and ICP-AES. These techniques can detect very low concentrations of metal ions and can analyze multiple elements at once. However, they are expensive and require skilled operators to manage the complex equipment (Jagaba *et al.*, 2020).

Health Risks Associated with the Presence of Heavy Metals in Water

Human Health Impacts

Heavy metals can accumulate in the human body when ingested or inhaled, making them potentially hazardous. While some heavy metals are essential for health and survival, as they contribute to bone formation, acid-base balance, and the maintenance of colloidal systems, they also support the production of crucial enzymes, structural proteins, and hormones. Despite their necessity, these metals have become pervasive in our environment, contaminating the air, water, and soil (Birniwa *et al.*, 2021).

Heavy metals are involved in the production of essential enzymes, structural proteins, and hormones but can be toxic in excessive amounts. Their toxicity varies, with the following metals listed in order of their harmfulness to living organisms: silver > copper > zinc > nickel > lead > cadmium > chromium > tin > iron > manganese > aluminum. These metals can inhibit biodegradation processes.

Each heavy metal has unique composition, physical properties, and chemical structures. Arsenic, manganese, nickel, titanium, molybdenum, vanadium, chromium, iron, zinc, silver, copper, cadmium, tin, cobalt, platinum, gold, mercury, and lead are heavy metals with densities greater than 5 g/cm^3 and are commonly encountered in daily life (Briffa *et al.*, 2020). The various types of toxicity caused by heavy metals affecting different organs are detailed in Table 2. The toxicity caused by different types of heavy metals is summarized in Table 3.

Table 2. Differeant types of toxicity related to organs.

Toxicological Effect	Description
Neurotoxicity	Adverse effects on the nervous system, leading to impairment of nerve function, cognition, or behavior.
Genotoxicity	Induction of damage to genetic material (DNA), potentially leading to mutations or other genetic abnormalities.
Reproductive and Developmental Toxicity	Harmful effects on reproductive organs or processes, as well as adverse impacts on fetal development during pregnancy.
Skin Toxicity	Harmful effects on the skin, which may include irritation, allergic reactions, or other dermatological problems.
Carcinogenicity	Ability to cause cancer or increase the risk of developing cancer.
Hepatotoxicity	Adverse effects on the liver, potentially leading to liver damage or impairment of liver function.
Nephrotoxicity	Adverse effects on the kidneys, leading to impaired kidney function or damage to kidney tissues.
Cardiovascular Toxicity	Adverse effects on the heart or circulatory system, potentially leading to cardiovascular diseases or dysfunction.
Immunological Toxicity	Harmful effects on the immune system, leading to altered immune response or increased susceptibility to infections or diseases.

Table 3. Toxicity produced by various types of heavy metals.

S. No.	Heavy Metals	Toxicity produced
1	Arsenic (As)	Neurotoxicity, Genotoxicity, Reproductive and Developmental Toxicity, Skin Toxicity, Carcinogenicity
2	Lead (Pb)	Hepatotoxicity, Carcinogenicity, Nephrotoxicity
3	Mercury (Hg)	Skin Toxicity, Cardiovascular Toxicity, Carcinogenicity, Nephrotoxicity
4	Cadmium (Cd)	Cardiovascular Toxicity, Hepatotoxicity, Nephrotoxicity, Nephrotoxicity, Neurotoxicity
5	Chromium (Cr)	Hepatotoxicity, Immunological Toxicity
6	Aluminium (Al)	Neurotoxicity

Table 4. Effects of heavy metals on human health

S. No.	Heavy Metals	Effects on Human Health
1	Arsenic (As)	Highly affects the dermal region (cancer), brain & cardiac problems. Bronchitis, dermatitis, poisoning
2	Lead (Pb)	Serious effect on mental health (Alzheimer's disease), mental retardation in children, development delay, chronic damage to nervous system, liver, kidney damage
3	Mercury (Hg)	Sclerosis, blindness, Minamata disease, deafness, gastric problems, renal disorders, tremors, gingivitis, damage to the nervous system
4	Cadmium (Cd)	Osteo-related problems, prostate cancer, lung diseases, renal issues, bone defects, kidney damage, bone marrow
5	Chromium (Cr)	Lung disorders (bronchitis, cancer), damage to renal and reproductive systems, irritability
6	Aluminium (Al)	Nervous system, Alzheimer's disease, Gehrig's disease
7	Iron (Fe)	Vomiting, diarrhoea, abdominal pain, dehydration & lethargy

Table 5. Various technology for removal of heavy metals and others pollutant from water

Technology/ treatment	Advantages	Disadvantages
Chemical precipitation	Simplicity of the method Low-priced capital costs Large quantity of metal sludge	Not metal discriminating High sludge disposal costs High maintenance expenses
Coagulation and flocculation	Inactivation capability of bacteria Fine sludge dewatering and settling attributes	Chemicals consumption Increased sludge volume production
Ion exchange	Metal selective High regeneration capability	Restricted pH tolerance High capital and maintenance costs

Flotation	Metal selective Removal of small particulates	High capital, operation, and maintenance costs Modest retention times
Membrane filtration	Less solid waste generation Less chemical consumption Modest space requirements Possible metal discrimination	High capital, operation, and maintenance costs Membrane fouling/clogging Limited flow rates
Electrochemical treatment	Moderately metal selective No chemicals required; can be engineered to tolerate suspended solids	High capital costs Production of hydrogen gas (in some processes)Filtration process for flocs
Adsorption	- Wide variety of target pollutants High capacity Fast kinetics Possibly selective depending on adsorbent	- Performance dependent on adsorbent type Physical or chemical activation needed to enhance sorption capacity

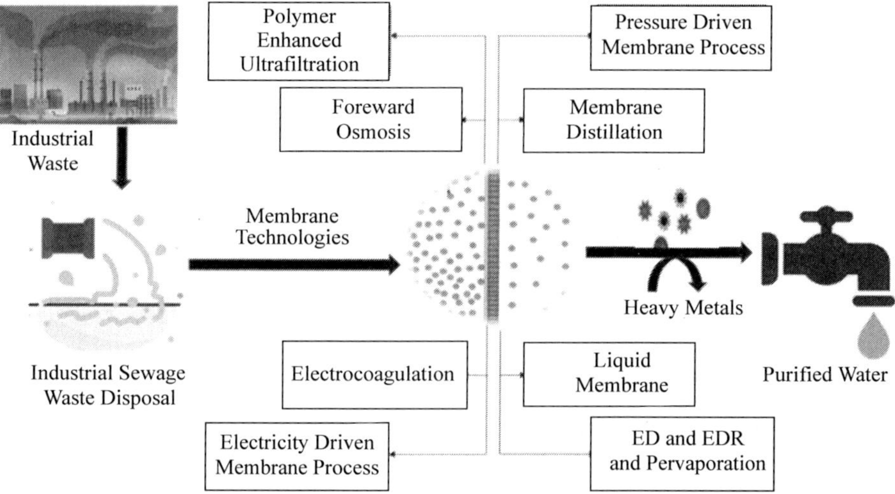

Membrane Technology For Wastewater Treatment

Over time, conventional wastewater treatment processes have made progress in handling overflows for discharge. However, further improvements are needed to make treated wastewater suitable for industrial and agricultural reuse. Membrane technology is increasingly favored for reclaiming water from various wastewater sources and ensuring its reuse (El Batouti *et al.*, 2021). In both developing and industrialized countries, substantial contaminants from human activities enter municipal water supply systems, raising public health and environmental concerns while also contributing to high volumes of wastewater (Tlili and Alkanhal, 2019).

Three emerging treatment technologies—membrane filtration, advanced oxidation processes, and UV irradiation—offer improved methods for protecting freshwater discharges. Research has explored various membrane materials for separating heavy metals from industrial effluents. Nanocomposite membranes, in particular, demonstrate superior removal efficiency compared to traditional polymeric membranes and can achieve complete removal of heavy metals (Alshahrani *et al.*, 2021).

Membrane technology offers a more efficient alternative to conventional separation methods. The membrane acts as a barrier between two phases, facilitating the separation and selective transport of chemical compounds based on their structure. The effectiveness of a membrane depends on its properties as a semi-permeable layer, which controls the movement of substances between phases. Optimizations, such as adjusting pyrolysis temperature and solution pH, can enhance the membrane's adsorption capabilities, improving its ability to remove organic contaminants (Aliyu *et al.*, 2018). This technology is particularly effective in removing heavy metals and oil droplets from industrial waste (see Figure 2.2).

Integrating membranes with photocatalysts helps in characterizing heavy metals by their composition and hydrophilic properties and reduces membrane fouling. Thorough physical and chemical characterization of membranes is essential to understand their basic structure and properties, and to evaluate their potential for various environmental applications (Yalcinkaya *et al.*, 2020). The applications of various membrane processes are detailed in Table 6.

Various membrane processes are used for the removal of heavy metals and other impurities, as described below.

- Pressure-driven membrane processes
- Polymer-Enhanced Ultrafiltration
- Forward Osmosis
- Electro-Dialysis (ED) and Electro-Dialysis Reversal (EDR)
- Pervaporation
- Electrically Driven Membrane Processes for Chemical Free Heavy Metal Ion Removal
- Electrocoagulation
- Heavy Metal Recovery by Membrane Distillation (MD)
- Removal of Heavy Metals by the Liquid Membrane
- FO-RO Hybrid Systems
- Membrane Bioreactors
- Membrane Distillation

Table 6. Application of various Membrane process

Membrane Process	Country	Year	Application
Microfiltration	Germany	1920	Laboratory use
Ultrafiltration	Germany	1930	Laboratory use
Hemodialysis	Holland	1950	Artificial kidney
Electrodialysis	USA	1955	Desalination
Reverse Osmosis	USA	1965	Desalination
X-flow Ultrafiltration	USA	1971	Concentration of macromolecules
Gas Separation	USA	1979	Hydrogen recovery
X-flow Microfiltration	Australia, USA	1980	Water treatment
Pervaporation	Germany, Holland	1982	Dehydration of organic solvents
Nanofiltration	USA	1986	Water softening
Electrodeionisation	USA	1987	Demineralisation
Submerged Membrane Filtration/ Bioreactor	Japan	1987	Water reclamation/sewage water treatment

Pressure-driven Membrane

Pressure-driven membrane processes are categorized into low-pressure and high-pressure types. High-pressure-driven processes include Reverse Osmosis (RO) and Nanofiltration (NF), while low-pressure processes encompass Ultrafiltration (UF) and Microfiltration (MF). These membrane technologies are increasingly vital in global water treatment systems, particularly for post-treatment stages (Pervez *et al.*, 2022).

High-pressure processes, such as NF and RO, are effective at removing dissolved impurities like small organic molecules and monovalent ions, but they require significant capital investment and high operating costs due to the pressures needed—ranging from 100 to 1,000 psi. In contrast, MF and UF membranes operate at much lower pressures, between 5 and 60 psi, but are less effective at removing dissolved ions and organic molecules (Diallo, 2014) (Figure 4).

Working Principle

In a generic membrane process, a feed stream is divided into two streams: retentate and permeate, using a membrane. In pressure-driven membrane techniques, the pressure difference between the feed and permeate sides drives the solvent through the membrane, which acts as a barrier to solid particles. The separation of particles and dissolved impurities depends on factors such as size, shape, and charge. The removal efficiency is quantified by the rejection coefficient (R), which ranges from 0 (indicating complete permeation) to 1 (indicating complete rejection). In full-scale installations, the recovery ratio—

the proportion of permeate flow relative to feed flow—typically ranges from 50% to 90%, with an average around 80% (Van Der Bruggen *et al.*, 2003).

Efficiency and Target Pollutants

Pressure-driven membranes are highly effective for removing pollutants from wastewater streams. Various types of pollutants that these membranes can remove are illustrated in Figure 5.6. Reverse Osmosis (RO) membranes are particularly valuable for extracting heavy metals from both water and wastewater. According to Regis *et al.* (2022), RO and Nanofiltration (NF) membranes can achieve approximately 95% removal of arsenic.The advantages and disadvantages of pressure-driven membranes are outlined in Table 7.

Table 7. Advantages and disadvantages of pressure-driven membranes (Ahmad *et al.*, 2018)

Membrane Process	**Advantages**	**Disadvantages**
Microfiltration (MF)	Compact modules Low capital costs No degradation due to heating Less pretreatment	High operational costs Fouling of low-mole cular-weight organics
Ultrafiltration (UF)	Effective in treating oily micro emulsions Low energy consumption and high efficiency Less pretreatment	Lower flux Increased fouling rates
Nanofiltration (NF)	Better removal efficiencies due to smaller pore size Can remove divalent ions including hardness Compact S/W modules	Higher energy consumption Needs good pretreatment More fouling rate than MF and UF
Reverse Osmosis (RO)	Can remove all solids and ions Best for heavy metal removal Reuse of product water for different purposes	Very high pressure requirements Increased fouling rates Needs very effective pretreatment

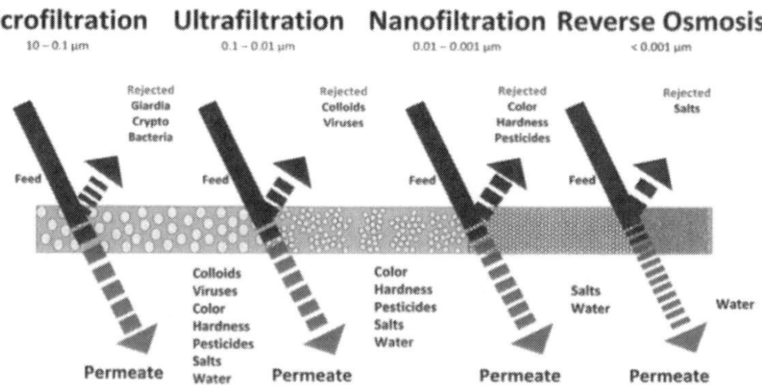

Figure 4. Pressure-driven membranes (Czarny *et al.*, 2017).

Membrane Modules and Selection

A membrane module is the hardware that houses the membrane and facilitates the separation of the incoming feed stream into retentate and permeate streams. Table 8 presents a comparison of different membrane modules.These modules contain extensive membrane areas and come in four main types:

1. Plate-and-Frame Module
2. Tubular Module
3. Spiral Wound Module
4. Hollow Fiber Module

Flat Sheet/Plate and Frame

Flat sheet modules consist of a flat sheet of membrane on the top and a flat plate on the bottom. In the middle, there is a web-like material that creates space for the removal of permeate (Scott, 1995).

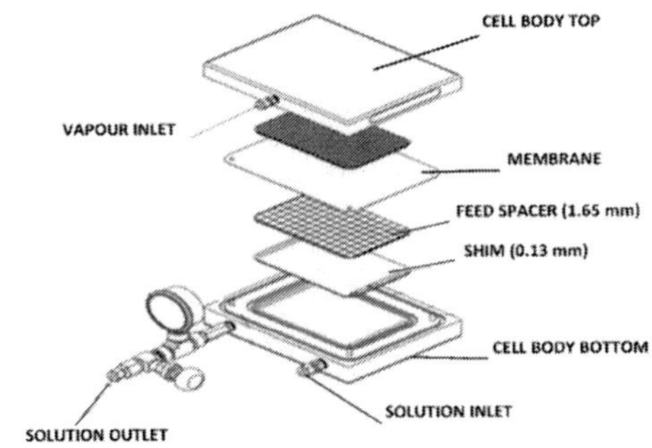

Figure 5. Flat sheet membrane module (Berdasco *et al.*, 2017)

Spiral Wound

This module features an envelope of membranes and spacers wrapped around a perforated collection tube and is encased within a tubular pressure vessel. These modules have high packing densities (surface area to volume ratio), which results in lower capital costs (Scott, 1995).

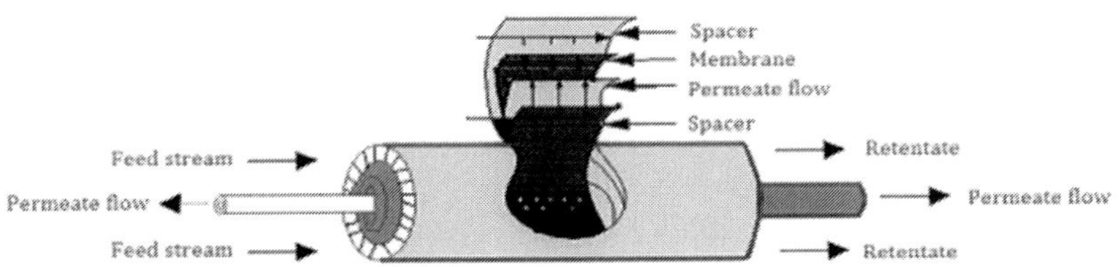

Figure 6. Spiral membrane module (Pal., 2017)

Tubular

A conventional tubular membrane system consists of numerous tubes arranged in a shell-and-tube configuration, where the feed liquid flows through all the tubes. The permeate, after passing through each tube, is collected in a single collection unit. The tubes typically have a porous fiberglass or paper support, with the membrane on the interior surface (see Figure 7). Due to their large core diameters, these modules are well-suited for handling feed streams containing relatively large particles (Berdasco *et al.*, 2017).

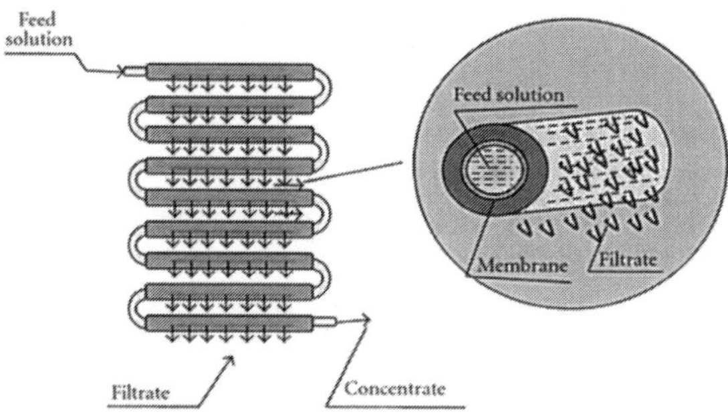

Figure 7. Tubular module (Allhathal Alanezi., 2012)

Hollow Fiber

In a hollow fiber membrane module, the feed flows either inside or outside the hollow fibers, and the permeate passes through the fiber wall to the opposite side. These fibers are made from an asymmetric membrane, with the active or skin layer facing the feed solution (see Figure 8). A bundle of hollow fibers is housed in a pressure vessel, and the open ends of the U-shaped fibers are sealed and connected to the head plate using epoxy (Ismail *et al.*, 2019).

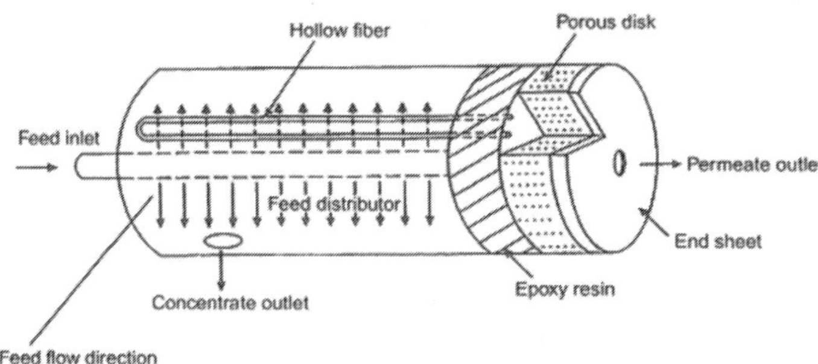

Figure 8. Hollow fiber module (Ismail., 2012)

Table 8. Comparison of membrane modules (Sridhar *et al.*, 2007)

Type of Module	Plate and Frame	Spiral Wound	Tubular	Capillary	Hollow Fiber
Diameter	-	50–200 mm	5–15 mm	0.5–5 mm	<0.5 mm
Packing Density (m^2/m^3)	30–500	200–800	30–200	600–1,200	500–9,000
Ease of Cleaning	Good	Fair	Excellent	Fair	Poor
Relative Cost	High	Low	High	Low	Low
Resistance to Fouling	Good	Moderate	Very Good	Moderate	Poor

Membrane Fouling

Membrane fouling occurs when solid particulates, microbes, and organic materials deposit on the membrane surface or within its pores, leading to a decrease in permeate flux. This fouling negatively impacts membrane performance, causing higher energy consumption and reduced filtration area. Fouling can take various forms, including colloidal fouling, bio-fouling, and both organic and inorganic fouling.

Table 10. Pressure-driven membrane processes for wastewater treatment (Obotey Ezugbe and Rathilal, 2020)

Pressure-Driven Membrane Process	Wastewater Treated	Reported Results	References
Ultrafiltration (UF)	Vegetable oil factory	COD (91%), TSS (100%), TOC (87%), PO4^3− (85%), Cl− (40%)	Mohammadi and Esmaeeli-far (2004)
Microfiltration (MF) coupled with Reverse Osmosis (RO)	Urban wastewater	Pesticides and pharmaceuticals removed to meet discharge limits	Rodriguez-Mozaz *et al.* (2015)
Microfiltration (MF)	Municipal wastewater (disinfection and phosphorus removal)	Contaminants removed to below detection limit	Dittrich *et al.* (1996)
Microfiltration (MF)	Synthetic emulsified oily wastewater	95% removal of organics	Wang *et al.* (2009)
Nanofiltration (NF) coupled with Reverse Osmosis (RO)	Dumpsite's leachate	95% recovery of water	Rautenbach *et al.* (2000)
Ultrafiltration (UF)	Poultry and slaughterhouse wastewater	COD and BOD removal of >94%, fats removal of 99%, TSS (98%)	Yordanov (2010)
Nanofiltration (NF)	Textile wastewater	COD (57%), salinity (30%), color (100%)	Ellouze *et al.* (2012)

Ultrafiltration (UF) coupled with Reverse Osmosis (RO)	Metal finishing industry	90%–99% removal of all contaminants	Petrinic *et al.* (2015)
Ultrafiltration (UF) coupled with Reverse Osmosis (RO)	Oily wastewater	Oil and grease (100%), COD and TOC (98%), TDS (95%), turbidity (100%)	Salahi *et al.* (2011)
Ultrafiltration (UF) coupled with Nanofiltration (NF) or Reverse Osmosis (RO)	Phenolic wastewater from paper mill	Phenol (94.9%), COD (95.5%)	Sun *et al.* (2015)

Polymeric Membranes

A polymer membrane is a thick film of polymeric material through which the permeate diffuses, driven by pressure, concentration, or other gradients (Jose, Kappen, and Alagar, 2018). Among the various polymers used in membrane technologies, biopolymers stand out for their environmental benefits and their natural ability to attract heavy metals. Biopolymers are derived directly or indirectly from natural sources such as plants, animals, and other organisms. Biodegradable materials, which are either entirely of biological origin or chemically synthesized from bio-origin molecules, are highly advantageous due to their abundance, non-toxicity, and biocompatibility (Hernández, Williams, and Cochran, 2014; Smith, Moxon, and Morris, 2016). These properties make biopolymers an excellent choice for a wide range of industrial applications (Bozell and Petersen, 2010; Mohan *et al.*, 2016).

Biopolymers can be classified in several ways based on their sources and properties. Generally, they are categorized into four main types:

1. Natural Biopolymers: These are derived entirely from biomass, including substances such as chitin, starch, and cellulose.
2. Microbially Synthesized Biopolymers: Produced through microbial fermentation, these include biopolymers like polyhydroxyalkanoates.
3. Chemically Synthesized Biopolymers from Microbial Fermentation: These are chemically synthesized from microbial fermentation, such as certain polyesters.
4. Chemically Synthesized Biopolymers from Non-Renewable Resources: Produced from petrochemicals, examples include polyethylene and nylon (Shankar and Rhim, 2018).

Biopolymers possess several advantageous properties, such as biodegradability, nontoxicity, biocompatibility, renewability, and natural abundance. Additionally, many biopolymers have a natural tendency to attract or adsorb heavy metals, making them highly suitable for water purification applications. As a result, biopolymers are widely used in diverse fields including food, packaging, medical, hygiene, textiles, cosmetics, pharmaceuticals, and water filtration (Sahana and Rekha, 2018; Hameed *et al.*, 2020; Raza, Khalil, and Abid, 2020; Gough *et al.*, 2021; Hussain, Ramakrishna, and Abid, 2021; Raza, Anwar, and Abid, 2021; Waqas *et al.*, 2022). With the increasing need for effective removal of heavy metals from drinking water and wastewater due to their toxicity, biopolymers offer a promising solution. Biosorption using biopolymer-based biosorbents is a particularly effective method for heavy metal removal and agricultural waste treatment, thanks to their natural abundance and cost-effectiveness.

Cellulose

Cellulose is a highly abundant natural material, with annual production estimates ranging between 10^{11} and 10^{12} tons due to photosynthesis (Rol *et al.*, 2019). Its polymer structure contains numerous hydroxyl groups that serve as active sites for binding metal ions (O'Connell, Birkinshaw, and O'Dwyer, 2008). The cellulose polymer is composed of microfibrils with nano-sized diameters, making it an excellent candidate for membrane manufacturing. Cellulose-based polymeric membranes are widely used in various industries, including water purification. The presence of hydroxyl groups in cellulose allows for chemical modifications, enhancing its suitability for different applications (Grigoray *et al.*, 2014; Lv *et al.*, 2018; Xu *et al.*, 2019).

Chitin

Chitin is a biopolymer sourced from the shells of crabs and shrimp. Chitosan, derived from chitin, contains about 20% residual sugar, including N-acetyl glucosamine, making it not entirely pure. The physical structure of chitosan is crucial in determining its biological properties. To activate its antibacterial and other biological functions, chitosan must be dissolved. Unlike chitin and cellulose, chitosan has amino groups in its chemical structure, which alters its properties. The presence of functional groups, such as amino linkages and hydroxyl groups, allows chitosan to attract metal ions with opposite charges due to free electrons (Sánchez, Butter, and Rivas, 2020).

Chitosan

Chitosan is a biopolymer known for its natural ability to adsorb heavy metals (Hudson and Smith, 1998). It is highly desirable for heavy metal removal from water due to its availability, low cost, eco-friendliness, and versatility in various applications. The specific binding functional groups in chitosan enhance its capacity to remove heavy metal ions from water. In particular, the presence of amine groups in the chitosan structure facilitates the binding of heavy metals.

Sodium Alginate

Sodium alginate contains multiple functional groups, including activc hydroxyl and carboxyl groups, which interact with heavy metals (Dong *et al.*, 2019). Modified sodium alginate, particularly with sulfonic functional groups, has proven effective for treating effluents (Fatin-Rouge *et al.*, 2006; Petrovič and Simonič, 2016). Sodium alginate nanocomposites are capable of capturing heavy metal ions such as Cd^{2+}, Pb^{2+}, Cu^{2+}, Cr^{2+}, and Co^{2+}.

Polylactic Acid

Polylactic acid (PLA) is a synthetic linear aliphatic thermoplastic polyester derived from the organic acid lactic acid (Castro-Aguirre, 2016). PLA is widely used in tissue engineering and drug delivery systems due to its biocompatibility (Chuan, 2020; Shujun Yuab *et al.*, 2021). Recently, it has also garnered attention for its ability to capture heavy metal ions through ultrafiltration membranes, owing to its hydrophilic and nontoxic properties.

Polycaprolactone

Polycaprolactone (PCL) is an excellent material for fabricating biocompatible membranes for heavy metal removal. Its advantageous properties include a low glass transition temperature and a melting point around 60°C, which make it suitable for various applications (Croisier *et al.*, 2012).

PCL and cellulose nanofiber membranes have demonstrated effective mechanical, adsorption, and filtration properties for removing contaminants from water (Carpenter, de Lannoy, and Wiesner, 2015). A composite membrane created by electrospinning agave cellulose and blending it with PCL exhibited enhanced properties (Yalcinkaya, 2019). This agave-based biomaterial, formed into a sustainable cellulose and PCL composite membrane, effectively filters heavy metals such as Fe^{3+} and Cr^{2+}. The electrospun membrane achieved removal rates of 99% for Cr^{2+} and 75% for Fe^{3+} (Hinestroza *et al.*, 2020).

Cellulose acetate (CA) and PCL nanostructured membranes have been developed for the adsorption of Pb^{2+} from water. CA's low cost and effectiveness make it a suitable material for use alone or in combination with other materials like PCL. The CA membrane features functional groups such as -COOH, $-SO_3H$, and $-NH_2$, which provide sites for the sequestration and adsorption of metal ions. The membrane's adsorption capabilities for electropositive ions have been analyzed using Langmuir and Freundlich isotherm models (Aquino *et al.*, 2018).

Gelatin

Gelatin is a sustainable, water-soluble biopolymer known for its nontoxicity, biodegradability, and cost-effectiveness. It contains various active groups, especially amino acids, which can adsorb metal ions and serve as a template for creating functional materials for heavy metal removal from contaminated water (Zhou *et al.*, 2017). Oxidation of the amino groups in gelatin macromolecules forms nitrite functional groups, enhancing its utility. Gelatin has been used as both a reducing and stabilizing agent in the preparation of gold-gelatin composites (Zhang *et al.*, 2009) and for the stable synthesis of manganese dioxide (MnO_2) (Wang and Li, 2002). Incorporating functional groups into gelatin generally increases its adsorption capacity (Wanga *et al.*, 2018a,b,c).

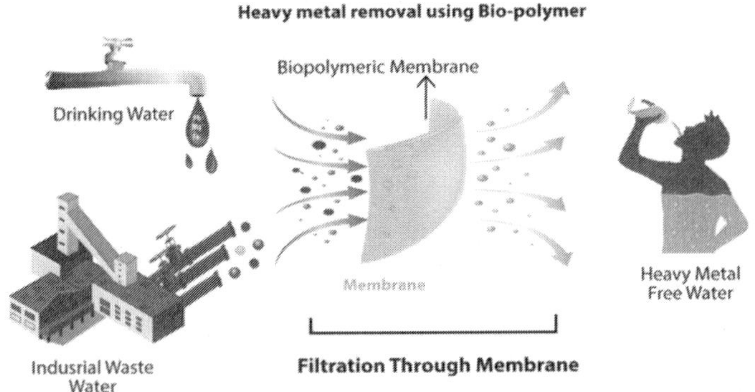

Figure 8. An illustration of heavy metal removal using membrane technology.

Photocatalytic Membrane

A significant number of toxic pollutants with low biodegradability have been detected in surface water, groundwater, and treated municipal wastewater (Lee *et al.*, 2017). These complex contaminants cannot be effectively removed by conventional treatment methods. One efficient approach to degrade organic pollutants is through advanced oxidation processes (AOPs) (Chuang and Shi, 2022). AOPs generate hydroxyl radicals (OH), which act as non-selective oxidizing agents capable of breaking down pollutants depending on their concentration. Among AOPs, heterogeneous photocatalysis holds promise for addressing low concentrations of pollutants. However, challenges such as short catalyst lifespan, low reusability, and poor separation efficiency have hindered its practical application in environmental remediation (Al-Maqdi *et al.*, 2021; Pan *et al.*, 2021).

To address these limitations, the use of photocatalytic membrane reactors (PMRs) offers a highly effective solution by enabling the separation of photocatalysts and photodecomposition products from reaction mixtures (Molinari *et al.*, 2021). PMRs integrate photocatalysis with pressure-driven membrane processes, such as nanofiltration (NF), ultrafiltration (UF), and microfiltration (MF), enhancing both contaminant degradation and separation efficiency (Mozia *et al.*, 2005).

In 1972, Fujishima and Honda pioneered the use of TiO_2 electrodes for water splitting via photoelectrochemical processes, marking a significant advancement in photocatalysis, one of the earliest forms of advanced oxidation processes (AOPs). Since then, photocatalysis has rapidly evolved for a variety of applications in environmental remediation and water treatment (Guo *et al.*, 2017). This technology has emerged as an alternative to conventional water treatment methods due to its ability to harness light energy to degrade pollutants. The principle behind photocatalysis involves the absorption of light by a photocatalyst, leading to the generation of electron-hole pairs that trigger redox reactions. The most widely used photocatalysts are semiconductors, which initiate reactions through light absorption (Mohd Kaus *et al.*, 2017).

Photocatalytic reactions are classified as either homogeneous or heterogeneous, depending on whether the photocatalyst and the reactants are in the same phase or different phases (Li *et al.*, 2021). Most traditional photocatalysts, including TiO_2, ZnO, and Nb_2O_5, are activated by ultraviolet (UV) light, which constitutes about 5% of the solar spectrum. Under UV irradiation, these photocatalysts can degrade both organic and inorganic contaminants in the liquid phase through selective reactions. However, UV light's high energy increases the formation of by-products, limiting its broader application (Molinari *et al.*, 2020).

To overcome this challenge, recent developments have focused on photocatalysts that are activated by visible light, which constitutes a much larger portion of the solar spectrum. Examples of visible-light photocatalysts include CdS, Bi_2MoO_6, $Bi2WO_6$, WO_3, Ag_2O, Cu_2O, $BiVO_4$, CdSe, and $RbPb_2Nb_3O_{10}$, all of which have shown promise in effectively degrading pollutants under visible light (Sang *et al.*, 2015; Meng *et al.*, 2017). This shift toward visible-light photocatalysis holds potential for more energy-efficient and environmentally friendly water treatment technologies.

Mechanism of Action

When a photocatalyst is illuminated with photon energy greater than its bandgap, electrons in the valence band (VB) are excited to the conduction band (CB), leaving behind holes in the VB. This process generates electron (e^-) and hole (h^+) pairs within the bulk phase of the photocatalyst (Albero and García, 2016). Once these charge carriers reach the surface of the photocatalyst, the holes (h^+) in the VB can oxidize adsorbed water (H_2O) molecules, producing highly reactive hydroxyl radicals (OH). Simultaneously, the electrons (e^-) in the CB reduce adsorbed oxygen (O_2), generating superoxide radicals ($\bullet O_2^-$).

These reactive radical species (hydroxyl and superoxide) are highly effective in breaking down pollutants adsorbed on the photocatalyst's surface into smaller, non-toxic molecules, such as water (H_2O) and carbon dioxide (CO_2). The mechanism for radical generation, particularly in TiO_2, a commonly used photocatalyst, can be summarized by the following reactions (Kumar and Pandey, 2017):

$$TiO_2 + hv(UV) \rightarrow TiO_2(e^-_{CB} + ^{h+}{}_{VB}) \quad (11.1)$$

$$O_2(h^+vB) + OH^- \rightarrow TiO_2 + OH^+ \quad (11.2)$$

$$TiO_2\ (e^-_{CB}) + O_2 \rightarrow TiO_2 + O_2^{\cdot-} \quad (11.3)$$

$$O_2^{\cdot-} + H + \rightarrow HO^{\cdot}_2 \quad (11.4)$$

$$\text{Pollutant} + OH^{\cdot} \rightarrow H_2O + CO_2 \quad (11.5)$$

$$\text{Pollutant} + O^{\cdot-}_2 \rightarrow H_2O + CO_2 \quad (11.6)$$

$$\text{Pollutant} + e^- \rightarrow \text{Reduction Product} \quad (11.7)$$

$$\text{Pollutant} + h^+ \rightarrow \text{Oxidation Product} \quad (11.8)$$

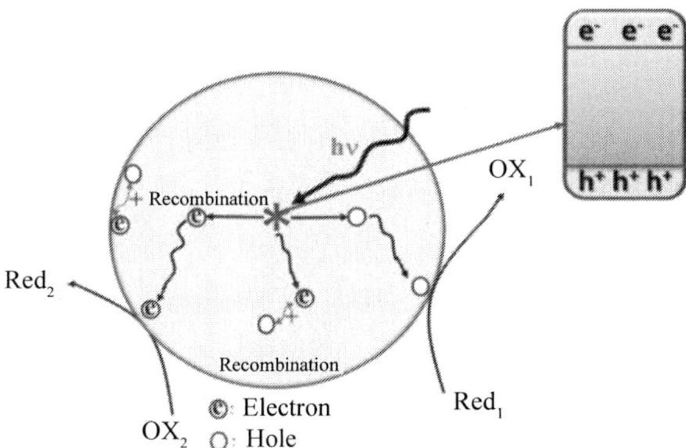

Figure 9. Mechanism of photocatalytic oxidation/reduction processes.

Factors Affecting the Photocatalytic Process

A suitable photocatalyst must exhibit high photoactivity under UV or visible light, resistance to photocorrosion, stability in the reaction medium, and cost-effectiveness (Molinari *et al.*, 2017). Photocatalytic efficiency is influenced by both intrinsic and extrinsic factors. Intrinsic parameters are related to the photocatalyst itself and include surface area, bandgap, particle size, surface defects, surface charge, functional groups, the recombination rate of charge carriers, and interfacial charge transfer. Extrinsic parameters, on the other hand, are external factors such as photocatalyst loading, pollutant concentration, solution pH, light intensity, and temperature, which govern the photocatalytic reaction.

Research indicates that reaction temperature affects the rate of photocatalytic reactions. Higher temperatures increase diffusion and mass transfer, thereby enhancing the reaction rate, while lower temperatures require a greater quantity of photocatalyst for effective mass transfer.

Photocatalytic removal of heavy metals from wastewater has gained significant attention. This process can operate at room temperature and atmospheric pressure, generating powerful oxidizing agents like hydroxyl radicals, which degrade both biodegradable and non-biodegradable contaminants without producing harmful intermediates. Additionally, photocatalysis is eco-friendly due to minimal chemical consumption and sludge production, and it is compatible with other physical and chemical methods while utilizing renewable solar energy.

Ionic Liquid Membrane

Ionic liquids (ILs), consisting of an organic cation such as imidazolium, ammonium, phosphonium, pyridinium, pyrrolidinium, or morpholinium, and an organic/inorganic anion like hexafluorophosphate, tetrafluoroborate, triflate, tosylate, or dicyanamide, have been widely used due to their high selectivity, polarity, ionic conductivity, thermal stability, near-zero vapor pressure, and non-flammability. These properties make ILs suitable as versatile solvents for various green processes (Welton *et al.*, 1999; Plechkova and Seddon, 2007; Wang *et al.*, 2016). ILs can interact with other molecules through different bonding mechanisms, such as hydrogen bonding, Van der Waals forces, ion pairing, and coulombic or electrostatic interactions. The ability to modify IL structures allows their properties to be tailored for specific applications, making them ideal for task-specific processes. This adaptability makes ILs useful in separation processes, electroanalytical and electrochemical applications, catalysis, and synthesis (Sawant *et al.*, 2011; Davis, 2004; Han and Armstrong, 2007) (see Figure 10).

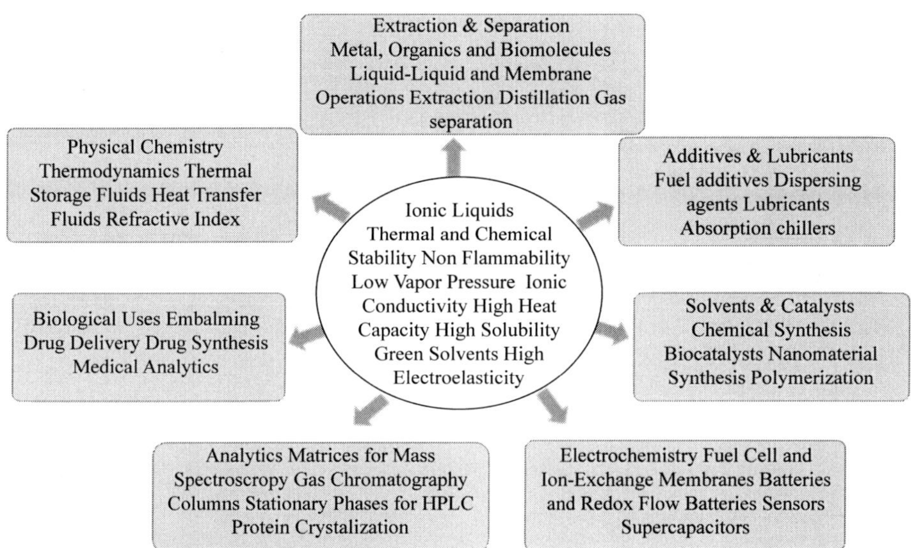

Figure 10. Most common of application of ILs.

Supported Ionic Liquid Membranes (SILMs)

Supported Ionic Liquid Membranes (SILMs) consist of a porous matrix where ionic liquids (ILs) are retained within the pores. The IL itself is responsible for the membrane's selectivity, while the supporting material serves primarily as a structural component. SILMs offer several advantages over traditional bulk IL membranes, such as requiring a smaller amount of IL, a simpler synthesis process, and the potential for recycling and reuse in multiple extraction cycles. SILMs can take various forms, including hollow fibers and flat sheets (see Figure 12.2). In this section, different types of SILMs employed for the removal of heavy metals from aqueous solutions are classified based on the type of cation present in the IL structure.

1. Phosphonium-Based Ils
2. Imidazolium-Based Ils
3. Ammonium-Based ILs

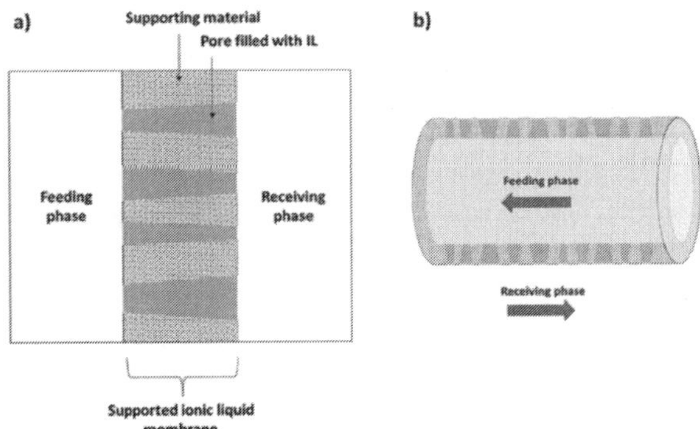

Figure 11. Scheme of different configurations of SLIM.

Polymer Ionic Liquid Inclusion Membranes (PILIMs)

Ammonium-Based Ils

Imidazolium-Based Ils

Phosphonium-Based Ils

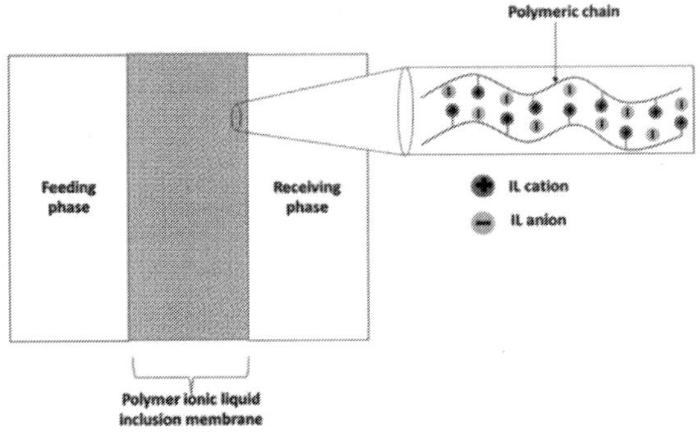

Figure 12. Generic scheme of structure of PILIMs.

Future Prospects

The distinctive properties of ionic liquids (ILs) have led to their widespread application in various research fields, including heavy metal removal, thanks to their selectivity and strong adsorption capacity. ILs are often used in bulk or emulsion form for this purpose; however, their limited reusability poses challenges for practical applications in wastewater treatment. To address this issue, IL solid membranes such as Supported Ionic Liquid Membranes (SILMs) have been developed, where the IL is retained within the pores of a porous matrix. Unfortunately, the stability of SILMs is compromised due to the potential leakage of ILs caused by degradation mechanisms.

A more promising alternative in terms of stability and reusability is the use of polymeric inclusion membranes (PIMs), where ILs are trapped within a polymeric matrix. These membranes provide greater stability compared to liquid membranes and SILMs, due to stronger interactions between the IL and the polymer. However, PIMs may exhibit slightly lower extraction capacities in solid/liquid extraction processes compared to liquid/liquid processes. Nonetheless, given the high cost of ILs, immobilizing them in membranes is an effective way to ensure their recycling and reuse in practical applications.

Life Cycle Assessment of Membranes for Removing Heavy Metals from Water and Wastewater

When selecting a process for removing heavy metals from water, it is crucial to ensure that the method chosen does not introduce new problems while addressing the original issue. The preferred method should have minimal impact on water flora and the environment, consume low energy, and offer high separation efficiency. Additionally, it should contribute to low environmental pollution and deliver high-quality final products.

To minimize the environmental impact and improve water purification, hybrid approaches that combine ceramic membranes with other processes such as photocatalysis, ozonation, coagulation, and activated carbon adsorption are increasingly being used. Although ceramic membranes come with a higher capital cost, their lower maintenance expenses and longer lifespan (up to 20 years) compared to polymeric membranes (8 years) make them a more cost-effective option over the entire product life cycle. The total cost of using both ceramic and polymeric membranes is nearly identical when accounting for the long-term benefits.

Ceramic membranes, despite their advantages, face challenges with membrane fouling, where colloidal particles or pore collapse can hinder performance over time, necessitating membrane replacement. Considering the total cost and longevity, ceramic membranes remain one of the best choices for water and wastewater treatment, though continued focus on reducing fouling is necessary for optimal performance.

Liquid Membrane

The emulsion liquid membrane (ELM) method offers several advantages over traditional liquid membrane technologies, such as bulk liquid membranes and supported liquid membranes, due to its cost-effectiveness, lower energy requirements, increased interfacial area, and higher mass transfer rate (Kumar, Thakur, and Panesar, 2019). ELM technology, particularly when stabilized with surfactants, has been applied successfully for separating toxic metal ions like lead, cadmium, mercury, and nickel.

However, despite its promise, the ELM method faces limitations that affect its stability and industrial viability. Issues such as leakage of internal reagents, coalescence of emulsion droplets, membrane dissolution, and the escape of water-soluble solutes into the emulsion hinder its large-scale application. These challenges result in reduced performance and increased operational risks, making it unsuitable for widespread industrial use (Abbassian and Kargari, 2016). For ELM to become a reliable method in industrial settings, further research is needed to address these stability issues.

In the emulsion liquid membrane (ELM) method, petroleum-based solvents such as heptane, hexane, chloroform, and kerosene are commonly used as the organic phase. These solvents contribute to significant environmental and cost impacts due to their toxicity, non-renewability, non-biodegradability, and flammability. Additionally, their high expense and limited availability further increase the overall cost of the life cycle of this method.

To mitigate these issues, research has explored the use of vegetable-based organic solvents, such as sunflower oil, palm oil, rice bran oil, and others. These alternative solvents offer several advantages: they are nontoxic, non-flammable, renewable, biodegradable, and generally cheaper than petroleum-based solvents. By using these vegetable-based solvents, the environmental impact and cost associated with the ELM method can be significantly reduced, making the process more sustainable and economically viable.

Economic Analysis of Membrane Technology

According to the United Nations, over 748 million people currently lack access to improved drinking water sources, and industrial water demands are expected to increase by 400% worldwide between 2000 and 2050 (UN-Water, 2015). Additionally, the global population is projected to rise from 7.6 billion today to 9.8 billion by 2050 (Luis and Moncayo, no date). This significant population growth heightens the need for freshwater and wastewater treatment, highlighting the urgency of addressing water shortages and potential conflicts.

In response to these challenges, effective and cost-efficient methods for removing contaminants such as pesticides, bacteria, organic particles, and heavy metals from water and wastewater are crucial. Membrane technologies have emerged as a preferred solution due to their environmentally friendly characteristics, relatively low energy consumption, and cost-effectiveness. The membrane market is expected to grow from USD 5.4 billion in 2019 to USD 8.3 billion by 2024, with a Compound Annual Growth Rate (CAGR) of 8.3%. Key drivers for this market growth include population expansion, increased awareness of wastewater reuse, and rapid industrialization.

In 2021, the market size for the Asia Pacific region, including China, India, Japan, Australia, and other Southeast Asian nations, was USD 2.47 billion. India and China are highlighted as particularly promising emerging markets, with a projected growth rate of 8.7% from 2022 to 2029. Europe, following Western European countries such as Germany, France, Spain, and Italy, is the second-largest region in terms of market size.

Reverse Osmosis (RO) membranes are the leading technology in the global membrane market. The Reverse Osmosis Membranes Market is expected to grow from USD 3.10 billion in 2022 to USD 6.00 billion by 2030, with a Compound Annual Growth Rate (CAGR) of 10.00% during this period. Initially used for desalinating seawater, RO membranes are now also employed to remove bacteria, pesticides, various organic wastes, and heavy metal ions (2023 Report by LG Chem Ltd., Toray Industries Inc., NX Filtration BV, Membranium (JSC RM Nanotech), Toyobo Co Ltd., Axeon Water Technologies, Report Code: FAF-2203).

Conclusion

Over the past decade, membrane filtration techniques have gained increasing importance in environmental engineering separations. However, several challenges are commonly associated with their use, including the need for pretreatment, fouling that leads to concentration polarization, energy requirements, and high capital and maintenance costs. Specifically, the cost of membrane modules, cleaning procedures such as backwashing and chemical cleaning, and overall maintenance can be significant (Akkoyunlu *et al.*, 2021).

To effectively manage membrane fouling, it is crucial to identify both the foulants and the fouling processes. A thorough understanding of these factors is essential for designing processes that minimize fouling and maintain high membrane productivity. This involves optimizing operating conditions, implementing effective pretreatment procedures, choosing appropriate cleaning solutions or methods, and selecting the right membrane for specific applications (Akkoyunlu *et al.*, 2021). Fouling remains a major obstacle to the widespread application of membrane separation techniques, as it can lead to increased operating and maintenance costs, reduced productivity and permeate quality, and frequent membrane replacement (Zularisam *et al.*, 2010).

References

2023 Report By LG Chem Ltd., Toray Industries Inc. NX Filtration BV Membranium (JSC RM Nanotech) Toyobo Co Ltd. Axeon Water Technologies, Report Code: FAF-220.

Abbassian, K., & Kargari, A. (2016). Modification of membrane formulation for stabilization of emulsion liquid membrane for extraction of phenol from aqueous solutions. Journal of Environmental Chemical Engineering, 4(4), 3926-3933.

Abdullah, N., Yusof, N., Lau, W. J., Jaafar, J., & Ismail, A. F. (2019). Recent trends of heavy metal removal from water/wastewater by membrane technologies. Journal of Industrial and Engineering Chemistry, 76, 17-38.

Abu, M., & Codjoe, S. N. A. (2018). Experience and future perceived risk of floods and diarrheal disease in urban poor communities in Accra, Ghana. International Journal of Environmental Research and Public Health, 15(12), 2830.

Ahmad, N. A., Goh, P. S., Abdul Karim, Z., & Ismail, A. F. (2018). Thin film composite membrane for oily waste water treatment: Recent advances and challenges. Membranes, 8(4), 86.

Al-Anezi, A. A. H., Sharif, A. O., Sanduk, M. I., & Khan, A. R. (2012). Experimental investigation of heat and mass transfer in tubular membrane distillation module for desalination. International Scholarly Research Notices, 2012(1), 738731.

Aliyu, U. M., Rathilal, S., & Isa, Y. M. (2018). Membrane desalination technologies in water treatment: A review. Water Practice & Technology, 13(4), 738-752.

Al-Maqdi, K. A., Elmerhi, N., Athamneh, K., Bilal, M., Alzamly, A., Ashraf, S. S., & Shah, I. (2021). Challenges and recent advances in enzyme-mediated wastewater remediation—a review. Nanomaterials, 11(11), 3124.

Alshahrani, A., Alharbi, A., Alnasser, S., Almihdar, M., Alsuhybani, M., & AlOtaibi, B. (2021). Enhanced heavy metals removal by a novel carbon nanotubes buckypaper membrane containing a mixture of two biopolymers: Chitosan and i-carrageenan. Separation and Purification Technology, 276, 119300.

Aquino, R. R., Tolentino, M. S., Elacion, R. M. P. D., Ladrillono, R., Laurenciana, T. R. C., & Basilia, B. A. (2018, October). Adsorptive removal of lead (Pb2+) ion from water using cellulose acetate/polycaprolactone reinforced nanostructured membrane. In IOP Conference Series: Earth and Environmental Science (Vol. 191, No. 1, p. 012139). IOP Publishing.

Berdasco, M., Coronas, A., & Vallès, M. (2017). Theoretical and experimental study of the ammonia/water absorption process using a flat sheet membrane module. Applied thermal engineering, 124, 477-485.

Birniwa, A. H., Abubakar, A. S., Huq, A. O., & Mahmud, H. N. M. E. (2021). Polypyrrole-polyethyleneimine (PPy-PEI) nanocomposite: An effective adsorbent for nickel ion adsorption from aqueous solution. Journal of Macromolecular Science, Part A, 58(3), 206-217.

Bozell, J. J., & Petersen, G. R. (2010). Technology development for the production of biobased products from biorefinery carbohydrates—the US Department of Energy's "Top 10" revisited. Green chemistry, 12(4), 539-554.

Briffa, J., Sinagra, E., & Blundell, R. (2020). Heavy metal pollution in the environment and their toxicological effects on humans. Heliyon, 6(9).

Carpenter, A. W., de Lannoy, C. F., & Wiesner, M. R. (2015). Cellulose nanomaterials in water treatment technologies. Environmental science & technology, 49(9), 5277-5287.

Choi, H. Y., Bae, J. H., Hasegawa, Y., An, S., Kim, I. S., Lee, H., & Kim, M. (2020). Thiol-functionalized cellulose nanofiber membranes for the effective adsorption of heavy metal ions in water. Carbohydrate Polymers, 234, 115881.

Chuan, D., Fan, R., Wang, Y., Ren, Y., Wang, C., Du, Y., ... & Guo, G. (2020). Stereocomplex poly (lactic acid)-based composite nanofiber membranes with highly dispersed hydroxyapatite for potential bone tissue engineering. Composites Science and Technology, 192, 108107.

Chuang, Y. H., & Shi, H. J. (2022). UV/chlorinated cyanurates as an emerging advanced oxidation process for drinking water and potable reuse treatments. Water Research, 211, 118075.

Croisier, F., Duwez, A. S., Jérôme, C., Léonard, A. F., Van Der Werf, K. O., Dijkstra, P. J., & Bennink, M. L. (2012). Mechanical testing of electrospun PCL fibers. Acta biomaterialia, 8(1), 218-224.

Cui, Z. F., Jiang, Y., & Field, R. W. (2010). Fundamentals of pressure-driven membrane separation processes. In Membrane technology (pp. 1-18). Butterworth-Heinemann.

Diallo, M. S. (2014). Water treatment by dendrimer-enhanced filtration: Principles and applications. In Nanotechnology applications for clean water (pp. 227-239). William Andrew Publishing.

Dittrich, J., Gnirss, R., Peter-Fröhlich, A., & Sarfert, F. (1996). Microfiltration of municipal wastewater for disinfection and advanced phosphorus removal. Water Science and Technology, 34(9), 125-131.

Dökmeci, A. H. (2020). Environmental impacts of heavy metals and their bioremediation. Heavy Metals-Their Environmental Impacts and Mitigation, 1-16.

Dong, Y., Sang, D., He, C., Sheng, X., & Lei, L. (2019). Mxene/alginate composites for lead and copper ion removal from aqueous solutions. RSC advances, 9(50), 29015-29022.

El Batouti, M., Al-Harby, N. F., & Elewa, M. M. (2021). A review on promising membrane technology approaches for heavy metal removal from water and wastewater to solve water crisis. Water, 13(22), 3241.

Ellouze, E., Tahri, N., & Amar, R. B. (2012). Enhancement of textile wastewater treatment process using nanofiltration. Desalination, 286, 16-23.

Grigoray, O., Wondraczek, H., Heikkilä, E., Fardim, P., & Heinze, T. (2014). Photoresponsive cellulose fibers by surface modification with multifunctional cellulose derivatives. Carbohydrate polymers, 111, 280-287.

Guo, Y., Qi, P. S., & Liu, Y. Z. (2017, May). A review on advanced treatment of pharmaceutical wastewater. In IOP Conference Series: Earth and Environmental Science (Vol. 63, No. 1, p. 012025). IOP Publishing.

H. Davis Jr, J. (2004). Task-specific ionic liquids. Chemistry letters, 33(9), 1072-1077.

Han, X., & Armstrong, D. W. (2007). Ionic liquids in separations. Accounts of Chemical Research, 40(11), 1079-1086.

Hernández, N., Williams, R. C., & Cochran, E. W. (2014). The battle for the "green" polymer. Different approaches for biopolymer synthesis: bioadvantaged vs. bioreplacement. Organic & biomolecular chemistry, 12(18), 2834-2849.

Ismail, A. F., Khulbe, K. C. & Matsuura, T. 2019. Chapter 5 - RO membrane module. In: Ismail, A. F., Khulbe, K. C. & Matsuura, T. (eds.) Reverse Osmosis (pp. 117–141). Elsevier.

Jagaba, A. H., Kutty, S. R. M., Hayder, G., Baloo, L., Ghaleb, A. A. S., Lawal, I. M., ... & Umaru, I. (2021). Degradation of Cd, Cu, Fe, Mn, Pb and Zn by Moringa-oleifera, zeolite, ferric-chloride, chitosan and alum in an industrial effluent. Ain Shams Engineering Journal, 12(1), 57-64.

Jagaba, A. H., Kutty, S. R. M., Khaw, S. G., Lai, C. L., Isa, M. H., Baloo, L., ... & Zango, Z. U. (2020). Derived hybrid biosorbent for zinc (II) removal from aqueous solution by continuous-flow activated sludge system. Journal of Water Process Engineering, 34, 101152.

Jose, A. J., Kappen, J., & Alagar, M. (2018). Polymeric membranes: Classification, preparation, structure physiochemical, and transport mechanisms. In Fundamental biomaterials: polymers (pp. 21-35). Woodhead Publishing.

Kapoor, D., & Singh, M. P. (2021). Heavy metal contamination in water and its possible sources. In Heavy metals in the environment (pp. 179-189). Elsevier.

Kumar, A., Thakur, A., & Panesar, P. S. (2019). A review on emulsion liquid membrane (ELM) for the treatment of various industrial effluent streams. Reviews in Environmental Science and Bio/Technology, 18, 153-182.

Landaburu-Aguirre, J., García, V., Pongrácz, E., & Keiski, R. (2006). Applicability of membrane technologies for the removal of heavy metals. Desalination, 1(200), 272-273.

Lee, C. M., Palaniandy, P., & Dahlan, I. (2017). Pharmaceutical residues in aquatic environment and water remediation by TiO 2 heterogeneous photocatalysis: a review. Environmental Earth Sciences, 76, 1-19.

Li, Y., Dong, H., Li, L., Tang, L., Tian, R., Li, R., ... & Zeng, G. (2021). Recent advances in waste water treatment through transition metal sulfides-based advanced oxidation processes. Water Research, 192, 116850.

主観的健康感を中心とした在宅高齢者における 健康関連指標に関する共分散構造分析

Lv, J., Zhang, G., Zhang, H., Zhao, C., & Yang, F. (2018). Improvement of antifouling performances for modified PVDF ultrafiltration membrane with hydrophilic cellulose nanocrystal. Applied Surface Science, 440, 1091-1100.

Mahmoudi, F., & Akbarzadeh, A. (2018). Sustainable desalination by permeate gap membrane distillation technology. In Emerging Technologies for Sustainable Desalination Handbook (pp. 157-204). Butterworth-Heinemann.

Masindi, V., & Muedi, K. L. (2018). Environmental contamination by heavy metals. Heavy metals, 10(4), 115-133.

Meng, X., Li, Z., & Zhang, Z. (2017). Pd-nanoparticle-decorated peanut-shaped BiVO4 with improved visible light-driven photocatalytic activity comparable to that of TiO2 under UV light. Journal of Catalysis, 356, 53-64.

Mohan, S., Oluwafemi, O. S., Kalarikkal, N., Thomas, S., & Songca, S. P. (2016). Biopolymers–application in nanoscience and nanotechnology. Recent advances in biopolymers, 1(1), 47-66.

Mohd Kaus, N. H., Rithwan, A. F., Adnan, R., Ibrahim, M. L., Thongmee, S., & Mohd Yusoff, S. F. (2021). Effective strategies, mechanisms, and photocatalytic efficiency of semiconductor nanomaterials incorporating rGO for environmental contaminant degradation. Catalysts, 11(3), 302.

Molinari, R., Lavorato, C., & Argurio, P. (2017). Recent progress of photocatalytic membrane reactors in water treatment and in synthesis of organic compounds. A review. Catalysis Today, 281, 144-164.

Molinari, R., Lavorato, C., & Argurio, P. (2020). Visible-light photocatalysts and their perspectives for building photocatalytic membrane reactors for various liquid phase chemical conversions. Catalysts, 10(11), 1334.

Molinari, R., Lavorato, C., & Argurio, P. (2021). The evolution of photocatalytic membrane reactors over the last 20 years: a state of the art perspective. Catalysts, 11(7), 775.

Mozia, S., Tomaszewska, M., & Morawski, A. W. (2005). A new photocatalytic membrane reactor (PMR) for removal of azo-dye Acid Red 18 from water. Applied Catalysis B: Environmental, 59(1-2), 131-137.

Murić, A., Petrinić, I., & Christensen, M. L. (2014). Comparison of ceramic and polymeric ultrafiltration membranes for treating wastewater from metalworking industry. Chemical Engineering Journal, 255, 403-410.

O'Connell, D. W., Birkinshaw, C., & O'Dwyer, T. F. (2008). Heavy metal adsorbents prepared from the modification of cellulose: A review. Bioresource technology, 99(15), 6709-6724.

Obotey Ezugbe, E., & Rathilal, S. (2020). Membrane technologies in wastewater treatment: a review. Membranes, 10(5), 89.

Pal, P. (2015). Chapter 4—Arsenic removal by membrane filtration. Groundwater Arsenic Remediation, 105-177.

Palacios Hinestroza, H., Urena-Saborio, H., Zurita, F., Guerrero de Leon, A. A., Sundaram, G., & Sulbarán-Rangel, B. (2020). Nanocellulose and polycaprolactone nanospun composite membranes and their potential for the removal of pollutants from water. Molecules, 25(3), 683.

Pan, X., Dong, W., Zhang, J., Xie, Z., Li, W., Zhang, H., ... & Lei, B. (2021). TiO2/chlorophyll S-scheme composite photocatalyst with improved photocatalytic bactericidal performance. ACS Applied Materials & Interfaces, 13(33), 39446-39457.

Pervez, M. N., Mahboubi, A., Uwineza, C., Zarra, T., Belgiorno, V., Naddeo, V., & Taherzadeh, M. J. (2022). Factors influencing pressure-driven membrane-assisted volatile fatty acids recovery and purification-A review. Science of the Total Environment, 817, 152993.

Petrinic, I., Korenak, J., Povodnik, D., & Hélix-Nielsen, C. (2015). A feasibility study of ultrafiltration/reverse osmosis (UF/RO)-based wastewater treatment and reuse in the metal finishing industry. Journal of cleaner production, 101, 292-300.

Plechkova, N. V., & Seddon, K. R. (2007). Ionic liquids:"designer" solvents for green chemistry. Methods and Reagents for Green Chemistry: An Introduction, 103-130.

Rautenbach, R., Linn, T., & Eilers, L. (2000). Treatment of severely contaminated waste water by a combination of RO, high-pressure RO and NF—potential and limits of the process. Journal of Membrane Science, 174(2), 231-241.

Regis, A. O., Vanneste, J., Acker, S., Martínez, G., Ticona, J., García, V., ... & Sharp, J. O. (2022). Pressure-driven membrane processes for boron and arsenic removal: pH and synergistic effects. Desalination, 522, 115441.

Rodriguez-Mozaz, S., Ricart, M., Köck-Schulmeyer, M., Guasch, H., Bonnineau, C., Proia, L., ... & Barceló, D. (2015). Pharmaceuticals and pesticides in reclaimed water: efficiency assessment of a microfiltration–reverse osmosis (MF–RO) pilot plant. Journal of hazardous materials, 282, 165-173.

Rol, F., Belgacem, M. N., Gandini, A., & Bras, J. (2019). Recent advances in surface-modified cellulose nanofibrils. Progress in Polymer Science, 88, 241-264.

Salahi, A., Badrnezhad, R., Abbasi, M., Mohammadi, T., & Rekabdar, F. (2011). Oily wastewater treatment using a hybrid UF/RO system. Desalination and water treatment, 28(1-3), 75-82.

Sánchez, J., Butter, B., & Rivas, B. L. (2020). Biopolymers applied to remove metal ions through ultrafiltration. A review. Journal of the Chilean Chemical Society, 65(4), 5004-5010.

Sang, Y., Liu, H., & Umar, A. (2015). Photocatalysis from UV/Vis to near-infrared light: towards full solar-light spectrum activity. ChemCatChem, 7(4), 559-573.

Sawant, A. D., Raut, D. G., Darvatkar, N. B., & Salunkhe, M. M. (2011). Recent developments of task-specific ionic liquids in organic synthesis. Green Chemistry Letters and Reviews, 4(1), 41-54.

Scott, K. (1995). Introduction to membrane separations. Handbook of Industrial Membranes, 3-185.

Shankar, S., & Rhim, J. W. (2018). Bionanocomposite films for food packaging applications. Reference module in food science, 1, 1-10.

Smith, A. M., Moxon, S., & Morris, G. A. (2016). Biopolymers as wound healing materials. In Wound healing biomaterials (pp. 261-287). Woodhead Publishing.

Sridhar, S., Smitha, B., & Aminabhavi, T. M. (2007). Separation of carbon dioxide from natural gas mixtures through polymeric membranes—a review. Separation & Purification Reviews, 36(2), 113-174.

Sun, X., Wang, C., Li, Y., Wang, W., & Wei, J. (2015). Treatment of phenolic wastewater by combined UF and NF/RO processes. Desalination, 355, 68-74.

Tlili, I., & Alkanhal, T. A. (2019). Nanotechnology for water purification: electrospun nanofibrous membrane in water and wastewater treatment. Journal of Water Reuse and Desalination, 9(3), 232-248.

Van der Bruggen, B., Vandecasteele, C., Van Gestel, T., Doyen, W., & Leysen, R. (2003). A review of pressure-driven membrane processes in wastewater treatment and drinking water production. Environmental progress, 22(1), 46-56.

Wang, J., Luo, J., Feng, S., Li, H., Wan, Y., & Zhang, X. (2016). Recent development of ionic liquid membranes. Green Energy & Environment, 1(1), 43-61.

Wang, P. *et al.* (2018a) 'Adsorption of hexavalent chromium by novel chitosan/poly(ethylene oxide)/permutit electrospun nanofibers', New Journal of Chemistry, 42(21), pp. 17740–17749. doi: 10.1039/C8NJ03899F.

Wang, X. and Li, Y. (2002) 'Rational synthesis of alpha-MnO2 single-crystal nanorods', Chemical Communications, pp. 764–765.

Wang, X., Huang, K., Chen, Y., Liu, J., Chen, S., Cao, J., ... & Jing, T. (2018). Preparation of dumbbell manganese dioxide/gelatin composites and their application in the removal of lead and cadmium ions. Journal of hazardous materials, 350, 46-54.

Wang, Y., Chen, X., Zhang, J., Yin, J., & Wang, H. (2009). Investigation of microfiltration for treatment of emulsified oily wastewater from the processing of petroleum products. Desalination, 249(3), 1223-1227.

Wang, Z., Wu, A., Colombi Ciacchi, L., & Wei, G. (2018). Recent advances in nanoporous membranes for water purification. Nanomaterials, 8(2), 65.

Water, U. N. (2015). World water development report—water for a sustainable world. United Nations.Available at: http://www.unesco. org/new/en/natural-sciences/environment/water/wwap/wwdr/2015-water-for-a-sustainable-world/.

Welton, T. (1999). Room-temperature ionic liquids. Solvents for synthesis and catalysis. Chemical reviews, 99(8), 2071-2084.

WHO, UNICEF . JMP; 2018. Data.https://washdata.org/data [WWW Document] accessed 5.18.18. [Google Scholar]WHO, UNICEF, 2018b. Data [WWW Document]. JMP. URL https://washdata.org/data (accessed 5.18.18).

World Health Organization, & UniCeF. (2010). Progress on sanitation and drinking-water: 2010 update. In Progress on sanitation and drinking-water: 2010 update (pp. 60-60).

Xu, X., Long, Y., Li, Q., Li, D., Mao, D., Chen, X., & Chen, Y. (2019). Modified cellulose membrane with good durability for effective oil-in-water emulsion treatment. Journal of Cleaner Production, 211, 1463-1470.

Yalcinkaya, F. (2019). Preparation of various nanofiber layers using wire electrospinning system. Arabian Journal of Chemistry, 12(8), 5162-5172.

Yalcinkaya, F., Boyraz, E., Maryska, J., & Kucerova, K. (2020). A review on membrane technology and chemical surface modification for the oily wastewater treatment. Materials, 13(2), 493.

Yordanov, D. J. B. J. O. A. S. (2010). Preliminary study of the efficiency of ultrafiltration treatment of poultry slaughterhouse wastewater. Bulgarian Journal of Agricultural Science, 16(6), 700-704.

Yu, S., Pang, H., Huang, S., Tang, H., Wang, S., Qiu, M., ... & Wang, X. (2021). Recent advances in metal-organic framework membranes for water treatment: A review. Science of the Total Environment, 800, 149662.

Zhou, Z., Liu, Y. G., Liu, S. B., Liu, H. Y., Zeng, G. M., Tan, X. F., ... & Cai, X. X. (2017). Sorption performance and mechanisms of arsenic (V) removal by magnetic gelatin-modified biochar. Chemical Engineering Journal, 314, 223-231

Annexure 1

Regulatory Standards

Standard values for computation of pollution indices.

Parameter	Weightage (K/MAC) upper permissible concentrations (MAC).	Standard Acceptable Limit (g/L)	Highest Permissible Limit (g/L)	Maximum Admissible Limit (g/L)
As (Arsenic)	0.02	50	10	50
Cadmium	0.37	5	3	3
Chromium	0.02	50	50	50
Copper	0.001	1,000	2,000	1,000
Iron	0.006	300	200	200
Manganese	0.02	100	500	50
Nickel	0.05	20	20	20
Lead	0.75	100	10	1.5
Zinc	0.0002	5,000	3,000	5,000

Standard of drinking water (WHO)

S. No.	Parameter	Maximum Permissible Limit
1	Al (Aluminum)	0.2 mg/L
2	Ca (Calcium)	75 mg/L
3	Fe (Iron)	0.3 mg/L
4	Mg (Magnesium)	50 mg/L
5	NO3 (Nitrate)	50 mg/L
6	pH	6.5–8
7	TDS (Total Dissolved Solids)	1000 mg/L
8	E. coli	0

7

Microbial Bioprocesses in Aquatic Environments: Applications and Innovations

Mayur Bhadarka[1*], D.T. Vaghela[2], Alwinpeter M.[3] and Hardik Sikotariya[4]

[1]*PG Scholar, Department of Aquatic Environment Management*

[2]*Associate Professor and Head, Department of sAquatic Environment Management*

[3]*PG Scholar, Department of Aquaculture*

[4]*PG Scholar, Department of Fisheries Resource Management*

College of Fisheries Science, Kamdhenu University, Veraval (362265), Gujarat, India

Abstract

Microbial bioprocesses are essential for maintaining the vitality and functionality of aquatic ecosystems, playing a key role in important biogeochemical cycles including those of carbon, nitrogen, and sulfur. Bioprocesses such as bioremediation, biofertilization, biofilms, biofloc technology, probiotics, bio-bleaching, bio-corrosion, and bio-fouling. these processes are applied in different domains including wastewater treatment, aquaculture, and environmental clean-up. In wastewater treatment, techniques like activated sludge and biofilms are utilized for the removal of nutrients and the breakdown of pollutants. In aquaculture, microbial management and probiotics help in sustaining water quality and improving the health of aquatic species. In bioremediation, microbes are employed to tackle contaminants like heavy metals, oil spills, and pesticides, with advancements in microbial consortia enhancing the effectiveness of pollutant-specific degradation. the role of microorganisms as bioindicators and biosensors. Recent developments in microbial biotechnology feature progress in genetically engineered microbes, the use of synthetic biology, and the implementation of microbial fuel cells (MFCs) for generating bioenergy. Challenges and limitations, including environmental constraints, the potential risks of genetically modified organisms, and the feasibility of large-scale applications.

Keywords: *Bioremediation, Biofertilization, Biofilms, Biofloc, Probiotics, Bio-Bleaching, Bio-Corrosion, Bio-Fouling*

1.0 Introduction

Microorganisms are fundamental to the well-being and functionality of aquatic ecosystems, playing a key role in managing water quality, nutrient cycling, and ecological equilibrium. They are crucial for the turnover of essential nutrients such as carbon, nitrogen, sulfur, and phosphorus, which are vital for the development of aquatic organisms (Sivakumar *et al.*, 2024). Microbial processes, including nitrogen fixation and methanogenesis, are central to the biogeochemical cycles that support aquatic environments (Shiky *et al.*, 2023). Through their metabolic functions, microbes regulate the availability of nutrients and influence primary productivity, thus sustaining the balance and efficiency of these ecosystems.

Microbes play a pivotal role in mitigating pollution and enhancing water quality. They are key agents in the bioremediation process, transforming toxic substances into less harmful forms, which helps in improving water quality (Shiky *et al.*, 2023). Techniques involving microbial indicators are valuable for assessing the health of aquatic systems and managing various contaminants (Savenko and Prysiazhniuk, 2022). In addition, aquatic plants and microbes form beneficial partnerships that boost nutrient absorption and promote plant health, which is essential for supporting the entire aquatic food web (Singh *et al.*, 2023). These relationships, shaped by environmental conditions, influence microbial diversity and their roles, thereby contributing significantly to plant growth and the overall stability of aquatic habitats.

Microbial processes play a crucial role in breaking down hazardous substances such as plastics and polycyclic aromatic hydrocarbons (PAHs), leading to improved soil health and reduced environmental toxins (Kaur and Vyas, 2024; Piotrowska and Boruszko, 2024). In the realm of sustainable agriculture, beneficial microorganisms enhance soil fertility and stimulate plant growth, even under challenging environmental conditions, thereby supporting more sustainable farming practices (Kaur and Vyas, 2024). The application of Effective Microorganisms (EM) has been effective in reducing soil pollutants and fostering healthier ecosystems (Piotrowska and Boruszko, 2024). Additionally, microbial processes are crucial for renewable energy production and bioremediation, particularly in handling mining waste by accumulating heavy metals and producing biofuels (Nair *et al.*, 2024). These processes also offer potential for enhancing waste management efficiency and reducing environmental contamination impacts. Despite their significant promise, there are ongoing challenges in optimizing these systems for practical use. Continued research is vital to fully realize their potential, advance sustainable development, and address the complexities associated with environmental management.

2.0 Principle And Applications Of Bioprocesses

2.1 Bioremediation

Bioremediation is an environmentally sustainable method that leverages microorganisms to cleanse pollutants from the environment. Its growing popularity stems from its cost-efficiency and minimal environmental impact. The core principles and practical applications of bioremediation can be divided into several key areas.

Principles of Bioremediation

- **Role of Microorganisms**: Microbes such as bacteria, fungi, and algae are essential in converting harmful pollutants into less toxic or harmless substances. These organisms break down contaminants like hydrocarbons, pesticides, and heavy metals through their natural metabolic activities (Ayilara and Babalola, 2023). In many cases, specific strains of microbes are chosen or genetically modified to efficiently target and degrade particular pollutants, making this approach highly suitable for addressing specific environmental problems. Their biological processes enable ecosystem restoration without introducing hazardous chemicals.
- **Bioremediation Mechanisms**: The process of bioremediation occurs through various techniques, such as bioaugmentation, biostimulation, and bioattenuation. Bioaugmentation involves adding specific microbial strains to accelerate the breakdown of pollutants, while biostimulation focuses on enhancing the growth and performance of existing microbes by providing nutrients or optimizing environmental conditions (Patel *et al.*, 2022). Bioattenuation, in contrast, allows contaminants to naturally degrade over time without direct human intervention, relying on the local microbial population's ability to gradually reduce pollution levels. The choice of mechanism depends on the characteristics of the contaminated site and the type of pollutants present.

Applications of Bioremediation

- **Oil Spill Cleanup**: Certain bacterial strains are particularly adept at degrading hydrocarbons, making them valuable for remediating oil spills. These bacteria decompose the complex molecules found in hydrocarbons, helping to rehabilitate affected marine ecosystems (Ron and Rosenberg, 2014). These microbes can also function in various environmental conditions, making them suitable for long-term remediation efforts.

- **Heavy Metal Removal**: Some bacteria can modify metal ions, making it easier to remove them from polluted environments through methods like biosorption, bioaccumulation, or conversion into less toxic forms. This is especially important for treating contaminated industrial and mining sites where heavy metal pollution poses significant risks (Sreerenjini *et al.*, 2024). Microbes have shown great potential in removing metals such as lead, mercury, and cadmium, aiding in large-scale environmental restoration.
- **Advancements in Genetic Engineering**: Modern genetic technologies like CRISPR are boosting the effectiveness of bioremediation by enhancing the capabilities of microorganisms. By modifying the genetic structure of microbes, researchers can create organisms that break down specific pollutants more efficiently, accelerating the overall remediation process (Sharma *et al.*, 2024). These innovations also enable the development of organisms that can thrive in extreme conditions, further expanding the possibilities for bioremediation.
- **Field-Scale Implementation**: The combination of molecular biological tools with conventional approaches has greatly improved the planning and oversight of bioremediation efforts at polluted sites (Madison *et al.*, 2023). These innovative tools offer more accurate detection and analysis of the microbial populations responsible for breaking down contaminants. Furthermore, continuous monitoring of microbial performance allows for timely modifications to bioremediation methods, enhancing pollutant degradation and supporting sustained recovery of the site.

2.2 Biofertilisation

Biofertilization is an eco-friendly agricultural method that uses microorganisms to enhance soil fertility and boost crop productivity. This approach not only reduces the reliance on chemical fertilizers but also promotes environmental sustainability by fostering healthier ecosystems. The principles and uses of biofertilization are grounded in various mechanisms and innovative technologies.

Principles of Biofertilization

- **Microbial Activity**: Biofertilizers consist of beneficial microorganisms like bacteria and fungi that assist in vital processes such as nitrogen fixation, phosphorus solubilization, and iron sequestration, making nutrients more readily available to plants (Akram *et al.*, 2020; Singh *et al.*, 2021). These microorganisms also help crops withstand stressors like drought and disease, improving overall plant health (Singh *et al.*, 2021).

Beneficial microbes, such as plant growth-promoting rhizobacteria (PGPRs), enhance nutrient uptake and boost plant resistance to various stress factors (Bhardwaj *et al.*, 2014). Furthermore, some microbial strains can produce hormones that stimulate plant growth.

- **Soil Health Enhancement**: Applying biofertilizers improves soil fertility by strengthening soil structure and boosting organic matter content, which fosters biodiversity and reduces erosion (Singh *et al.*, 2021). By decreasing the use of chemical fertilizers, biofertilization also lowers the risk of nutrient runoff into waterways, which helps prevent eutrophication and supports more sustainable agricultural systems. This contributes to a more balanced nutrient cycle and healthier soil ecosystems.

Applications of Biofertilization

- **Increasing Crop Yields**: Biofertilizers have been found to raise crop yields by 20%–35% while reducing the need for synthetic fertilizers by as much as 60% (Singh *et al.*, 2021). This not only enhances productivity but also cuts costs and minimizes environmental impact. Additionally, biofertilizers improve nutrient uptake, enabling plants to more efficiently absorb available nutrients.
- **Effluent Utilization**: Biological effluents, derived from organic waste, can be applied via advanced irrigation systems, conserving freshwater and improving soil nutrient levels (Vivaldi *et al.*, 2024). This practice supports a circular economy by converting organic waste into valuable agricultural inputs. Moreover, using bio-based effluents lowers the environmental impact of farming, aiding in water conservation and soil improvement.
- **Soil Remediation**: When paired with fertilizers, specific microbial consortia can effectively decompose pollutants like hydrocarbons, demonstrating biofertilization's contribution to bioremediation (Silva-Castro *et al.*, 2012). This integrated method not only speeds up the breakdown of contaminants but also improves soil health and fertility. By combining microbial action with nutrient enrichment, this approach provides a more efficient and sustainable solution for cleaning up polluted areas.

2.3 Biofilms

Biofilms are complex microbial communities that have important roles in both environmental and healthcare contexts.

Principle of Biofilms

- **Mechanism of Biofilm Formation:** The process of biofilm formation begins with bacteria adhering to a surface, where they then develop into microcolonies that mature into intricate three-dimensional structures (Paraje, 2023). The biofilm's extracellular polymeric substance (EPS) matrix, composed of polysaccharides, proteins, and extracellular DNA, forms a robust barrier that protects the bacteria from both antibiotics and the immune system (Shree *et al.*, 2023). This matrix not only anchors the bacteria but also supports their growth and enhances their resistance by shielding them from various environmental pressures.
- **Resistance Mechanisms:** Biofilms exhibit increased resistance due to several key factors: the lower metabolic activity of the bacteria within the biofilm, the presence of persister cells capable of surviving harsh conditions, and the protective EPS matrix (Joshi *et al.*, 2023; Shree *et al.*, 2023). Additionally, biofilms facilitate horizontal gene transfer, enabling the exchange of antibiotic resistance genes between bacteria, which complicates treatment and accelerates the spread of resistance (Bello *et al.*, 2023). This gene transfer contributes to the rapid development of multi-drug-resistant bacterial strains, posing significant challenges to managing infections and effective treatment.
- Implications for Treatment
- The enduring nature and robustness of biofilms require the development of new treatment strategies, including specialized antibiofilm agents and combination therapies, to disrupt and dismantle biofilm structures effectively (Paraje, 2023; Joshi *et al.*, 2023). These strategies are designed to eliminate infections caused by biofilms and prevent their reformation by addressing various stages of biofilm development and stability. Additionally, research is focused on creating treatments that can break down the EPS matrix and block the horizontal transfer of resistance genes, which is crucial for improving the success rate of treatments and combating antibiotic resistance.

Applications of Biofilm

Environmental Applications

- **Pollution Control**: Biofilms are particularly effective at degrading pollutants such as heavy metals and organic substances, which makes them crucial for processes like wastewater treatment and soil cleanup (Shen *et al.*, 2024). Their structure, which provides an extensive surface area, facilitates more efficient microbial activity and pollutant breakdown. Additionally, biofilms can be engineered to address specific contaminants, enhancing the overall effectiveness of environmental remediation efforts.

- **Bioremediation**: In addition to tackling traditional pollutants, biofilms are also capable of addressing newer contaminants such as microplastics and antibiotic resistance genes (ARGs), showcasing their adaptability across different environmental challenges (Shen *et al.*, 2024). This flexibility is beneficial for developing new strategies to combat pollutants that are resistant to standard remediation techniques.

Industrial and Clinical Implications

- **Industrial Use**: Biofilms are utilized in bioreactors for the production of important biochemicals and in wastewater treatment processes. However, they can also lead to issues such as corrosion of industrial equipment and pipelines, which results in higher maintenance costs and operational difficulties (Rao, 2023). The tendency of biofilms to form on surfaces can disrupt industrial processes, requiring careful management to minimize their adverse effects.
- **Healthcare Challenges**: In healthcare settings, biofilms are a significant cause of chronic infections and contribute to antibiotic resistance, making treatment more complicated. They frequently develop on medical devices, including catheters and implants, leading to persistent infections that are difficult to manage with conventional antibiotics (Ali *et al.*, 2023; Joshi *et al.*, 2023). Addressing the resilience of biofilms to antimicrobial treatments is crucial for developing alternative therapies and strategies to control and reduce their impact.

2.4 Biofloc

Biofloc Technology (BFT) is an advanced aquaculture approach aimed at improving water quality and nutrient efficiency by converting inorganic nitrogen into microbial biomass using heterotrophic bacteria. This biomass, rich in nutrients, becomes a valuable food source for aquatic species, promoting more sustainable practices in aquaculture.

Principles of Biofloc Technology

- **Nutrient Conversion**: BFT effectively manages excess nutrients, especially ammonia, by transforming them into microbial protein that can be utilized by fish and shrimp (Ekasari, 2009). This process helps in lowering ammonia levels in the water and enhances the nutrient profile of the aquaculture environment. By recycling these nutrients, BFT supports high water quality and contributes to the health and growth of aquatic life.

- **Microbial Community**: The creation of bioflocs relies on a diverse microbial community crucial for maintaining water quality and minimizing disease (Choo and Caipang, 2015). These microorganisms play a key role in biological filtration, reducing pathogen levels, and fostering a balanced ecosystem. The interactions among different microbes help stabilize the aquatic environment, which in turn promotes healthier and more vigorous growth of the cultured species.

Applications in Aquaculture

- **Improved Production**: BFT has shown significant advantages in freshwater tilapia farming, including faster growth rates and more efficient feed conversion ratios (Choo and Caipang, 2015). This approach enhances feed utilization and can lead to increased productivity and profitability. Additionally, it results in more consistent and reliable production outcomes due to the improved growth performance.
- **Environmental Sustainability**: By reducing nutrient discharge into the environment and decreasing dependence on artificial feeds, BFT supports more sustainable aquaculture practices (Ekasari, 2009; Choo and Caipang, 2015). This method helps lower the ecological footprint of aquaculture by recycling nutrients and reducing the need for external feed inputs, thus promoting a more eco-friendly and resource-efficient aquaculture system.
- **Sustainable Development**: BFT supports sustainable aquaculture by enhancing food security and reducing resource inputs, aligning with global development goals (Bossier and Ekasari, 2017). This approach minimizes environmental impact by reducing waste and nutrient discharge, and it helps lower dependence on external feed sources. Additionally, by fostering efficient nutrient recycling and improving overall system resilience, BFT contributes to the long-term viability and sustainability of aquaculture operations.

2.5 Probiotics

Probiotics are live microorganisms that offer health benefits when ingested in adequate amounts, primarily by restoring and maintaining a balanced gut microbiota. They have wide-ranging applications, especially in the realms of food and healthcare, where they enhance health outcomes and ensure food safety.

Principles of Probiotics

- **Microbial Composition**: Probiotics consist mainly of beneficial bacteria such as *Lactobacillus* and *Bifidobacterium*, which support nutrient absorption and help regulate immune responses (Pawar *et al.*, 2022). These microorganisms play a crucial role in maintaining a healthy gut by outcompeting harmful bacteria and fostering a balanced microbiome. Their presence is essential for promoting overall gastrointestinal health and enhancing immune system function.

- **Mechanisms of Action**: Probiotics operate through various mechanisms, such as inhibiting the growth of pathogenic organisms, producing antimicrobial substances, and modulating immune responses (Pawar *et al.*, 2022; Arunavarsini *et al.*, 2024). They also contribute to preserving the gut barrier's integrity and reducing inflammation, which further enhances their health benefits.

Applications in Food Industry

- **Food Products**: Probiotics are incorporated into a range of products, including dairy products, fermented foods, and beverages, to support gut health and improve food safety by inhibiting harmful bacteria (Zambelli and De Mendonça, 2024). Their inclusion not only extends the shelf life of these products but also boosts their nutritional profile. Moreover, probiotic-enriched foods can help create a more diverse and beneficial gut microbiota.
- **Consumer Acceptance**: Successful integration of probiotics into food products relies on educating consumers about their health benefits and ensuring high product quality (Zambelli and De Mendonça, 2024). Effective labeling and clear communication about the benefits of probiotics can enhance consumer trust and increase the demand for probiotic-rich foods.

Healthcare Applications

- **Disease Prevention**: Probiotics are effective in preventing and managing a range of health issues, such as gastrointestinal disorders, metabolic conditions, and even mental health disorders, by promoting a healthy gut flora (Arunavarsini *et al.*, 2024). They can aid in digestion, alleviate symptoms of irritable bowel syndrome (IBS), and improve overall gut health.
- **Therapeutic Potential**: Probiotics offer therapeutic benefits for specific conditions, including antibiotic-associated diarrhea and immune system support (Arunavarsini *et al.*, 2024; Singh *et al.*, 2022). They can help restore gut microbiota balance after antibiotic use, reduce gastrointestinal symptoms, and bolster immune responses, making them a valuable component of therapeutic strategies.

2.6 Bio-Bleaching

Bio-bleaching is a biotechnology-based process that uses microorganisms to extract metals from ores and polluted environments, providing a more environmentally friendly alternativc to traditional mining and remediation techniques. It is particularly valuable in sectors like mining and environmental restoration, where microbial activity helps dissolve and recover valuable metals.

Principles of Bio-Bleaching

- **Microbial Action**: Bio-bleaching relies on acidophilic microorganisms, such as *Acidithiobacillus ferrooxidans*, which oxidize iron and sulfur compounds, producing acids that dissolve metals from sulfide minerals (Jones and Santini, 2023). These microorganisms thrive in highly acidic conditions, making them effective in breaking down metal sulfides. Their metabolic processes release metals like copper, zinc, and gold, providing a more sustainable and less chemical-intensive alternative to traditional leaching methods.
- **Mechanisms**: The bio-bleaching process involves complex enzymatic reactions, including the thiosulfate and polysulfide pathways, which break down mineral surfaces and release the metals within (Jones and Santini, 2023). These pathways convert metal sulfides into soluble forms, making metal extraction easier and more efficient. By reducing the need for harsh chemicals, bio-bleaching minimizes environmental harm and lowers the ecological impact of metal extraction.

Applications of Bio-Bleaching

- **Mining Industry**: Bio-bleaching is widely used for metal recovery from ores, offering a more eco-friendly alternative to traditional mining techniques (Mishra *et al.*, 2022). It reduces the dependence on toxic chemicals such as cyanide and lowers the energy requirements of metal extraction processes. Furthermore, bio-bleaching is particularly useful for extracting metals from low-grade ores, which are not economically viable using conventional methods, making it a cost-effective solution for the mining industry.
- **Environmental Remediation**: Bio-bleaching also plays a significant role in environmental cleanup, particularly in reducing heavy metal contamination in sediments and aquatic ecosystems (Fonti *et al.*, 2016). This technique can be used to remove harmful metals like arsenic, lead, and mercury from polluted water bodies, decreasing their ecological impact and improving overall water quality. As a result, bio-bleaching

is a valuable tool for mitigating pollution and supporting ecosystem restoration efforts.

2.7 Bio-Corrosion

Bio-Corrosion or Microbiologically influenced corrosion (MIC) is a significant issue across numerous industries due to the damaging impact of microorganisms on metals and other materials. This process is largely driven by biofilms produced by certain bacteria, which can accelerate the corrosion of metals. Gaining an understanding of how bio-corrosion works is essential for developing ways to prevent or mitigate its effects.

Mechanisms of Bio-Corrosion

- **Microbial Activity**: Certain microorganisms, including sulfate-reducing bacteria (SRB), accelerate corrosion by generating corrosive substances such as biogenic sulfuric acid. These bacteria thrive in environments lacking oxygen, like pipelines, oil rigs, and storage tanks, where they release by-products that degrade materials. Additionally, the acidic microenvironments they create further intensify the corrosion process (Oyewole *et al.*, 2022).
- **Biofilm Formation**: Biofilms play a critical role in localized metal corrosion. These microbial layers not only allow bacteria to grow but also act as barriers to oxygen diffusion, creating anaerobic conditions that promote the activity of SRBs. The biofilms also trap corrosive elements, forming electrochemical gradients that lead to pitting corrosion, which, if uncontrolled, can result in serious structural damage (Oyewole *et al.*, 2022).

Applications and Mitigation Strategies

- **Mitigation Strategies**: An effective method to manage bio-corrosion involves using bio-based materials like *Bacillus subtilis*, which enhance resistance to corrosion by consuming dissolved oxygen and filling in cracks within concrete structures. These bacteria form protective layers that shield the material from further degradation and reinforce its structural integrity. This approach is particularly useful in environments susceptible to microbial-induced damage, such as sewage systems and pipelines, where corrosion can be a significant issue (Kawaai, 2022).
- **Environmental Monitoring**: Research on bio-corrosion is vital for environmental monitoring, particularly in industries like water treatment, pipelines, and sewage systems. By analyzing microbial activity, early signs of corrosion can be detected, allowing preventive measures to be taken before serious structural damage occurs. This research supports

the development of advanced sensors and real-time monitoring systems designed to track microbial corrosion in tough, industrial environments.

- **Innovative Materials**: Bio-corrosion research has paved the way for the creation of advanced materials such as bio-concrete, which includes anaerobic granular sludge. This material has been shown to significantly reduce corrosion rates and lower sulfide absorption, making it highly suitable for sewer systems and other harsh environments. Furthermore, bio-concrete contributes to long-term infrastructure stability by utilizing microbial self-healing mechanisms that repair small cracks, thereby enhancing the durability and lifespan of structures over time (Song *et al.*, 2021).

2.8 Bio-Fouling

Bio-fouling refers to the accumulation of microorganisms, plants, algae, or small aquatic organisms on submerged surfaces, presenting significant challenges in areas such as biomedicine and environmental management. Understanding bio-fouling is crucial for devising effective methods to prevent and manage its impact.

Principles of Bio-fouling

- **Mechanisms**: The process of bio-fouling starts with microorganisms adhering to a surface, which leads to the development of biofilms. These biofilms modify the surface properties, facilitating the further colonization by additional organisms. As biofilms mature, they become increasingly difficult to remove, resulting in persistent fouling that can disrupt the performance of equipment and systems. Over time, this can lead to issues such as mechanical obstructions, corrosion, and diminished efficiency in submerged systems (Masters, 2013).
- **Factors Influencing Bio-fouling**: The extent of bio-fouling is influenced by various environmental factors. Key elements include changes in temperature, the availability of nutrients, and the characteristics of the submerged surface, such as its texture and chemical composition. Rough or hydrophobic surfaces generally encourage faster attachment and colonization, while smoother or hydrophilic surfaces may slow down the process. Additionally, seasonal variations and water flow dynamics can affect the rate and distribution of bio-fouling (Bright, 1988).

Applications of Bio-fouling Research

- **Biomedical**: Bio-fouling is a serious concern in medical devices, as the buildup of microorganisms on surfaces can cause infections and other complications. By studying the mechanisms of bio-fouling, researchers can create anti-fouling surfaces that prevent microbial attachment. This is especially important for devices like catheters, implants, and surgical tools, where biofilm formation can compromise both patient health and the performance of the equipment. Developing materials that resist bacterial growth helps significantly lower the risk of infections and device-related failures, enhancing patient safety (Ansari, 2012).

- **Environmental**: In marine settings, bio-fouling has a notable impact on both the efficiency of ships and the well-being of marine ecosystems. The accumulation of organisms on ship hulls increases drag, resulting in greater fuel consumption and higher operational costs. Additionally, bio-fouling can lead to the spread of invasive species, which can upset the balance of local ecosystems. Research into bio-fouling processes supports the creation of environmentally friendly anti-fouling coatings that not only boost ship performance but also reduce the environmental footprint. These coatings aim to eliminate the need for harmful chemicals that could negatively affect marine organisms (Kikuchi, 2010).

3.0 Utilization of Microbial Bioprocesses

3.1 Wastewater treatment

Microbial bioprocesses are essential in wastewater treatment, harnessing the natural ability of microorganisms to break down pollutants and improve water quality. These processes are valued for their efficiency, cost-effectiveness, and environmental sustainability. Microbial technologies work by utilizing microorganisms to decompose organic matter, transforming it into non-toxic byproducts such as carbon dioxide and water (Singh and Shyu, 2024). Common biological treatment methods include anaerobic digestion, aerobic systems, and constructed wetlands, all of which are highly effective at purifying wastewater by removing contaminants (Sultana *et al.*, 2024). Additionally, advancements in microbial bioprocessing are enhancing the removal of emerging pollutants, such as pharmaceuticals and microplastics, making these methods increasingly crucial for modern wastewater management. The integration of these technologies into treatment plants has also contributed to reducing energy consumption and greenhouse gas emissions, further promoting sustainability.

3.2 Aquaculture

Microbial bioprocesses are essential for enhancing sustainability and efficiency in aquaculture by utilizing beneficial microorganisms to improve water quality, provide eco-friendly feed options, and strengthen disease resistance in aquatic organisms. Marine bacteria are utilized in biological filters to eliminate harmful compounds like nitrate and ammonia, thereby enhancing the water quality in recirculating aquaculture systems (Alaa *et al.*, 2023). Probiotics are increasingly applied to ensure water safety and foster healthier aquatic habitats (Midhun and Arun, 2023; Alaa *et al.*, 2023). Microbial protein feed (MPF), derived from agricultural and industrial byproducts, offers a sustainable and resource-efficient substitute for conventional fish feeds, cutting down on carbon emissions while delivering high-quality protein for fish and shrimp (Zhou *et al.*, 2023). Incorporating microalgae and other microorganisms into aquafeeds not only boosts their nutritional value but also reduces dependence on wild-caught fish, addressing important ecological issues (Sarker, 2023). Moreover, advances in biotechnology have led to the creation of vaccines and the improvement of fish growth factors, which enhance disease resistance and overall productivity (Labh, 2023). These developments collectively make aquaculture practices more effective and environmentally friendly.

3.3 Waste Treatment and Resource Recovery

Microbial biomass plays a crucial role in managing waste from the agro-food sector, such as potato wastewater and whey, effectively lowering disposal expenses while generating valuable byproducts like glucans and vitamins (Yang, 2024). Moreover, nutrient-rich dairy waste can be transformed through microbial processes into biofertilizers and biofuels using both aerobic and anaerobic methods (Usmani *et al.*, 2021). These approaches not only minimize waste but also contribute to a circular economy by converting waste into beneficial resources, thereby enhancing sustainable waste management and mitigating environmental impact.

3.4 Biorefineries and Value-Added Products

In biorefineries, microorganisms act as biocatalysts, converting waste materials into valuable chemicals and polymers, thus improving resource recovery and decreasing dependence on finite resources (Sar *et al.*, 2023). Additionally, microbial processes that utilize food and forestry by-products can produce organic fertilizers and other valuable chemicals, supporting sustainable agriculture and enhancing soil health (Kivelä and Hakalehto, 2022). These innovations facilitate a more efficient and sustainable use of resources, aligning with broader objectives of environmental conservation and resource management.

4.0 Microorganisms as Bioindicators and Biosensors.

4.1 Microorganisms as Bioindicators

Microorganisms are essential bioindicators for evaluating environmental health and detecting pollution. Their variety and abundance offer key information about alterations in ecosystem conditions, making them crucial for monitoring ecosystems on land and in water. Bacteria, fungi, and algae are particularly responsive to environmental changes, indicating the presence of pollutants like heavy metals, hydrocarbons, and organic substances (Sangwan *et al.*, 2024; Ogidi *et al.*, 2024). Due to their sensitivity to contaminants, microorganisms serve as early warning indicators, supplying important data for environmental monitoring and preservation efforts.

Fungi as Environmental Indicators: Fungi are vital to ecosystems due to their wide-ranging ecological roles and high sensitivity to environmental changes. Their presence or absence can serve as key indicators of ecosystem health, making them valuable tools for both environmental monitoring and pollution remediation. Fungi are useful for assessing air, water, and soil quality, often detecting pollutants that may not be identified by other methods (Warnasuriya *et al.*, 2023). Moreover, certain fungal species have the ability to degrade pollutants, contributing to the remediation of contaminated environments.

Microbial Communities in Soil: Microorganisms in soil, including bacteria and fungi, reflect the health of the soil, with changes in their populations signaling environmental disturbances or contamination. Shifts in these communities can act as early indicators of declining soil health, enabling timely corrective actions. The advent of metabarcoding technology has allowed for more precise detection of these microbial shifts, particularly in soils that are degraded or nutrient-deficient (Joos *et al.*, 2023). This technology has significantly improved soil health monitoring, offering more accurate assessments of ecological balance.

Microbes in Sediment: Microbial communities in sediment play an essential role in assessing environmental quality, as they respond to changes in their chemical and physical surroundings. Certain microbial taxa, such as *Rhodanobacteraceae*, have been identified as indicators of low nitrogen levels, while other groups, like microeukaryotes, are highly sensitive to heavy metal contamination. These microorganisms provide valuable insights into nutrient cycling and pollution levels in aquatic environments (Wei *et al.*, 2023), making them crucial for long-term monitoring and potential remediation efforts.

Microbial Indicators in Aquatic Ecosystems: In coastal and aquatic environments, microbial communities serve as important indicators of pollution, particularly in the interconnected soil-water-sediment systems. The identification of specific microbial genera that are correlated with particular pollutants has enhanced the precision of microbial-based environmental monitoring (Bourhane *et al.*, 2022). These communities provide early warnings of ecological stress, such as nutrient imbalances or toxic substances, guiding environmental management efforts to safeguard marine and coastal ecosystems.

4.2 Microorganisms as Biosensors

Microorganisms serve as powerful biosensors in environmental monitoring, relying on their natural biochemical and physiological responses to identify pollutants. These microbial-based sensors are developed to detect and measure environmental contaminants with advantages such as high specificity, fast detection, and cost-effectiveness. By utilizing the chemical sensitivity of microbial species, these systems offer an efficient way to evaluate environmental conditions with greater precision and in real time.

Mechanisms of Microbial Biosensors: Microbial biosensors work by exploiting the biochemical interactions between microorganisms and pollutants, converting these chemical signals into measurable outputs. These outputs are typically detected through electrochemical or optical devices, enabling the real-time monitoring of contamination in various environments (Sharma *et al.*, 2024; Carducci *et al.*, 2024). Advances in genetic engineering have enhanced the accuracy and adaptability of these biosensors, allowing them to be customized for specific pollutants such as heavy metals, pesticides, and organic compounds. This increased precision allows for the creation of highly specialized sensors that are capable of addressing unique environmental issues (Bala *et al.*, 2022). These biosensors can detect extremely low concentrations of contaminants, making them ideal for monitoring critical areas such as industrial discharge sites and polluted water bodies.

Applications in Environmental Monitoring: Microbial biosensors are widely applied in real-time environmental monitoring, particularly for water quality assessments. These biosensors can identify a broad range of pollutants, such as heavy metals, pesticides, and other harmful chemicals, offering a quick and accurate method for detecting environmental hazards (Gautam, 2024; Aiyer *et al.*, 2023). Their integration into portable devices enhances their usability, enabling field testing and continuous environmental monitoring without the need for time-consuming laboratory procedures. This portability and efficiency make microbial biosensors valuable for environmental regulators and industries, helping them meet environmental standards and regulations (Bala *et al.*, 2022). Additionally, their ability to operate in remote or inaccessible locations makes them a versatile solution for large-scale environmental assessments.

5.0 Innovations in Microbial Biotechnology for Aquatic Systems

5.1 Genetically engineered microbes

Advancements in microbial biotechnology, particularly through genetically modified microbes, are transforming aquatic systems by improving water quality, promoting sustainable aquaculture, and enhancing biofuel production. These developments tap into the unique abilities of microorganisms to tackle both environmental and economic issues. For example, engineered microbes like compound microbial agents (CMA), which include *Bacillus subtilis* and lactic acid bacteria, have demonstrated notable improvements in water quality for aquaculture species such as *Micropterus salmoides*. These microbes boost oxygen levels and decrease harmful nitrogen compounds, fostering healthier aquatic environments (Li *et al.*, 2024). In the realm of sustainable aquaculture, genetic modifications of fish species are being employed to optimize growth rates and bolster disease resistance, thereby increasing production efficiency and sustainability (Kumar *et al.*, 2024). Additionally, marine microbes are being genetically engineered to produce bioethanol, leveraging their resilience in saline conditions. This innovation not only provides a sustainable energy source but also helps manage waste in aquatic environments (Immanuel *et al.*, 2024). However, while these advancements hold great promise, concerns about their ecological impacts and the potential long-term effects of genetically modified organisms on natural ecosystems remain important considerations as microbial biotechnology evolves.

5.2 Synthetic Biology

Advancements in microbial biotechnology are profoundly reshaping aquatic systems through synthetic biology, leading to enhancements in water quality and the overall health of aquatic organisms. These innovations utilize microorganisms to develop sustainable solutions for both aquaculture and wastewater treatment. For example, compound microbial agents (CMAs) have been effective in improving water quality for species like *Micropterus salmoides* by elevating dissolved oxygen levels and lowering harmful nitrogen compounds (Li *et al.*, 2024). They also enhance fish health by diversifying gut microbiota, which benefits both the aquatic environment and the species within it (Li *et al.*, 2024). In wastewater management, microbial technologies offer a cost-effective and efficient alternative to traditional methods, employing microorganisms to break down pollutants. These technologies are adaptable and can be implemented on various scales, from household to industrial settings, providing versatile solutions for aquatic ecosystems (Singh and Shyu, 2024). Additionally, recent advances in biotechnological tools are refining processes in aquatic food production, such as fermentation and quality control, which supports more sustainable practices. By incorporating microbial systems, these advancements boost productivity and help address food security issues in response to increasing global demand for aquatic protein (Kumar *et al.*, 2023).

5.3 Microbial Fuel Cells (MFCs)

Microbial fuel cells (MFCs) represent a cutting-edge advancement in microbial biotechnology, with significant promise for enhancing both wastewater treatment and renewable energy production. These systems use electrogenic bacteria to convert organic waste into electrical energy, thus achieving both energy generation and pollutant removal. MFCs function through bacterial electron transport mechanisms, which are akin to those found in the mitochondria of eukaryotic cells, highlighting an evolutionary connection between these biological processes (Garimella *et al.*, 2024). The performance of MFCs hinges on the choice of electroactive bacteria and substrates, as specific microbial strains have demonstrated high power outputs and coulombic efficiencies (Indriyani *et al.*, 2023). MFCs are effective in processing various types of wastewaters, including those with low organic content, while concurrently producing electricity, aligning with goals for sustainable development (Gupta, 2024). Additionally, the application of microbial technologies within MFCs is both cost-effective and versatile, allowing for their deployment in a range of settings and scales (Singh and Shyu, 2024).

6.0 Challenges and Limitations of Microbial Bioprocesses

6.1 Environmental and Ecological Constraints

Nutrient and Environmental Factors: The effectiveness of microbial growth and bioremediation largely relies on the availability of nutrients and environmental conditions such as pH and temperature (Thakur *et al.*, 2024). However, fermentation processes can be hindered by substrate and product toxicity, which may impede microbial metabolism (Jarboe *et al.*, 2022).

Regulatory and Technological Challenges: Regulatory constraints, particularly those concerning genetically modified organisms (GMOs), can restrict the application of bioremediation technologies for certain persistent pollutants (Thakur *et al.*, 2024). Furthermore, the high costs of raw materials and the low yields observed in downstream processing present challenges for scaling up fermentation processes, particularly for the production of biosurfactants (Barbosa *et al.*, 2023).

Economic and Process Design Issues: Large-scale bioleaching operations often face engineering constraints that limit the adoption of promising microbial technologies, primarily due to inefficiencies in the slowest subprocesses

(Petersen, 2023). Achieving optimal conditions for microbial cultures remains difficult, leading to inconsistent results that can affect the technology's acceptance within the scientific community (Aguilar *et al.*, 2019).

6.2 Risks of Introducing Genetically Modified Microbes

Environmental Risks: The transfer of modified genes to native microbial populations could potentially disrupt local ecosystems (Sharma *et al.*, 2024). Additionally, the effectiveness of genetically engineered microorganisms (GEMs) can be compromised by pollutant toxicity and bioavailability, posing challenges for remediation efforts (Phian *et al.*, 2022).

Regulatory Challenges: Stringent regulations surrounding the release of GEMs can limit their use in bioremediation projects (Thakur *et al.*, 2024). Public concerns about safety and ethical implications can also lead to opposition against the use of these organisms (Meruvu, 2021).

Technological Limitations: The physiological traits of GEMs may constrain their effectiveness in real-world conditions, impacting their commercial viability (Montaño López *et al.*, 2022).

6.3 Sustainability of Large-Scale Applications in Aquatic Environments

Physiological Limitations: Microbial metabolic engineering often struggles to achieve the desired production levels due to limitations in the host organisms' physiological characteristics (Montaño López *et al.*, 2021). Moreover, the ability of microbes to adapt to new pollutants is crucial, yet environmental toxicity and bioavailability issues can hinder their survival and effectiveness (Phian *et al.*, 2022).

Environmental Dynamics: The complexity of aquatic ecosystems complicates the implementation of microbial remediation strategies. The success of these strategies depends on factors such as nutrient availability and the interactions with native microbial communities (Phian *et al.*, 2022).

Economic Factors: The high cost of raw materials and the complexity of scaling up fermentation processes for biosurfactant production pose significant barriers (Barbosa *et al.*, 2023). Additionally, the need for advanced technologies to monitor and optimize these processes further increases the financial challenges (Saravanan *et al.*, 2020).

Conclusion

Microbial bioprocesses are crucial for the vitality and proper functioning of aquatic ecosystems, significantly impacting the regulation of biogeochemical cycles and promoting environmental sustainability. Their broad range of applications in wastewater treatment, aquaculture, and bioremediation demonstrates their capacity for effective nutrient removal, pollutant degradation, and overall aquatic health maintenance. Innovations such as biofilms, biofloc technology, probiotics, and advanced microbial systems reveal the profound potential of microbes in enhancing water quality and tackling environmental contaminants. Future prospects include significant advancements through genetic engineering, synthetic biology, and microbial fuel cells, which are expected to drive further innovations and sustainable applications. Investigating microbial diversity in extreme habitats and employing cutting-edge tools like metagenomics and proteomics present promising opportunities to refine microbial systems. Ongoing research and development are vital to addressing existing challenges and limitations, such as environmental constraints and the risks posed by genetically modified organisms. By merging microbial technologies with emerging fields such as nanotechnology and renewable energy, we can create more effective and sustainable solutions for managing aquatic environments, which is essential for confronting future environmental challenges and improving our ability to protect and restore aquatic ecosystems.

References

Aguilar, C. N., Hernández-Almanza, A., Sabu, A., Loredo, A., Sudhakaran, S., Davila-Medina, D., ... & Teixeira, J. A. (2019). Microbial technology: advances and challenges. Advances in Food Bioproducts and Bioprocessing Technologies, 3-24.

Aiyer, K., Mukherjee, D., & Doyle, L. E. (2023). A weak electricigen-based bioelectrochemical sensor for real-time monitoring of chemical pollutants in water. ACS Applied Bio Materials, 6(10), 4105-4110.

Akram, M. S., Cheema, M. A., Waqas, M., Bilal, M., & Saeed, M. (2020). Role of bio-fertilizers in sustainable agriculture. Mediterr. J. Basic Appl. Sci.

Alaa, E., Abou, G. M., Hassan, S. M., Abdelatif, H. H., El-Din, N. G. S., Abdelnaby, H. M., & Ibrahim, H. A. (2023). Applications of Marine Bacteria in the Aquaculture Industry for Improving Water Quality and Treating Microbial Attack. Egyptian Journal of Aquatic Biology & Fisheries, 27(4).

Ali, A., Zahra, A., Kamthan, M., Husain, F. M., Albalawi, T., Zubair, M., ... & Noorani, M. S. (2023). Microbial biofilms: applications, clinical consequences, and alternative therapies. Microorganisms, 11(8), 1934.

Ansari, R. R. (2012). Biomedical Imaging: Principles and Applications.

Arunavarsini, K., Yuvansasi, V., Sekar, M., & Mahenthiran, R. (2024). Probiotics and its Application in Humans: An Overview. Journal of Pharmaceutical Research International, 36(7), 220-237.

Arunavarsini, K., Yuvansasi, V., Sekar, M., & Mahenthiran, R. (2024). Probiotics and its Application in Humans: An Overview. Journal of Pharmaceutical Research International, 36(7), 220-237.

Bala, R., Somal, M. K., Kumar, M., Karnwal, A., & Sachan, R. S. K. (2022). Microbial biosensors for environmental monitoring. Environmental Microbiology: Advanced Research and Multidisciplinary Applications, 1, 105. https://doi.org/10.2174/9781681089584122010008

Barbosa, F. G., Castro-Alonso, M. J., Rocha, T. M., Sánchez-Muñoz, S., de Arruda, G. L., Viana, M. C. A., ... & Da Silva, S. S. (2023). An Overview of Developments and Challenges in the Production of Biosurfactant by Fermentation Processes. Biosurfactants and Sustainability: From Biorefineries Production to Versatile Applications, 117-142.

Bello, O. O., Martins, F. T., Bello, T. K., Anuoluwa, I. A., Anyakudo, M. M., Amolegbe, O. A., & Ilemobayo, A. M. Occurrence and Role of Bacterial Biofilms in Different Systems.

Bhardwaj, D., Ansari, M. W., Sahoo, R. K., & Tuteja, N. (2014). Biofertilizers function as key player in sustainable agriculture by improving soil fertility, plant tolerance and crop productivity. Microbial cell factories, 13, 1-10.

Bossier, P., & Ekasari, J. (2017). Biofloc technology application in aquaculture to support sustainable development goals. Microbial biotechnology, 10(5), 1012-1016.

Bourhane, Z., Lanzén, A., Cagnon, C., Said, O. B., Mahmoudi, E., Coulon, F., ... & Duran, R. (2022). Microbial diversity alteration reveals biomarkers of contamination in soil-river-lake continuum. Journal of Hazardous Materials, 421, 126789.

Bright, F.V. (1988). Bioanalytical applications of fluorescence spectroscopy. Analytical chemistry, 60 18, 1031A-1039A .

Carducci, N. G. G., Dey, S., & Hickey, D. P. (2024). Recent Developments and Applications of Microbial Electrochemical Biosensors.

Choo, H. X., & Caipang, C. M. A. (2015). Biofloc technology (BFT) and its application towards improved production in freshwater tilapia culture. Aquaculture, Aquarium, Conservation & Legislation, 8(3), 362-366.

Ekasari, J. (2009). Teknologi biotlok: teori dan aplikasi dalam perikanan budidaya sistem intensif bioflocs technology: theory and application in intensive aquaculture system. Jumal Akuakultur Indonesia, 8(2), 117-126.

Fonti, V., Dell'Anno, A., & Beolchini, F. (2016). Does bioleaching represent a biotechnological strategy for remediation of contaminated sediments?. Science of the Total Environment, 563, 302-319.

Garimella, S. S. S., Rachakonda, S. V., Pratapa, S. S., Mannem, G. D., & Mahidhara, G. (2024). From cells to power cells: harnessing bacterial electron transport for microbial fuel cells (MFCs). Annals of Microbiology, 74(1), 19.

Gautam, R. (2024). Biosensors for Environmental Monitoring. In Sensors for Environmental Monitoring, Identification, and Assessment (pp. 204-217). IGI Global.

Gupta, A. D. (2024). Microbial fuel cells (MFCs) for waste recycling and energy production. Biodegradable Waste Processing for Sustainable Developments, 220.

Immanuel, S. J., Yogesh, P., Andrew, P. M., & Ramanathan, S. (2024). Genetically engineered marine microbes for the production of bioethanol. In Microbial Biotechnology for Bioenergy (pp. 419-428). Elsevier.

Indriyani, Y. A., Rusmana, I., Anwar, S., Djajakirana, G., & Santosa, D. A. (2023). Harvesting bioelectricity from microbial fuel cells (MFCs) powered by electroactive microbes. Jurnal Teknik Pertanian Lampung (Journal of Agricultural Engineering), 12(3), 583-596.

Jarboe, L. R. (2022). Progress and challenges for microbial fermentation processes within the biorefinery context. AZ of Biorefinery, 447-471.

Jones, S., & Santini, J. M. (2023). Mechanisms of bioleaching: iron and sulfur oxidation by acidophilic microorganisms. Essays in Biochemistry, 67(4), 685-699.

Joos, L., De Tender, C., Holderbeke, A., Clement, L., Vandecasteele, B., & Debode, J. (2023). Exploring the microbial response as a potential bio-indicator for soil health: Insights from a controlled incubator experiment. Agriculture, Ecosystems & Environment, 356, 108634.

Joshi, S., Lahiri, D., Ray, R. R., & Davoodbasha, M. (Eds.). (2023). Microbial Biofilms: Challenges and Advances in Metabolomic Study.

Kaur, K., & Vyas, P. (2024). Microbes for Sustainable Development of Environmental Issues. In Microbial Applications for Environmental Sustainability (pp. 1-20). Singapore: Springer Nature Singapore.

Kawaai, K., Nishida, T., Saito, A., & Hayashi, T. (2022). Application of bio-based materials to crack and patch repair methods in concrete. Construction and Building Materials, 340, 127718.

Kikuchi, K. (2010). Design, synthesis and biological application of chemical probes for bio-imaging. Chemical Society Reviews, 39(6), 2048-2053.

Kivelä, J., & Hakalehto, E. (2022, May). Utilization of microbial strains and communities as a precept of recycled organic fertilization. In EGU General Assembly Conference Abstracts (pp. EGU22-10211).

Kumar, V., Parida, S. N., Roy, S., Dhar, S., Bisai, K., Behera, B. K., & Das, B. K. (2024). An Overview of Modern Biotechnological Tools in Aquatic Food Production–A Review. Annals of Animal Science, 24(1), 13-25.

Labh, S. N. (2023). Applications of biotechnology in aquaculture nutrition and fisheries. In Frontiers in aquaculture biotechnology (pp. 163-170). Academic Press.

Li, Q., Sun, L. H., Yang, B. C., Jiang, J. H., Chen, J. M., Zheng, G., ... & Guo, J. L. (2024). Effectiveness of a novel compound microbial agent for water quality and the gut microbiota of Micropterus salmoides. Frontiers in Marine Science, 11, 1379711.

Madison, A. S., Sorsby, S. J., Wang, Y., & Key, T. A. (2023). Increasing in situ bioremediation effectiveness through field-scale application of molecular biological tools. Frontiers in Microbiology, 13, 1005871.

Masters, B. R. (2014). Fluorescence microscopy: from principles to biological applications.

Meruvu, H. (2021). Genetically modified microbes for nanobiotechnology. In Nanobiotechnology (pp. 123-132). Elsevier.

Midhun, S. J., & Arun, D. (2023). History and development of microbial technology in aquaculture. In Recent Advances in Aquaculture Microbial Technology (pp. 1-13). Academic Press.

Mishra, S., Panda, S., Akcil, A., & Dembele, S. (2023). Biotechnological avenues in mineral processing: Fundamentals, applications and advances in bioleaching and bio-beneficiation. Mineral Processing and Extractive Metallurgy Review, 44(1), 22-51.

Montaño López, J., Duran, L., & Avalos, J. L. (2022). Physiological limitations and opportunities in microbial metabolic engineering. Nature Reviews Microbiology, 20(1), 35-48.

Nair, V., Singh, A., & Gowda, N. K. Bioprocesses for Sustainable Management of Mine Waste in Contaminated Environmental Matrices. In Sustainable Management of Mining Waste and Tailings (pp. 146-178). CRC Press.

Ogidi, O. I., Angaye, T. C., & Carbom, H. E. (2024). Microorganisms in the Monitoring of Terrestrial and Aquatic Ecosystems: Applications, Challenges, and Future Prospects. Biomonitoring of Pollutants in the Global South, 639-665.

Oyewole, O. A., Yakubu, J. G., & Raji, R. O. (2022). The Roles of Biofilms in Corrosion. In Microbial Biofilms (pp. 123-145). CRC Press.

Paraje, M. G. (2023). Biofilms: The Achilles' heel of antimicrobial resistance. Advanced Pharmacy, 1, 225. https://doi.org/10.2174/9789815049428123010011

Patel, A. K., Singhania, R. R., Albarico, F. P. J. B., Pandey, A., Chen, C. W., & Dong, C. D. (2022). Organic wastes bioremediation and its changing prospects. Science of the Total Environment, 824, 153889.

Pawar, R., Dhawal, P., Nabar, B., Barve, S., & Zambare, V. (2022). Mechanisms and applications of probiotics in healthcare industry. In Biotechnology in Healthcare (pp. 225-257). Academic Press.

Petersen, J. (2023). From understanding the rate limitations of bioleaching mechanisms to improved bioleach process design. Hydrometallurgy, 221, 106148.

Phian, S., Nagar, S., Kaur, J., & Rawat, C. D. (2022). Emerging issues and challenges for microbes-assisted remediation. In Microbes and Microbial Biotechnology for Green Remediation (pp. 47-89). Elsevier.

Piotrowska, A., & Boruszko, D. (2024). Potential of Effective Microorganisms in the Aspect of Sustainable Development.

Rao, T. S. (2023). Industrial applications and implications of biofilms. In Understanding Microbial Biofilms (pp. 713-738). Academic Press.

Ayilara, M. S., & Babalola, O. O. (2023). Bioremediation of environmental wastes: the role of microorganisms. Frontiers in Agronomy, 5, 1183691.

Ron, E. Z., & Rosenberg, E. (2014). Enhanced bioremediation of oil spills in the sea. Current Opinion in biotechnology, 27, 191-194.

Sangwan, S., Kumar, M., Lamba, R., Singh, S., Kumari, A., & Wati, L. Bioindicators: Natural Biotic Sensors of Environmental Pollution and Ecological Disturbance. In Environmental Nexus Approach (pp. 311-337). CRC Press.

Sar, T., Harirchi, S., Yazdian, F., & Taherzadeh, M. J. (2023). Microorganisms Used in Biorefineries. In Biotic Resources (pp. 21-58). CRC Press.

Saravanan, A., Karishma, S., Kumar, P. S., Yaashikaa, P. R., Jeevanantham, S., & Gayathri, B. (2020). Microbial electrolysis cells and microbial fuel cells for biohydrogen production: Current advances and emerging challenges. Biomass Conversion and Biorefinery, 1-21.

Sarker, P. K. (2023). Microorganisms in fish feeds, technological innovations, and key strategies for sustainable aquaculture. Microorganisms, 11(2), 439.

Savenko, N., & Prysiazhnyuk, N. M. (2022). The role of microorganisms of the aquatic environment in the formation of the ecological and sanitary state of water bodies. Animal Husbandry Products Production and Processing: збірник наукових праць, (2), 175.

Sharma, D., Parashar, A., Sharma, D., Attri, S., & Chaudhary, A. (2024). Microbial Sensors: A Tool for Accelerating Sustainable Green Technologies. In Microbial Approaches for Sustainable Green Technologies (pp. 1-20). CRC Press.

Sharma, P., Bano, A., Tong, Y. W., & Singh, S. (2024). Monitoring of microbes and possible remediation through genetically engineered microbes (pp. 181-192). Elsevier. https://doi.org/10.1016/b978-0-443-13659-7.00004-7

Sharma, V., Sehgal, R., Angra, V., & Gupta, R. (2024). Applications of genome editing in bioremediation. In Genome editing in bacteria (Part 2) (Vol. 1, p. 150). https://doi.org/10.2174/9789815223798124010007

Shen, D. S., Xing, X. J., Long, Y. Y., & Hui, C. (2024). Applications of biofilm in environmental pollution control and the related challenges. Ying yong sheng tai xue bao= The journal of applied ecology, 35(5), 1426-1434.

Shilky, Patra, S., Harshvardhan, A., Kumar, A., & Saikia, P. (2023). Role of microbes in controlling the geochemical composition of aquatic ecosystems. Hydrogeochemistry of Aquatic Ecosystems, 265-281.

Shree, P., Singh, C. K., Sodhi, K. K., Surya, J. N., & Singh, D. K. (2023). Biofilms: Understanding the structure and contribution towards bacterial resistance in antibiotics. Medicine in Microecology, 16, 100084.

Silva-Castro, G. A., Uad, I., Gónzalez-López, J., Fandiño, C. G., Toledo, F. L., & Calvo, C. (2012). Application of selected microbial consortia combined with inorganic and oleophilic fertilizers to recuperate oil-polluted soil using land farming technology. Clean Technologies and Environmental Policy, 14, 719-726.

Singh, D., Thapa, S., Geat, N., Mehriya, M. L., & Rajawat, M. V. S. (2021). Biofertilizers: Mechanisms and application. In Biofertilizers (pp. 151-166). Woodhead Publishing.

Singh, M., Chaudhary, P., Bhutani, S., Bhasin, S., Mehra, A., & Tripathi, K. (2023). Overview of Microbial Associations and Their Role Under Aquatic Ecosystems. In Current Status of Fresh Water Microbiology (pp. 77-115). Singapore: Springer Nature Singapore.

Singh, R. P., Shadan, A., & Ma, Y. (2022). Biotechnological applications of probiotics: a multifarious weapon to disease and metabolic abnormality. Probiotics and Antimicrobial Proteins, 14(6), 1184-1210.

Singh, S., & Shyu, D. J. Microbial Technologies: Effectiveness and Feasibility Viewpoint. In Microbial Nexus for Sustainable Wastewater Treatment (pp. 237-269). CRC Press.

Sivakumar, H., Gupta, J., & Verma, A. (2024). Revolutionizing Water Quality Through Microbial Interactions in Aquatic Ecosystems for Pollution Abatement. In Microbiology-2.0 Update for a Sustainable Future (pp. 17-38). Singapore: Springer Nature Singapore.

Song, Y., Chetty, K., Garbe, U., Wei, J., Bu, H., O'moore, L., ... & Jiang, G. (2021). A novel granular sludge-based and highly corrosion-resistant bio-concrete in sewers. Science of The Total Environment, 791, 148270.

Sreerenjini, R., Nair, S. R., Hemalatha, M., Ganesh, S. K., Mohanasrinivasan, V., & Devi, C. S. (2024). Applications of potential microbes in bioremediation. In Emerging Innovative Trends in the Application of Biological Processes for Industrial Wastewater Treatment (pp. 263-290). Elsevier.

Sultana, M., Hossain, A., Nizhum, A. H., & Hossain, M. A. Unveiling the Prospects of High-throughput Microbial Biotechnology in the Effective Management of Wastewater. In Microbial Nexus for Sustainable Wastewater Treatment (pp. 339-363). CRC Press.

Thakur, N. (2024). Krishnendu Mondal, Santosh Kumar 2, Ashish Kumar Singh 2, Ishfaq Nabi Najar 2.

Usmani, Z., Sharma, M., Gaffey, J., Sharma, M., Dewhurst, R. J., Moreau, B., ... & Gupta, V. K. (2022). Valorization of dairy waste and by-products through microbial bioprocesses. Bioresource technology, 346, 126444.

Vivaldi, G. A., Salcedo, F. P., & Tallou, A. (2024). Application of biological effluent. In J. E. Ayars, D. Zaccaria, & K. M. Bali (Eds.), Microirrigation for crop production (2nd ed., pp. 393–410). Elsevier Science. https://doi.org/10.1016/B978-0-323-99719-5.00012-5

Warnasuriya, S. D., Udayanga, D., Manamgoda, D. S., & Biles, C. (2023). Fungi as environmental bioindicators. Science of the Total Environment, 892, 164583.

Wei, H., Liu, C., Cui, X., Shen, Z., & Wang, J. (2024). Distribution characteristics of microorganisms in sediments of Dagu River and their biological indicator function for evaluating eco-environmental quality of rural river. Environmental Research, 245, 118032.

Yang, D. (2024). Research on the application of microbial biomass in resource utilization of agricultural product processing waste. Applied and Computational Engineering. https://doi.org/10.54254/2755-2721/58/20240728

Zambelli, R. A., & de Mendonça, L. G. Application of Probiotics in Food Industry: A Promising Scope for Improving Food Quality. In Probiotics (pp. 282-304). CRC Press.

Zhou, Y. M., Liu, Y., Liu, W., & Shen, Y. (2023). Generation of microbial protein feed (MPF) from waste and its application in aquaculture in China. Journal of Environmental Chemical Engineering, 11(2), 109297.

8

Microplastics in Water: Environmental Impacts, Detection, and Mitigation

D.T. Vaghela [1*], Mayur Bhadarka [2] and Milan Ram [3]

[1] *Department of Aquatic Environment Management*

[2] *Department of Aquatic Environment Management*

[3] *College of Fisheries Science, Kamdhenu University, Veraval (362265), Gujarat, India*

Abstract

Microplastics, or plastic fragments smaller than 5 millimeters, have become prevalent contaminants in aquatic systems, arising from the breakdown of larger plastics, personal care products, and synthetic fabrics. Their presence disrupts marine and freshwater ecosystems by being consumed by wildlife, disturbing food webs, and polluting habitats. Additionally, microplastics can transport harmful pollutants and release toxic chemicals, worsening their environmental impact. Identifying these particles involves sophisticated methods such as microscopy, spectroscopy, and thermal analysis, although differentiating them from other particulates remains challenging. Major sources include direct releases, the fragmentation of larger plastic waste, and atmospheric deposition, with pathways like stormwater runoff and wastewater treatment aiding their dispersal. Addressing microplastic pollution requires a combination of regulatory measures, including bans on microbeads, technological advancements such as enhanced filtration systems, and increased public and industry engagement. Continued research is essential for developing effective solutions, understanding microplastic pathways in the environment, and assessing their long-term effects on health.

Keywords: *Microplastic, Environmental impacts, Detection, Sources, Mitigation strategies*

1.0 Introduction

Human activities have led to the accumulation of a diverse range of materials, including synthetic polymers like plastics, in lake sediments. Plastics have long been recognized as a significant element of pollution in rivers (Balas *et al.*, 2001; Tudor and Williams, 2004). The term 'microplastics' usually refers plastic particles with a diameter of less than 5 mm, which is the standard definition used by most experts. However, there is a proposal to redefine 'microplastics' as particles under 1 mm to specifically address those in the micrometer range (Andrady ,2011; Brownw *et al.*, 2011). 'mesoplastics' would refer to particles ranging from 1 mm to 2,500 mm (GESAMP, 2015). Lambert *et al.* (2014) defined plastics as macroplastics (>5 mm), mesoplastics (5 mm to >1 mm), microplastics (1 mm to >0.1 μm), and nanoplastics (≤0.1 μm). Despite these distinctions, the 5 mm limit remains commonly used because it includes particles small enough to be ingested by organisms (GESAMP, 2015). Approximately 54.5% of the microplastics found floating in the ocean are made of polyethylene, while 16.5% are composed of polypropylene. The remaining microplastics consist of polyvinyl chloride, polystyrene, polyester, and polyamides. Due to their lower density compared to seawater, polyethylene and polypropylene remain buoyant and impact the ocean's surface. In contrast, materials with higher density sink to the seafloor, where they affect the benthic environment (Issac and Kandasubramanian, 2021). The accumulation of microplastics in the oceans is largely due to land-based human activities, including industrial manufacturing, agricultural operations, and the disposal of municipal solid waste in landfills. These sources play a major role in marine microplastic pollution through mechanisms such as runoff, atmospheric deposition, and inadequate waste management practices (Driedger *et al.*, 2015; Vandermeersch *et al.*, 2015).

Microplastics have been found in 87% of tap water samples across 34 countries, with concentrations ranging significantly from 0.028 to 728.105 items per liter (Sun *et al.*, 2024). These particles are also commonly found in marine protected areas (MPAs), with concentrations varying between 0 and 809,000 items per cubic meter (Nunes *et al.*, 2023). Polyethylene (PE) and polypropylene (PP) are the most frequently detected types of microplastics, which raises serious concerns about their environmental impact (Nunes *et al.*, 2023). These particles are particularly problematic because they can carry toxic substances and pathogens, posing potential health risks to both humans and wildlife (Waghmare and Dar, 2024; Chandra and Walsh, 2024).

The widespread presence of microplastics in water has emerged as a significant health concern due to their potential toxic effects on humans. These small particles can contain harmful chemicals and pathogens, posing risks of both acute and chronic health issues, including toxicity, cancer, and endocrine disruption (Madiraju *et al.*, 2024; Meshram & Mhatre, 2024). Microplastics pose a significant threat to aquatic ecosystems by disrupting marine life and contributing to the bioaccumulation of toxins. When microplastics accumulate in aquatic ecosystems, they can be ingested by fish and other marine organisms, which are then consumed by people, thereby amplifying health risks (Bhusare *et al.*, 2024; Sinitsyna *et al.*, 2023). Microplastic pollution presents a serious threat to marine ecosystems, originating from various human activities such as aquaculture, fishing, tourism, and the discharge from industrial and household wastewater systems (Bhadarka *et al.*, 2024). Moreover, microplastics can disrupt nutrient absorption and cause physical damage to marine life, leading to potential disruptions in food chains. Addressing the impacts of microplastics is essential for protecting human health and maintaining the balance of ecosystems.

2.0 Environmental Impacts of Microplastics in Water

2.1 Ecological Impacts

Microplastics (MPs) pose significant ecological risks to aquatic systems, affecting both individual species and the broader environment. Their pervasive presence in water bodies results in several detrimental effects, including the accumulation of toxins and disruption of food chains. MPs undermine the health of fish and other marine organisms by causing immune system problems, reproductive issues, and digestive tract damage (Bhat *et al.*, 2024; Bhusare *et al.*, 2024). Moreover, they function as carriers for toxic substances, contributing to conditions like "Plasticosis" in seabirds, characterized by persistent inflammation (Siwach *et al.*, 2024). When smaller aquatic creatures ingest MPs, these particles can accumulate up the food chain, impacting larger predators and ultimately posing risks to human health (Megha *et al.*, 2024; Bhusare *et al.*, 2024). The long-term presence of MPs can lead to a decline in biodiversity and ecosystem stability, disrupting the natural equilibrium of aquatic environments. The enduring nature of these particles means that the damage they cause is not only immediate but also continuous, necessitating urgent global efforts to address their effects on marine life and human health.

2.2 Chemical Impact

Microplastics (MPs) in aquatic environments present serious chemical threats to ecosystems and human health, as they release harmful additives and pollutants. These small plastic fragments, originating from industrial waste, consumer products, or the breakdown of larger plastics, introduce a variety of toxic chemicals into water systems, lowering water quality and harming marine life. MPs can release toxic chemicals like phthalic acid esters and bisphenol A into aquatic ecosystems, disrupting endocrine systems and causing developmental, reproductive, and immune issues in marine species (Ziembowicz and Kida, 2024). In addition to these direct effects, MPs act as carriers for pathogens, persistent organic pollutants (POPs), and heavy metals such as lead and cadmium. These harmful substances attach to the surface of

MPs and are ingested by marine life, leading to health problems like impaired reproduction, stunted growth, developmental abnormalities, and increased mortality rates (Debnath *et al.*, 2024). As MPs move through the food chain, these toxins bioaccumulate, threatening higher organisms, including humans who consume contaminated seafood. Furthermore, MPs and their associated pollutants can alter water chemistry, affecting pH levels, oxygen availability, and nutrient balance, which exacerbates the damage to aquatic ecosystems. The long-lasting nature of MPs means their harmful chemical effects persist, posing ongoing risks to the health of water systems and the species that depend on them.

2.3 Impact on Biodiversity

Microplastics in aquatic environments represent a serious threat to biodiversity, affecting various ecosystems and species. In freshwater systems, these particles are commonly found, with levels reaching as high as 3.5 particles per liter in water and 56.9 particles per gram in sediments, predominantly due to urban land use (Gonzalez-Saldias *et al.*, 2024). They disturb benthic communities by impairing feeding abilities, stunting growth, and causing abnormal behaviors in organisms like gastropods and bivalves (Hussain *et al.*, 2024). Additionally, microplastics can lead to the buildup of harmful substances within the food chain, intensifying adverse effects on aquatic life. Moreover, they can boost the success of invasive species in fluctuating water conditions, allowing these species to outcompete native ones and causing shifts in community structure and loss of biodiversity (Zhang *et al.*, 2024). This disruption can create cascading impacts throughout the ecosystem. In karst systems, microplastics endanger both protected species and water quality, with concentrations between 37 and 86 items per liter, highlighting the urgent need for thorough monitoring and management (Balestra *et al.*, 2024). These issues underscore the critical importance of tackling microplastic pollution to protect freshwater ecosystems and the diverse species they sustain.

2.4 Human Health Concerns

Microplastics in water pose significant human health concerns due to their widespread presence and potential toxicological effects. Research has revealed that microplastics are commonly found in drinking water, with contamination levels reaching up to 83% in tap water samples worldwide (Chandra & Walsh, 2024). The ingestion of these particles has been associated with various health risks, including inflammatory responses and endocrine disruption, which could potentially lead to serious conditions such as cancer and reproductive issues (Madiraju *et al.*, 2024; Haleem *et al.*, 2023). Microplastics are frequently detected in urban tap water, with an average concentration of 35.33

particles per liter in Bangladesh, indicating shortcomings in water treatment processes (Hossain *et al.*, 2024). They can enter the human body through drinking water, consumption of seafood, and inhalation, raising concerns about cumulative exposure (Madiraju *et al.*, 2024). Additionally, microplastics can serve as carriers for harmful chemicals and viruses, potentially enhancing their transmission and infectivity (Wang *et al.*, 2024). The leaching of toxic substances from microplastics during water treatment further compounds these health risks, affecting multiple organ systems. The persistence of microplastics in the environment also means that exposure is ongoing, adding to the urgency of addressing this issue to safeguard public health. Improved water treatment technologies and stricter regulations are crucial to mitigating these risks and ensuring safer water supplies.

3.0 Detection and Characterization of Microplastics in Water

3.1 Sampling Techniques

To effectively assess and monitor microplastics in aquatic environments, various sampling methods are employed. These techniques are crucial for obtaining precise data on microplastic contamination. Surface water samples are commonly collected using methods such as net tows or filtration, which involve either dragging fine mesh nets through the water or filtering water samples to capture microplastic particles. Sediment samples are obtained using sediment cores or grab samples, which help researchers examine microplastics that have settled on the lake or riverbeds. Additionally, samples from aquatic organisms, such as fish and invertebrates, are collected to evaluate the extent of microplastic accumulation within these species. These approaches collectively offer a thorough understanding of microplastic distribution and levels across different environmental compartments, which is vital for assessing their ecological and health effects. To enhance the precision of these assessments, advanced techniques like microscopy, spectroscopy, and chemical analysis are used alongside these sampling methods. These technologies help in accurately identifying and characterizing the size, shape, and composition of microplastic particles, which is essential for better understanding their impact and developing effective mitigation strategies. The detection and characterization of microplastics (MPs) in water are essential for understanding their environmental effects and potential health risks. Several analytical methods have been developed, each with its own strengths and limitations.

3.2 Analytical Techniques

3.2.1 Pyrolysis-GC-MS Gas (Chromatography-Mass Spectrometry)

Pyrolysis coupled with gas chromatography/mass spectrometry (Py-GC/MS) has been employed to analyze polymer compositions, including for microplastic (MP) identification. This method is highly effective for identifying a range of polymer types in water samples, with detection limits as low as 11.3 µg/L. Pyrolysis-Gas Chromatography-Mass Spectrometry (GC-MS) works by breaking down plastic polymers into their individual monomers, which are then analyzed to determine the types and amounts of microplastics present. The method is appreciated for its rapid analysis, taking about 20 minutes, which makes it ideal for processing large numbers of samples quickly. However, it involves complex sample preparation and may not be as effective for detecting very small particles or those with similar thermal degradation characteristics (Dalmau-Soler *et al.*, 2024; Hermabessiere *et al.*, 2018)).

3.2.2 Raman Spectroscopy

Raman spectroscopy is a valuable tool for analyzing microplastics due to its ability to address challenges associated with diverse background media. It is effective for analyzing a broad range of particle sizes down to 1 µm with modern instruments and provides reliable identification of chemical compositions. Additionally, Raman spectroscopy-based microscopy techniques enable the detection of microplastics within living organisms, even at the subcellular level (Ribeiro-Claro *et al.*, 2017). When used in conjunction with convolutional neural networks (CNNs), Raman Spectroscopy provides improved quantitative analysis of microplastics, achieving high precision in concentration measurement with an R^2 value of 0.9972. This technique uses laser light to interact with microplastics, producing distinctive spectral fingerprints for different polymer types. The use of CNNs enhances data analysis and improves the detection of microplastics in complex samples. While it excels in distinguishing various types of plastics and offers detailed molecular data, it may require costly equipment and thorough calibration (Luo *et al.*, 2024).

3.2.3 UV Imaging

This innovative method uses DSLR cameras to capture microplastics that fluoresce under ultraviolet light, allowing for efficient counting and size distribution analysis. UV imaging takes advantage of the fluorescence emitted by microplastics when exposed to UV light, and this fluorescence is recorded by high-resolution cameras. This approach enables rapid and non-destructive analysis of large water volumes, making it effective for real-time monitoring of microplastic pollution. However, its effectiveness can be limited by the need for specific fluorescent dyes or coatings, and it may not detect microplastics that do not fluoresce strongly under UV light (Jaruda and Gabayno, 2024).

3.2.4 Digital Image Analysis

This method integrates light microscopy with digital image processing to assess microplastics (MPs) based on their two-dimensional and three-dimensional shapes. It effectively captures around 90% of the variation in MP characteristics, making it a robust tool for detailed morphological studies (Dacewicz *et al.*, 2024). This technique enables high-resolution imaging and accurate measurement of MP features, which enhances our understanding of their distribution and ecological impact.

3.2.5 Smartphone Microscopy

A novel and affordable smartphone-based system has been created to detect microplastics as small as 20 μm. This system delivers performance comparable to more costly commercial microscopes, offering a practical and accessible method for monitoring and detecting microplastics in diverse environments (Karki *et al.*, 2024). Its compact design and low cost make it a valuable tool for both field and laboratory applications.

3.2.6 SEM and FTIR

Scanning Electron Microscopy (SEM) and Fourier Transform Infrared Spectroscopy (FTIR) are essential for examining the chemical composition and surface structure of microplastics. SEM provides detailed images of MP surfaces, highlighting structural details and potential sources of contamination, while FTIR analyzes chemical bonds and identifies polymer types. Together, these techniques offer a thorough understanding of the origins and potential hazards of microplastics, supporting more effective environmental monitoring and assessment (Özgenç, 2024; Li *et al.*, 2024).

3.3 Size and Shape Distribution

Microplastics (MPs) are pervasive contaminants in aquatic systems, characterized by their diminutive size and diverse forms. These particles usually range from 1 μm to 5 mm, and their varying sizes and shapes significantly impact their environmental effects (Sunil *et al.*, 2024). In the Delaware Bay estuary, microplastic concentrations have been recorded from below detectable levels up to 4.12 MPs per cubic meter, with smaller particles being more prevalent in coastal areas (Nitzberg *et al.*, 2024). Research using static light scattering has shown average particle sizes of 3-3.5 μm, highlighting the prevalence of smaller microplastics in marine environments (Gouse Sandhani *et al.*, 2024).

The dominance of these smaller particles is particularly concerning as they are more prone to being ingested by marine organisms.

fibrous microplastics are the most frequently identified, making up 83.1% of MPs found in South African tap water (Ramaremisa *et al.*, 2024). These fibers often have higher aspect ratios, which are useful for estimating their volume and understanding their potential to entangle or be consumed by marine life (Nitzberg *et al.*, 2024). The variability in microplastic shapes and sizes influences their behavior in the environment and their interactions with aquatic organisms. Thorough characterization of these characteristics is crucial for assessing the full scope of microplastics ecological impact and for devising effective measures to combat this widespread pollution challenge.

3.4 Challenges in Detection

Detecting microplastics in aquatic environments presents significant challenges due to their tiny size, low concentrations, and the limitations of current detection methods. A major issue is the absence of standardized protocols in traditional sampling techniques, which can lead to inconsistent data collection and analysis across different studies (Viaroli *et al.*, 2024). The minute size of microplastics, which can be as small as 1 μm, makes detection particularly difficult, requiring advanced imaging technologies like high-resolution microscopy and complex analytical tools (Jaruda and Gabayno, 2024; Karki *et al.*, 2024). While these methods are effective, they are often costly and necessitate specialized equipment, which can restrict their use and availability.

Environmental factors further complicate microplastic detection. The density and composition of microplastics can vary widely, influencing their distribution in water bodies. For instance, microplastics made from polyethylene and polypropylene tend to float, while those made from denser materials like polystyrene and certain polyesters sink (Harpreet Kaur *et al.*, 2024). This variability can make sampling and data interpretation more complex. Additionally, contamination from sampling equipment, such as PVC components, can skew microplastic concentration measurements (Viaroli *et al.*, 2024). This highlights the need for careful selection and handling of sampling tools to prevent contamination and ensure accurate data. Overcoming these methodological and environmental hurdles is essential for enhancing the reliability of microplastic detection and improving our understanding of their effects on aquatic ecosystems.

4.0 Sources and Pathways of Microplastics in Water Bodies

4.1 Primary Sources

The direct release of microplastics, particularly from synthetic textiles during washing, represents a significant environmental challenge. Studies show that household laundering is a major contributor to microplastic fiber (MPF) pollution, with approximately 5.69 million tons being released globally each year, primarily through washing machines (Wang *et al.*, 2024).

4.1.1 Direct Release of Microplastics

A. **Microbeads and Industrial Abrasives:** While specific research on microbeads is limited, industrial abrasives are known to release microplastics directly into water systems during their usage. This discharge contributes to the overall microplastic pollution in aquatic environments, with these particles persisting and accumulating over time, thereby intensifying the contamination issue.

B. **Synthetic Textiles**: The washing of synthetic textiles is a notable source of MPF pollution. Studies reveal that the quantity of microplastics released during washing is significantly affected by various factors, including the type of fabric. For instance, fleece garments, which feature a raised texture, release similar amounts of MPFs as non-raised textiles. This indicates that fabric texture and type can influence the volume of fibers released (Klinkhammer *et al.*, 2024). Additionally, the type of detergent used and the settings on washing machines also impact the extent of microplastic release.

4.1.2 Impact of Washing Practices

A. **Washing Parameters:** The conditions under which textiles are washed can greatly influence MPF emissions. Variables such as the size of the wash load, washing duration, and water temperature play critical roles. Research has shown that larger wash loads and shorter wash cycles can reduce the amount of microplastic fibers released. This suggests that optimizing washing practices may help mitigate fiber discharge into wastewater (Klinkhammer *et al.*, 2024).

B. **Material Influence:** Different fabric types affect microplastic emissions differently. Studies indicate that recycled polyester tends to release more microplastics compared to virgin polyester, likely due to variations in fiber quality and manufacturing processes. This underscores the importance of considering the environmental impact of fabric materials when choosing textiles (Akyildiz *et al.*, 2024). Understanding these factors is crucial for developing effective strategies to reduce microplastic pollution from textiles.

4.2 Secondary Source

4.1.1 Fragmentation of Larger Plastics

A. **Degradation Process**: Large plastic items like bottles, bags, and packaging materials gradually break down into microplastics duc to physical wear and exposure to ultraviolet (UV) radiation. Environmental factors such as temperature changes and mechanical wear further accelerate this fragmentation (Kumar *et al.*, 2024; Adetuyi *et al.*, 2024).

B. **Long-Term Impact**: Once broken down, these microplastics can persist in the environment for extended periods. They continue to degrade into even smaller particles over time, perpetuating their environmental presence and contributing to ongoing pollution.

4.1.2 Wastewater Treatment Plants

A. **Filtration Limitations**: Many wastewater treatment facilities face challenges in effectively removing microplastics from treated water due to the limitations of current filtration technologies. Consequently, microplastics often escape into natural water bodies through effluent discharge (Bhusare *et al.*, 2024; Adetuyi *et al.*, 2024).

B. **Environmental Distribution**: Once released, these microplastics can spread across rivers, lakes, and coastal areas, contaminating aquatic ecosystems and posing threats to marine life and habitats.

4.1.3 Atmospheric Deposition

A. **Transport Mechanism**: Microplastics can be carried into the atmosphere through various activities, such as industrial emissions and the use of plastic products. These airborne particles eventually fall into water bodies through processes like precipitation or atmospheric deposition (Adetuyi *et al.*, 2024).

B. **Wide Reach**: This atmospheric transport allows microplastics to reach even remote and otherwise untouched water bodies, significantly expanding their environmental footprint.

4.1.4 Soil Runoff

A. **Agricultural Contributions**: The use of plastic materials in agriculture, such as plastic mulch and containers, can lead to soil erosion and runoff. This runoff can carry microplastics into nearby rivers and lakes, worsening water pollution (Niu *et al.*, 2024; Adetuyi *et al.*, 2024).

B. **Continued Pollution**: As agricultural practices continue to evolve, the potential for microplastic contamination through soil runoff remains a pressing issue. Effective management and mitigation strategies are needed to address this ongoing environmental challenge.

Despite the well-documented negative impacts of microplastics, some research suggests they may have specific uses, such as enhancing wastewater filtration systems. Nonetheless, the benefits of such applications do not outweigh the considerable risks associated with microplastics, particularly their role in carrying harmful toxins and pollutants (Kumar *et al.*, 2024; Bhusare *et al.*, 2024).

4.3 Pathways into Aquatic Systems

Microplastics infiltrate aquatic ecosystems through several pathways, each contributing to their widespread environmental impact. Key sources include stormwater runoff, wastewater treatment plants, atmospheric deposition, and agricultural runoff, particularly from the use of biosolids.

4.3.1 Stormwater Runoff

Stormwater runoff plays a major role in transporting microplastics into water bodies, with concentrations differing by location. For instance, research has shown an average of 4.5 microplastic particles per liter in stormwater samples, primarily consisting of fibers (Wolfand *et al.*, 2023). The volume of microplastics carried by stormwater is strongly influenced by the intensity of rainfall, with heavy rain events leading to substantial influxes of microplastics into aquatic systems. This highlights the need for improved stormwater management strategies to mitigate the effects of heavy rains on microplastic pollution (Ochoa *et al.*, 2024).

4.3.2 Wastewater Treatment Plants

Wastewater treatment plants are a significant source of microplastic pollution, as their effluents often contain substantial amounts of microplastics, with varying shapes and types detected in different areas (Rochman *et al.*, 2022). Despite improvements in treatment technologies, many facilities still fail to completely remove microplastics from wastewater, resulting in their release into rivers and oceans. This issue points to the necessity for enhanced treatment methods and more advanced filtration technologies to reduce microplastic discharge into natural water bodies.

4.3.3 Atmospheric Deposition

Atmospheric deposition also contributes to microplastic pollution by depositing airborne particles onto land and water surfaces. Studies have found that microplastics from the atmosphere can accumulate in agricultural soils at a rate of approximately 15 particles per kilogram of dry soil per year (Adhikari *et al.*, 2023). This deposition adds to soil contamination and poses a risk of microplastics being washed into nearby water bodies through soil erosion and runoff. The pervasive nature of atmospheric deposition calls for further research into its sources and the impact on both land and water environments.

4.3.4 Agricultural Runoff

Agricultural runoff, particularly from the application of biosolids, is a notable contributor to microplastic pollution. The prolonged use of biosolids has led to significant microplastic accumulation in soils, which can subsequently be transported into water bodies through runoff (Adhikari *et al.*, 2023). This runoff not only contaminates aquatic systems but also affects soil quality and crop health. Improving waste management practices and seeking alternatives to biosolids are essential steps to address the issue of microplastic pollution from agricultural sources.

5.0 Mitigation Strategies for Microplastic Pollution

Effectively addressing microplastic pollution requires a multi-pronged strategy, encompassing measures such as banning microbeads, establishing global agreements, and implementing Extended Producer Responsibility (EPR) schemes. Each of these methods is essential for tackling the pervasive issue of microplastics.

5.1 Banning Microbeads

Implementing bans on microbeads found in personal care products has proven to be an effective strategy for reducing microplastic pollution. Microbeads are a major contributor to microplastic contamination in aquatic environments, and their removal from products like cosmetics and cleaning agents has significantly improved water quality (Das *et al.*, 2024). Regulatory measures that restrict these harmful particles and promote the use of eco-friendly alternatives can mitigate the impact of microplastics on ecosystems. Furthermore, such bans can stimulate the creation of innovative, sustainable product formulations that minimize environmental harm.

5.2 Global and Regional Agreements

Developing comprehensive international frameworks, such as a potential global plastics treaty, is vital for a coordinated global response to microplastic pollution (Kurniawan *et al.*, 2024). These agreements facilitate international cooperation, ensuring that countries adopt and maintain sustainable practices and benefit from shared advancements in technology (Nikpay and Roodsari, 2024). By promoting cross-border collaboration, these agreements can implement more effective and uniform measures to address microplastic pollution on a global scale.

5.3 Extended Producer Responsibility (EPR) Schemes

EPR schemes assign manufacturers responsibility for the entire lifecycle of their products, from design through disposal. This approach incentivizes the creation of recyclable and biodegradable materials, which can help reduce the generation of microplastics (Samuel *et al.*, 2024). Integrating EPR into waste management systems can significantly cut down on microplastic waste, foster sustainable product design, and improve recycling practices (Thacharodi *et al.*, 2024). Moreover, this strategy encourages manufacturers to adopt more environmentally friendly practices throughout their production processes, further mitigating the impact of microplastics.

5.4 Advanced Filtration Systems

Modern treatment technologies like membrane bioreactors and electrocoagulation significantly boost the efficiency of microplastic removal in wastewater treatment plants (WWTPs). These advanced systems can remove between 69-86% of total suspended solids (TSS), effectively tackling microplastic contamination (Ajao, 2024; Tsamoutsoglou *et al.*, 2024). Moreover, the Advanced Primary Filtration (APF) method has been shown to outperform traditional approaches by dramatically lowering both biological oxygen demand (BOD) and chemical oxygen demand (COD) (Tsamoutsoglou *et al.*, 2024). This advancement not only improves the quality of treated water but also reduces the environmental footprint of wastewater discharge.

5.5 Biodegradable Plastics and Alternatives

The development of biodegradable plastics is essential for mitigating microplastic pollution. These plastics are designed to decompose more swiftly in natural environments, thus reducing long-term pollution and the buildup of microplastics (Samuel *et al.*, 2024; Thacharodi *et al.*, 2024). Focusing on biodegradable alternatives lessens the ecological impact of plastic products and fosters more sustainable waste management practices. Progress in creating these materials is critical for addressing the wider challenges posed by plastic waste and its environmental effects.

5.6 Innovations in Plastic Recycling and Waste Management

Innovative recycling techniques, such as those for converting discarded fishing nets into valuable composite materials, play a crucial role in promoting a circular economy and decreasing marine debris (Liotta *et al.*, 2023). Effective waste management approaches are vital for transforming plastic waste into useful products, thereby mitigating the accumulation of microplastics in natural habitats (Thacharodi *et al.*, 2024). By enhancing recycling processes and advocating for sustainable waste practices, we can substantially reduce the environmental impact of plastics and improve the sustainability of waste management systems.

5.7 Public Awareness and Education Campaigns

Public education campaigns should focus on raising awareness about the origins and impacts of microplastics and promoting eco-friendly practices. Effective outreach can significantly improve public knowledge and engagement. For example, a study indicates that only 12% of individuals are aware of microplastics, and greater awareness is associated with increased willingness to take action (Al Masud *et al.*, 2024). Through targeted community engagement, these campaigns can drive substantial behavioral changes, such as decreased plastic consumption and enhanced waste management (Al Masud *et al.*, 2024). By improving public understanding, such initiatives can encourage more informed consumer decisions and cultivate a culture of environmental responsibility.

5.8 Industry Responsibility

Industries must adopt sustainable practices, such as designing products for durability, ease of recycling, and developing biodegradable options (Samuel *et al.*, 2024). A comprehensive framework is needed to support companies in reducing plastic use and enhancing packaging design. Implementing these measures could significantly cut down on microplastic pollution and lessen environmental damage (Nikpay and Roodsari, 2024). Embracing these sustainable practices not only helps in reducing pollution but also fosters long-term sustainability in product design and production. By taking proactive environmental responsibility, industries can play a pivotal role in addressing the microplastic issue and improving ecological health.

5.9 Development of Biodegradable Polymers

Biodegradable plastics are vital for addressing microplastic pollution as they break down under specific environmental conditions, reducing their long-term presence in ecosystems (Samuel *et al.*, 2024; Angela, 2024). Recent advancements in green chemistry focus on creating these materials to be both functional and environmentally friendly, promoting increased reusability and recyclability (Samuel *et al.*, 2024). These developments support a shift toward more sustainable materials and offer a solution to the challenges presented by conventional plastics. As biodegradable polymers continue to improve, they hold significant promise for reducing the accumulation of microplastics in the environment.

5.10 Understanding Microplastic Pathways

Microplastics originate from a variety of sources, such as synthetic fabrics and tire abrasion, and are dispersed throughout marine and freshwater systems, affecting diverse ecosystems (Debnath *et al.*, 2024; Madiraju *et al.*, 2024). Research highlights the urgent need for improved waste management and advanced technologies to intercept microplastics before they enter aquatic environments (Samuel *et al.*, 2024; Anamika Singh *et al.*, 2024). By enhancing our knowledge of how microplastics spread, we can develop more effective strategies to prevent their release and mitigate their impact, thereby safeguarding both ecological systems and human health.

5.11 Long-term Studies on Human Health Impacts

There is growing concern about the health risks linked to microplastic exposure through seafood and drinking water, with potential connections to inflammatory and endocrine disorders (Madiraju *et al.*, 2024; Debnath *et al.*, 2024). Conducting long-term studies is crucial for thoroughly assessing the chronic health effects of microplastics and informing public health policies (Madiraju *et al.*, 2024). These investigations will provide critical insights into the prolonged impacts of microplastic exposure, which will be essential for developing effective health guidelines and regulatory measures to address this pressing issue.

Conclusion

Microplastics, which originate from the breakdown of larger plastics, personal care products, and synthetic fabrics, have emerged as a pervasive contaminant in aquatic ecosystems, impacting both marine and freshwater environments. Their broad environmental effects include ingestion by various organisms, disruptions to food webs, contamination of habitats, and the capacity to carry and release harmful pollutants. Detecting and analyzing microplastics present significant challenges, though advances in microscopy, spectroscopy, and new technologies offer promising solutions. Addressing this issue effectively requires a combination of regulatory actions, technological advancements, and community involvement, such as implementing microbead bans, enhancing filtration systems, and raising public awareness. Given the serious threats posed to both environmental and human health, coordinated global efforts are essential. Future research should prioritize developing biodegradable alternatives, improving understanding of microplastic movement, and assessing long-term health effects. Updated policies and international cooperation are vital to managing plastic waste and mitigating the microplastic issue, thereby protecting ecological and human health for future generations.

References

Adetuyi, B. O., Mathew, J. T., Inobeme, A., Falana, Y. O., Adetunji, C. O., Shahnawaz, M., & Yerima, M. B. (2024). Sources, Uses and Transport of Secondary Microplastics from Land to Marine Environment. In Microplastic Pollution (pp. 35-49). Singapore: Springer Nature Singapore.

Adhikari, K., Pearce, C. I., Sanguinet, K. A., Bary, A. I., Chowdhury, I., Eggleston, I., ... & Flury, M. (2024). Accumulation of microplastics in soil after long-term application of biosolids and atmospheric deposition. Science of the Total Environment, 912, 168883.

Ajao, A. (2024). Conventional and Advanced Treatment Technologies for Microplastics in Water Treatment Facilities.

Akyildiz, S. H., Fiore, S., Bruno, M., Sezgin, H., Yalcin-Enis, I., Yalcin, B., & Bellopede, R. (2024). Release of microplastic fibers from synthetic textiles during household washing. Environmental Pollution, 357, 124455.

Al Masud, A., Islam, A. R. M. T., Al Mamun, A., Alam, G. M., Arabameri, A., Bari, A. M., ... & Malafaia, G. (2024). A new approach from public behavioral attitudes and perceptions towards microplastics: Influencing factors, and policy proposals. Ocean & Coastal Management, 253, 107144.

Anamika Singh, Dr. N. Nagabhooshanam, Kushagra Kulshreshtha, Kuldeep Sharma and Padala Ramesh Kumar (2024), Exploring the Ecotoxicological Effects and Mitigation Strategies for Microplastics Pollution in Aquatic Ecosystems, Global NEST Journal, 26(7), 05938.

Andrady, A. L. (2011). Microplastics in the marine environment. Marine pollution bulletin, 62(8), 1596-1605.

Angela, Cui. (2024). 3. Plastic Degradation and its Effects. doi: 10.58445/rars.1220

Balas, C. E., Williams, A. T., Simmons, S. L., & Ergin, A. (2001). A statistical riverine litter propagation model. Marine Pollution Bulletin, 42(11), 1169-1176.

Balestra, V., Galbiati, M., Lapadula, S., Zampieri, V., Cassarino, F., Gajdošová, M., ... & Bellopede, R. (2024). Microplastic pollution calls for urgent investigations in stygobiont habitats: A case study from Classical karst. Journal of Environmental Management, 356, 120672.

Bhat, R. A. H., Sidiq, M. J., & Altinok, I. (2024). Impact of microplastics and nanoplastics on fish health and reproduction. Aquaculture, 741037.

Bhadarka, M., Vaghela, D. T., Sikotariya, H., Biswas, S., & Makwana, K. (2024). Effect of microplastic pollution on marine organisms. AgriGate: An International Multidisciplinary e-Magazine, 4(2), 139-145.

Bhusare, S., Satkar, S. G., Sahu, A., Savaliya, B., Karale, T., & Gautam, R. (2024). Microplastic (MP) Pollution in Aquatic Ecosystems and Environmental Impact on Aquatic Animals. Uttar Pradesh Journal of Zoology, 45(5), 59-68.

Browne, M. A., Crump, P., Niven, S. J., Teuten, E., Tonkin, A., Galloway, T., & Thompson, R. (2011). Accumulation of microplastic on shorelines woldwide: sources and sinks. Environmental science & technology, 45(21), 9175-9179.

Chandra, S., & Walsh, K. B. (2024). Microplastics in water: Occurrence, fate and removal. Journal of Contaminant Hydrology, 104360.

Dacewicz, E., Łobos-Moysa, E., & Chmielowski, K. (2024). Identification Tools of Microplastics from Surface Water Integrating Digital Image Processing and Statistical Techniques. Materials, 17(15), 3701.

Dalmau-Soler, J., Boleda, M. R., & Lacorte, S. (2024). Routine method for the analysis of microplastics in natural and drinking water by pyrolysis coupled to gas chromatography-mass spectrometry. Journal of Chromatography A, 1730, 465153.

Das, B., Sarkar, S., Paul, S., Zafar, A., Biswas, D., Panda, J. R., & Chakraborty, T. (2024). Mitigation Strategies of Emerging Pollutants from Cosmetic Industry Effluent: A Review. European Journal of Theoretical and Applied Sciences, 2(3), 886-899.

Debnath, R., Prasad, G. S., Amin, A., Malik, M. M., Ahmad, I., Abubakr, A., ... & Faggio, C. (2024). Understanding and addressing microplastic pollution: Impacts, mitigation, and future perspectives. Journal of Contaminant Hydrology, 104399.

Driedger, A. G., Dürr, H. H., Mitchell, K., & Van Cappellen, P. (2015). Plastic debris in the Laurentian Great Lakes: A review. Journal of Great Lakes Research, 41(1), 9-19.

Gesamp (2015) Sources, fate and effects of microplastics in the marine environment: a global assessment. Joint Group of Experts on the Scientific Aspects of Marine Environmental Protection Reports and studies 90

Gonzalez-Saldias, F., Sabater, F., & Gomà, J. (2024). Microplastic distribution and their abundance along rivers are determined by land uses and sediment granulometry. Science of The Total Environment, 933, 173165.

Gouse Sandhani, C., Shanmugam, P., & Annamalaisamy Sannasiraj, S. (2024, April). Comprehensive Investigation of Microplastics size distribution in Marine Environment: Detection, Quantification, and Optical Analysis Using Static Light Scattering (SLS). In EGU General Assembly Conference Abstracts (p. 18753).

Haleem, N., Kumar, P., Zhang, C., Jamal, Y., Hua, G., Yao, B., & Yang, X. (2023). Microplastics and associated chemicals in drinking water: A review of their occurrence and human health implications. Science of The Total Environment, 169594.

Harpreet Kaur, Taranveer Kaur, Uday Karanbir Singh, Richa Rastogi, Pushpinder Kaur,"Microplastics: Challenges, Solutions and Detection Techniques", Futuristic Trends in Chemical Material Sciences & Nano Technology Volume 3 Book 24,IIP Series, Volume 3, May, 2024, Page no.105-114, e-ISBN: 978-93-5747-617-1, DOI/ Link: https://www.doi.org/10.58532/V3BECS24P2CH3

Hermabessiere, L., Himber, C., Boricaud, B., Kazour, M., Amara, R., Cassone, A. L., ... & Duflos, G. (2018). Optimization, performance, and application of a pyrolysis-GC/ MS method for the identification of microplastics. Analytical and bioanalytical chemistry, 410, 6663-6676.

Hossain, M. B., Yu, J., Sarker, P. K., Banik, P., Sultana, S., Nur, A. A. U., ... & Arai, T. (2024). Microplastic accumulation, morpho-polymer characterization, and dietary exposure in urban tap water of a developing nation. Frontiers in Sustainable Food Systems, 8, 1397348.

Hussain, M., Blache, D., & Maloney, S. K. (2024). Impact of Microplastics on the Physiology of Benthos. In Microplastic Pollution (pp. 363-372). Singapore: Springer Nature Singapore.

Issac, M. N., & Kandasubramanian, B. (2021). Effect of microplastics in water and aquatic systems. Environmental Science and Pollution Research, 28, 19544-19562.

Jaruda, R. J., & Gabayno, J. F. (2024, July). Detection of microplastics in surface water using a DSLR lens-based UV imaging system. In Journal of Physics: Conference Series (Vol. 2793, No. 1, p. 012018). IOP Publishing.

Jaruda, R. J., & Gabayno, J. F. (2024, July). Detection of microplastics in surface water using a DSLR lens-based UV imaging system. In Journal of Physics: Conference Series (Vol. 2793, No. 1, p. 012018). IOP Publishing.

Karki, A., Thaiba, B. M., Shishir Acharya, K. C., Sedai, T., Kandel, B., Paudyal, H., ... & Neupane, B. B. (2024). Smartphone microscopic method for imaging and quantification of microplastics in drinking water. Microscopy Research and Technique, 87(10), 2266-2274.

Karki, A., Thaiba, B. M., Shishir Acharya, K. C., Sedai, T., Kandel, B., Paudyal, H., ... & Neupane, B. B. (2024). Smartphone microscopic method for imaging and quantification of microplastics in drinking water. Microscopy Research and Technique, 87(10), 2266-2274.

Klinkhammer, K., Kolbe, S., Brandt, S., Meyer, J., Ratovo, K., Bendt, E., & Rabe, M. (2024). Release of fibrous microplastics from functional polyester garments through household washing. Frontiers in Environmental Science, 12, 1330922.

Kumar, P., Chaudhary, S., & Bhalla, A. (2024). Occurrence and Fate of Emerging Contaminants with Microplastics Current Scenario, Sources and Effects. In Occurrence, Distribution and Toxic Effects of Emerging Contaminantsx (pp. 58-100). CRC Press.

Kurniawan, T. A., Mohyuddin, A., Othman, M. H. D., Goh, H. H., Zhang, D., Anouzla, A., ... & Pasaribu, B. (2024). Beyond surface: Unveiling ecological and economic ramifications of microplastic pollution in the oceans. Water Environment Research, 96(7), e11070.

Lambert S, Sinclair CJ, Boxall ABA (2014) Occurrence, degradation and effects of polymerbased materials in the environment. Rev Environ Contam Toxicol 227:1–53. doi:10.1007/978-3-319-01327-5_1

Li, Y., Chen, P., Tang, Y., Yang, Y., Zhou, C., Bu, J., & Zhong, S. (2024). Microplastics in Water: A Review of Characterization and Removal Methods. Sustainability, 16(10), 4033.

Liotta, I., Avolio, R., Castaldo, R., Gentile, G., Ambrogi, V., Errico, M. E., & Cocca, M. (2024). Mitigation approach of plastic and microplastic pollution through recycling of fishing nets at the end of life. Process Safety and Environmental Protection, 182, 1143-1152.

Luo, Y., Su, W., Rabbi, M. F., Wan, Q., Xu, D., Wang, Z., ... & Wu, J. (2024). Quantitative analysis of microplastics in water environments based on Raman spectroscopy and convolutional neural network. Science of The Total Environment, 926, 171925.

Madiraju, S. V. H., Pamula, A. S. P., & Darsi, B. P. (2024). Theoretical Review on Microplastic Pollution: A Multifaceted Threat to Marine Ecosystems, Human Health, and Environment. Igmin Research, 2(6), 460-468.

Megha, K. B., Anvitha, D., Parvathi, S., Neeraj, A., Sonia, J., & Mohanan, P. V. (2024). Environmental impact of microplastics and potential health hazards. Critical Reviews in Biotechnology, 1-31.

Meshram, L. N., & Mhatre, K. J. (2024). Microplastics: Impacts on Environment and Human Health Hazards. Uttar Pradesh Journal of Zoology, 45(4), 22-33.

Nikpay, M., & Toorchi Roodsari, S. (2024). Crafting a Scientific Framework to Mitigate Microplastic Impact on Ecosystems. Microplastics, 3(1), 165-183.

Nitzberg, E. J., Parmar, S., Arbuckle-Keil, G., Saba, G. K., Chant, R. J., & Fahrenfeld, N. L. (2024). Microplastic concentration, characterization, and size distribution in the Delaware Bay estuary. Chemosphere, 142523.

Niu, J., Xu, D., Wu, W., & Gao, B. (2024). Tracing microplastic sources in urban water bodies combining their diversity, fragmentation and stability. npj Clean Water, 7(1), 37.

Nunes, B. Z., Huang, Y., Ribeiro, V. V., Wu, S., Holbech, H., Moreira, L. B., ... & Castro, I. B. (2023). Microplastic contamination in seawater across global marine protected areas boundaries. Environmental Pollution, 316, 120692.

Ochoa, L., Chan, J., Auguste, C., Arbuckle-Keil, G., & Fahrenfeld, N. L. (2024). Stormwater runoff microplastics: Polymer types, particle size, and factors controlling loading rates. Science of The Total Environment, 929, 172485.

Özgenç, E. (2024). Advanced analytical techniques for assessing and detecting microplastic pollution in water and wastewater systems. Environmental Quality Management.

Ramaremisa, G., Tutu, H., & Saad, D. (2024). Detection and characterisation of microplastics in tap water from Gauteng, South Africa. Chemosphere, 356, 141903.

Ribeiro-Claro, P., Nolasco, M. M., & Araújo, C. (2017). Characterization of microplastics by Raman spectroscopy. Comprehensive Analytical Chemistry, 75, 119-151.

Rochman, C. M., Grbic, J., Earn, A., Helm, P. A., Hasenmueller, E. A., Trice, M., ... & Teh, S. (2022). Local monitoring should inform local solutions: morphological assemblages of microplastics are similar within a pathway, but relative total concentrations vary regionally. Environmental Science & Technology, 56(13), 9367-9378.

Samuel, H., Makong, F. D., & Ori, M. (2024). Green Chemistry Strategies for Mitigating Microplastic Pollution in Aquatic Environments. Asian Journal of Environmental Research, 1(2), 73-82.

Sinitsyna, O. O., Yeremin, G. B., Turbinskii, V. V., Pushkareva, M. V., Shiryaeva, M. A., Markova, O. L., & Borisova, D. S. (2023). Microplastics pollution in water is a threat for human health and the environment (literature review). Health Risk Anal, 3, 158-165.

Siwach, S., Bharti, M., Yadav, S., Dolkar, P., Modeel, S., Yadav, P., & Negi, R. K. (2024). Unveiling the ecotoxicological impact of microplastics on organisms-the persistent organic pollutant (POP): A comprehensive review. Journal of Contaminant Hydrology, 104397.

Sun, T., Teng, Y., Ji, C., Li, F., Shan, X., & Wu, H. (2024). Global prevalence of microplastics in tap water systems: Abundance, characteristics, drivers and knowledge gaps. Science of The Total Environment, 929, 172662.

Sunil, M., Mithun, N., Charles, M., Chidangil, S., Kumar, S., & Lukose, J. (2024). Visualization and characterisation of microplastics in aquatic environment using a home-built micro-Raman spectroscopic set up. Journal of Environmental Management, 354, 120351.

Thacharodi, A., Hassan, S., Meenatchi, R., Bhat, M. A., Hussain, N., Arockiaraj, J., & Pugazhendhi, A. (2024). Mitigating microplastic pollution: A critical review on the effects, remediation, and utilization strategies of microplastics. Journal of Environmental Management, 351, 119988.

Tsamoutsoglou, K., Katzourakis, V. E., Chrysikopoulos, C. V., & Gikas, P. (2024). Performance evaluation and parameter estimation of the advanced primary filtration (APF) process in wastewater treatment plants. Separation and Purification Technology, 344, 127106.

Tudor, D. T., & Williams, A. T. (2004). Development of a 'Matrix Scoring Technique' to determine litter sources at a Bristol Channel beach. Journal of Coastal Conservation, 10(1), 119-127.

Vandermeersch, G., Van Cauwenberghe, L., Janssen, C. R., Marques, A., Granby, K., Fait, G., ... & Devriese, L. (2015). A critical view on microplastic quantification in aquatic organisms. Environmental Research, 143, 46-55.

Viaroli, S., Lancia, M., Lee, J. Y., Ben, Y., Giannecchini, R., Castelvetro, V., ... & Re, V. (2024). Limits, challenges, and opportunities of sampling groundwater wells with plastic casings for microplastic investigations. Science of The Total Environment, 174259.

Waghmare, P. R., & Dar, M. A. (2024). Report of Microplastic Presence in Drinking Water. In Microplastic Pollution (pp. 225-244). Singapore: Springer Nature Singapore.

Wang, C., Song, J., Nunes, L. M., Zhao, H., Wang, P., Liang, Z., & Xing, B. (2024). Global microplastic fiber pollution from domestic laundry. Journal of Hazardous Materials, 477, 135290.

Wang, X., Zheng, K., Wang, Y., Hou, X., He, Y., Wang, Z., ... & Liu, X. (2024). Microplastics and viruses in the aquatic environment: a mini review. Frontiers in Microbiology, 15, 1433724.

Wolfand, J., Poor, C. J., Taylor, B. L., Morrow, E., Radke, A., & Diaz-Gunning, E. (2023). Microplastics: The occurrence in stormwater runoff and the effectiveness of bioretention systems for removal. Journal of Environmental Engineering, 149(11), 04023078.

Zhang, R. F., Guo, Y. J., Lai, Y. W., Zhao, T. T., Li, G. L., Yan, Z. G., ... & Rillig, M. C. (2024). Microplastics promote the invasiveness of invasive alien species under fluctuating water regime. Journal of Applied Ecology.

Ziembowicz, S., & Kida, M. (2024). The effect of water ozonation in the presence of microplastics on water quality and microplastics degradation. Science of The Total Environment, 929, 172595.

9

Toxicants and Their Effects on Aquatic Systems Bioaccumulation, Biomagnification and Detoxification

Mayur Bhadarka[1], D.T. Vaghela[2], Rushikesh Valmik[1] and Pinak Bamaniya[3]

[1]PG Scholar, Department of Aquatic Environment Management
[2]Department of Aquatic Environment Management
[3]College of Fisheries Science, Kamdhenu University, Veraval (362265), Gujarat, India

---

Abstract

Toxicants, whether naturally occurring or synthetic, present significant threats to aquatic ecosystems, leading to severe ecological damage and health hazards. These pollutants—such as pesticides, heavy metals, microplastics, and pharmaceuticals—contaminate water bodies through various routes, including industrial discharges, agricultural runoff, and insufficient wastewater treatment. They interfere with biological processes, causing problems like neurological damage, reproductive issues, and changes in water chemistry. Understanding the processes of bioaccumulation, biomagnification, bioconcentration, bioavailability and detoxification is essential for grasping how these contaminants impact aquatic organisms and human health. The rise of emerging pollutants, like microplastics and pharmaceuticals, adds further complexity, posing new challenges to ecosystem stability. Addressing these issues requires effective management and mitigation strategies, including reducing pollutant sources, improving waste treatment, and utilizing advanced detection techniques. Through research on these toxicants and their interactions, it is possible to better evaluate their effects, craft effective environmental policies, and implement measures to safeguard both aquatic ecosystems and public health.

Keywords: *Toxicants, Bioaccumulation, Biomagnification, Bioconcentration, Bioavailability, Detoxification*

1.0 Introduction

A toxicant is any chemical, whether naturally occurring or man-made, that can harm a living organism. A toxin, on the other hand, is a specific type of toxicant that is produced by a living organism. The two terms are not interchangeable—while all toxins are toxicants, not all toxicants are toxins (Cope *et al.*, 2004). Toxicants significantly affect aquatic systems, leading to ecological damage and posing considerable health risks to both aquatic organisms and humans. Pollutants such as pesticides, heavy metals, microplastics, and pharmaceuticals deteriorate water quality and interfere with biological processes, resulting in imbalances within aquatic ecosystems. For example, pesticides can cause neurological issues and biochemical alterations in aquatic species, leading to stress, illness, and death (Saha and Dutta, 2024). These substances also enter the food chain through biomagnification, affecting higher-level organisms and potentially jeopardizing human health (Saha and Dutta, 2024; AbuQamar *et al.*, 2024). Likewise, heavy metals and other toxic substances modify the physical and chemical properties of water, hindering the growth and development of aquatic plants and animals (Rajak *et al.*, 2024). Emerging pollutants, like microplastics and pharmaceuticals, add to these challenges by introducing new threats to aquatic ecosystems, disturbing reproductive and feeding patterns, and creating long-term risks for biodiversity and human well-being (Thanigaivel *et al.*, 2023). These adverse effects underscore the critical need for robust management and mitigation strategies to protect both environmental and public health. Pharmaceuticals and personal care products, even at low levels, pose a threat to aquatic organisms by disrupting endocrine systems and causing reproductive and developmental issues. These pollutants can remain in aquatic environments, building up over time and compromising the health of entire ecosystems (Zhu *et al.*, 2023; Shanmuganathan *et al.*, 2023). Continuous exposure to these substances not only alters biological processes but also puts vulnerable species at risk, making their long-term consequences a major concern for the sustainability of aquatic environments. Studying toxicants in aquatic systems is crucial for understanding their ecological effects, evaluating biodiversity risks, and shaping regulatory policies. Research on the interactions between toxicants and aquatic organisms sheds light on the intricate dynamics involved, emphasizing the need for thorough toxicity testing and ecological evaluations. Toxicants can have serious outcomes, such as population extinction or bistability, where species survival depends on initial conditions and toxicant concentrations (Liu and Huang, 2020). Chronic exposure studies are vital in establishing water quality standards that protect aquatic life from harmful substances (Baldisserotto, 2021). Moreover, developing advanced detection methods for human-made pollutants is essential for effective environmental monitoring (Thanigaivel *et al.*, 2023). The persistent release of emerging contaminants, including microplastics and pharmaceuticals, adds complexity to aquatic pollution and requires ongoing toxicological studies (Lu *et al.*, 2021). Understanding how foundational species like microalgae respond to these toxicants is key to predicting ecosystem health and protecting human safety (Lu *et al.*, 2021).

Toxicants enter aquatic systems from various sources and pathways, significantly impacting aquatic life and ecosystems. Heavy metals, such as lead and mercury, commonly enter through industrial discharges and accumulate in organisms, causing harmful effects (Jeong *et al.*, 2023). Pesticides from agricultural runoff contaminate water bodies, affecting aquatic species and leading to neurological issues (Saha and Dutta, 2024). Pharmaceuticals, including psychotropic drugs, persist in water due to insufficient wastewater treatment, posing risks even at low levels (Zhu *et al.*, 2023). Microplastics and other emerging pollutants disrupt ecosystems and may interact with other toxicants, amplifying their effects (Thanigaivel *et al.*, 2023). These toxicants enter aquatic environments through runoff from agriculture, inadequate wastewater discharge, and atmospheric deposition, highlighting the need for effective mitigation strategies.

Understanding bioaccumulation, biomagnification, and detoxification is crucial for assessing the environmental impact of heavy metals and engineered nanomaterials. Bioaccumulation is the buildup of pollutants in organisms, biomagnification is the increase in pollutant concentration up the food chain, and detoxification is how organisms break down and remove harmful substances. These processes highlight how pollutants affect ecosystems and human health. To mitigate these effects, it's essential to reduce pollutant sources, enhance waste management, and strengthen regulations. Implementing best practices in industry and agriculture, utilizing advanced treatment technologies, and ensuring regular monitoring and enforcement are key strategies. Additionally, public education on reducing chemical use and proper disposal plays a critical role in protecting aquatic ecosystems.

2.0 Types of toxicants in aquatic system

Aquatic ecosystems are under significant threat from a range of toxic substances. Pollutants such as heavy metals, pesticides, pharmaceuticals, and synthetic chemicals are entering water bodies through industrial processes, agricultural runoff, and other sources. These contaminants not only harm aquatic organisms but also disrupt food chains and present risks to human health. To protect these ecosystems and ensure the well-being of aquatic life, it is crucial to implement effective pollution control strategies and maintain continuous monitoring.

2.1 Chemical Toxicants

Heavy Metals: Metals like lead and mercury are often released into water bodies from industrial activities. These metals can contaminate the water, accumulate in aquatic organisms, and cause severe damage (Rajak *et al.*, 2024).

Pesticides: Pesticides used in agriculture can enter aquatic ecosystems via runoff, harming non-target species and disrupting natural food chains (Rajak *et al.*, 2024; Lu *et al.*, 2021).

Pharmaceuticals and Personal Care Products (PPCPs): Even in trace amounts, pharmaceuticals and personal care products can have detrimental effects on aquatic life. These pollutants are becoming an increasing concern in water environments (Lu *et al.*, 2021; Erzinger *et al.*, 2021).

Industrial Chemicals: Industrial chemicals have severe impacts on aquatic ecosystems and pose long-lasting, often irreversible, risks to both aquatic systems and human health.

The combined effects of various pollutants can be more damaging than individual contaminants. Assessing the overall health of aquatic ecosystems is challenging because the cumulative impact of multiple pollutants can intensify their harmful effects (Erzinger *et al.*, 2021).

2.2 Biological toxicants

The effects of biological toxicants can vary depending on the specific contaminant, its concentration, and the sensitivity of the affected organism. The important biological toxicants are described below:

Harmful Algal Blooms (HABs): Excessive growth of algae can produce toxins that are harmful to aquatic organisms and humans. These toxins can cause skin irritation, respiratory problems, and even death.

Pathogens

Bacteria: As bacteria are ubiquitous, they cause wide degree of spread and contamination. Bacteria like E. coli and Salmonella can contaminate water sources, leading to waterborne diseases in humans and animals.

Viruses: While viruses are often associated with human and animal health, they can also pose significant threats to aquatic ecosystems. they cause behaviour change in organisms, having role in nutrient cycling, cause infection in aquatic organism and leads to mortality of host also. It leads to some water borne disease like viral infections such as hepatitis A and polio can be transmitted through contaminated water.

Parasites: Parasites are organisms that live on or inside another organism (the host) and derive nourishment from it. They directly can cause mortality to their host, leading to population declines, affect food web and influences nutrient cycling indirectly. they also cause disease like Giardia and Cryptosporidium can cause gastrointestinal illness (Moore *et al.*, 2000).

2.3 Physical toxicants

Physical toxicants also have detrimental effects on aquatic ecosystem. In aquatic systems it includes heat (thermal pollution) and radiation, Sedimentation, plastics and microplastic etc.

Thermal pollution: Thermal pollution is the degradation of water quality as a result of changes to the ambient water temperature. Warm water increases respiration rates, reduces the amount of oxygen water can hold, promote harmful algal blooms, and shifts species composition. A common cause of thermal pollution is the use of water as a coolant by power plants and industrial manufacturers (Huertos, 2020).

Radiation: Radiation from nuclear waste disposal, nuclear and power station, from electronic devices etc., are the sources of radiation cause devastating effects on aquatic environment. Radiation pollution can cause to genetic damage, diversity loss,, impaired immune response, and lead to bioaccumulation also (Huertos, 2020).

Sedimentation: Excessive sedimentation can smother aquatic habitats, reducing light penetration and affecting primary productivity. This can lead to (i) reduced food availability for higher trophic levels (ii) Habitat degradation for benthic organisms (iii) Impaired water quality due to increased turbidity.

Plastics and microplastics: Plastics have become ubiquitous in modern society, leading to widespread pollution of aquatic environments. These pollutants can have detrimental effects on aquatic ecosystems and organisms.

Although the defined size of these microparticles has varied between different studies over the past decade, the National Oceanic and Atmospheric Administration (NOAA) now defines the term "microplastic" as fragments smaller than 5 mm in diameter (Arthur *et al.*, 2009). Microplastics are so small that they can be easily absorbed by organisms in the food chain. Because of their shape and size, they can attract pollutants from the water and release harmful chemicals. When animals eat microplastics, these toxins can enter the food chain and build up in larger organisms (Teuten *et al.*, 2009).

3.0 Pathways of Toxicants Entry into Aquatic Systems

Agricultural Runoff: Agricultural runoff is a significant source of pesticide contamination in water bodies. Common pesticides such as DEET, DDT, metolachlor, and malathion have been found in water systems, where they can harm organisms, even at low concentrations. Many of these pesticides belong to chemical families that persist in the environment for extended periods. Specialized treatment methods are required to clean water contaminated by pesticides, though several traditional techniques have been used for this purpose in the past (Yadav *et al.*, 2022).

Industrial Waste Discharge: Variations in governance, enforcement, and local policies across regions lead to major differences in wastewater management for industries, ranging from minimal regulation in underdeveloped or developing nations to strict control in some developed countries. One of the major challenges industries face is the disposal of wastewater. When this untreated wastewater is released into rivers or other bodies of water, it causes environmental pollution and poses risks to human health (Tong *et al.*, 2022).

Atmospheric Deposition: Atmospheric deposition is the process through which pollutants from the air, such as dust, heavy metals, polycyclic aromatic hydrocarbons, dioxins, furans, sulfates, and nitrates, are deposited on land or water. This process is a growing environmental concern, as pollutants are carried by the atmosphere and eventually settle on the ground or into water bodies through rain, snow, or other means.

Acid Rain: Acid rain, caused by the combination of sulfuric and nitric acids formed from sulfur and nitrogen oxides in the atmosphere, is one of the most serious environmental issues with transboundary effects. Acid rain can acidify water bodies, leading to harmful effects on aquatic life, including fish. Lakes and streams are particularly vulnerable as they lack the buffering capacity of soil and plants. Acidification can alter species composition in affected lakes and rivers, disrupting food webs and reducing the number of benthic organisms, which in turn decreases species like flies, mosquitoes, and mayflies (Likens, 1985).

Urban waste water: Wastewater from households and industry can harm aquatic environments. When this wastewater is dumped into rivers, it can pollute the water with organic matter and nutrients. This pollution can use up the oxygen in the water, making it difficult for fish and other creatures to survive. Additionally, too much nitrogen and phosphorus can cause plants and algae to grow too much, blocking sunlight and using up even more oxygen (EEA, 2019).

Oil spill: Petroleum contamination is a growing problem that affects both land and water. While there is a lot of focus on oil spills in the ocean, these spills can also harm terrestrial and aquatic ecosystems. Oil spills in the ocean can have a serious impact on coastal businesses and people who rely on the sea for resources (saadoun *et al.*, 2015). When oil is spilled, the thick slick spreads with the wind and currents, undergoing chemical and physical changes. The contaminants it contains, such as polycyclic aromatic hydrocarbons (PAHs), have oxidative and carcinogenic properties, binding to proteins and DNA. Inhaling these hydrocarbons can irritate the respiratory tract and cause narcosis in mammals and birds.

In the short term, oil spills can create hypoxic zones near the ocean surface, causing marine life to die or flee. These events also result in economic consequences, such as the closure of fishing areas, which impacts surrounding communities. Additionally, human health is affected, with symptoms like respiratory, eye, and skin irritation. Species that were already endangered before the spill face long-term risks, as their populations may not recover to pre-spill levels.

4.0 Effects of Toxicants on Aquatic System

Impacts on Water Quality: Lakes play crucial environmental, social, and economic roles, serving as sources of drinking water, recharging groundwater, controlling floods, supporting biodiversity, and providing livelihoods. However, the discharge of effluents containing heavy metals can degrade these water bodies, making them unhealthy for the organisms they support. In aquatic environments, the toxicity of metals is influenced by various abiotic factors such as oxygen levels, water hardness, pH, alkalinity, and temperature.

Eutrophication: The Cultural eutrophication excessive nutrient pollution of coastal waters, is driven by population growth, food production (agriculture, animal husbandry, aquaculture), and energy consumption. It is one of the most widespread pollution issues globally (Howarth *et al.*, 2002). Nutrients entering coastal waters include both inorganic forms (ammonium, nitrate, nitrite, phosphate, silicate) and organic forms (dissolved organic and particulate matter). About half of the dissolved nitrogen carried by rivers into coastal waters is in the form of dissolved organic nitrogen (DON) (Meybeck, 1982). The primary environmental impacts of eutrophication include increased suspended particles from extensive macroalgal blooms, reduced water clarity, and increased precipitation rates that destroy benthic habitats by shading submerged vegetation. Additionally, eutrophication leads to bottom-water hypoxia, elevated CO2 levels from the decomposition of excessive organic matter, and water acidification. This alters biogeochemical processes, causing sediment anoxia, accumulation of harmful hydrogen sulfide, and disruptions in nutrient cycling.

Impacts on Aquatic Organisms: Pollutants like petroleum hydrocarbons, heavy metals, and pesticides cause direct toxic effects in aquatic environments. Sensitive species may suffer sublethal impacts or even die, leading to ecological disruptions, such as trophic cascades or reduced competition, which can alter species dynamics. Pollutants can also directly affect keystone or foundational species, and changes in nutrient and oxygen levels can impair ecosystem functions (Fleeger *et al.*, 2003). Contaminants may cause changes in predator-prey interactions or competitive relationships, potentially intensifying their impact. These behavioral alterations can indirectly affect ecosystems, amplifying the harm caused by contaminants.

Impacts on Ecosystem Structure: Toxicants can have severe consequences for ecosystems (Hunt *et al.*, 2020). They can kill organisms in contaminated lakes and rivers, reducing species diversity and causing biodiversity loss (Milan *et al.*, 2024). Toxicants also contribute to reproductive issues in animals and impair the resilience of ecosystems. By disrupting population dynamics, they reduce ecosystem productivity and stability, limiting the ecosystem's capacity to sustain itself.

5.0 Bioaccumulation

Bioaccumulation is defined as the increase of contaminant concentrations in aquatic organisms following uptake from the ambient environmental medium. Different sources of exposure contribute to contaminant bioaccumulation (Wang *et al.*, 2014). Bioaccumulation is a growth-dependent remediating mechanism, where the toxic heavy metals have been up taken and transported into cells by live yeast cells (Wang, 2016). The process takes place in two steps including rapid passive biosorption and slows ion penetration through the cell membrane (Kieliszek *et al.*, 2015). Passive biosorption occurs through ion exchange, electrostatic interaction, complexation, and coordination, whereas ion penetration through cell membrane happens via an active cell transport mechanism. Bioaccumulation in aquatic organisms is influenced by several mechanisms, including the physicochemical properties of pollutants, biological traits of the organisms, and environmental factors. Understanding these mechanisms is essential for evaluating ecological risks.

5.1 Key Mechanisms of Bioaccumulation:

Physicochemical Properties: The potential for substances such as antibiotics and polychlorinated diphenyl ethers (PCDEs) to bioaccumulate is largely determined by their chemical structure, like the octanol-water partition coefficient (Zhu *et al.*, 2023; Zhang *et al.*, 2023). For instance, higher chlorination levels in PCDEs are linked to greater bioaccumulation factors (BCFs) (Zhang *et al.*, 2023).

Biological Factors: Characteristics unique to organisms, such as lipid content and metabolic pathways, significantly impact bioaccumulation. For example, different species exhibit varying abilities to biotransform arsenic, influencing its accumulation (Zhang *et al.*, 2022). Pharmaceuticals also show differing accumulation patterns across aquatic species (Del Carmen Gómez-Regalado, 2023).

Environmental Influences: Environmental conditions such as pH, salinity, and organic matter content can modify the rates at which contaminants are absorbed and assimilated by organisms (Zhang *et al.*, 2022; Del Carmen Gómez-Regalado, 2023).

5.2 Examples of Bioaccumulation

Bioaccumulation in aquatic organisms is a major environmental issue, particularly concerning heavy metals and organic pollutants. Various studies show how aquatic species accumulate these harmful substances, impacting both their health and the broader food web.

Heavy Metals in Fish: Research has found that fish species such as *Clarias gariepinus* and *Tilapia zilli* accumulate significant amounts of heavy metals like cadmium, lead, and zinc in their gills and muscles. For example, *T. zilli* exhibited the highest levels of cadmium, raising concerns about the potential risks to human health from consuming contaminated fish (IKWUEMESI *et al.*, 2023).

Antibiotics and Aquatic Organisms: Antibiotics like enrofloxacin and azithromycin also pose a threat, showing high bioaccumulation potential in fish. The accumulation of these substances is influenced by factors such as the organism's lipid content and environmental conditions (Zhu *et al.*, 2023).

Polycyclic Aromatic Hydrocarbons (PAHs): PAHs tend to accumulate in aquatic organisms, particularly at higher trophic levels, impacting fish reproduction and overall health. Studies have shown that the bioaccumulation factor for PAHs is notably higher in zooplankton compared to fish, highlighting the complex interactions within the aquatic food web (Mai *et al.*, 2024).

Pharmaceuticals: Antibiotics like enrofloxacin and doxycycline show a high potential for bioaccumulation in aquatic organisms, with bioconcentration factors exceeding 3.3 L/kg (Zhu *et al.*, 2023). Other pharmaceuticals, such as antidepressants and anti-inflammatory drugs like diclofenac, have also been found to accumulate in fish and mollusks. The extent of bioaccumulation varies significantly across different species and environmental conditions (Del Carmen Gómez-Regalado, 2023).

Heavy Metals: Trace metals like arsenic, cadmium, and lead tend to bioaccumulate in aquatic organisms, particularly in the muscle and liver tissues of invertebrates and fish. The bioaccumulation of these metals is influenced by factors like food assimilation efficiency and environmental conditions, including pH and salinity (Radomyski *et al.*, 2018).

Organic Pollutants: Polychlorinated diphenyl ethers (PCDEs) exhibit significant bioaccumulation within aquatic food chains, with bioaccumulation factors ranging from 2.42 to 4.03 L/kg, depending on the species (Zhang *et al.*, 2023). The degree of bioaccumulation is closely linked to the number of chlorine substitutions in the compounds (Zhang *et al.*, 2023).

6.0 Biomagnification

Biomagnification is the phenomenon where the concentration of harmful substances increases in organisms higher up in the food chain. This occurs because toxins, such as pesticides, heavy metals, or industrial pollutants like mercury or PCBs (Polychlorinated Biphenyls), are ingested by organisms at lower levels, including plants and small aquatic animals. As predators consume these organisms, the toxic substances accumulate and intensify in concentration with each step up the food chain.

6.1 The Process of Biomagnification

Introduction into the Environment: Pollutants, including heavy metals like mercury, pesticides such as DDT, and various industrial chemicals, infiltrate ecosystems through several routes. These Elements are discharged into the air, water, and soil due to various geological processes and human activities. The contamination of water sources with heavy metals is a significant environmental concern, as it negatively impacts plants, animals, and human health (Rezania *et al.*, 2016).

Absorption by Primary Producers: Primary producers, including phytoplankton and benthic diatoms, take up heavy metals from both sediments and water. These metals can accumulate in concentrations that are notably higher than those found in particulate organic matter (JAINGAM, 2018).

Bioaccumulation: Toxic substances are transferred from food into an organism, resulting in higher concentrations of these substances within the organism compared to their levels in the food source (Gray, 2002).

Consumption by Higher Trophic Levels: When predators eat primary producers or smaller organisms that have absorbed these pollutants, the toxins enter the food chain. This feeding process transfers the contaminants to higher trophic levels. In higher trophic level the concentration of toxicants is higher than lower trophic levels (Carpenter *et al.*, 1985; Díaz-Delgado *et al.*, 2024).

6.2 Example of Biomagnification

Deep-Sea Ecosystems: In the Aviles submarine Canyon, polychlorinated biphenyls (PCBs) exhibited significant biomagnification, with a trophic magnification factor (TMF) of 6.2 for top predators such as cetaceans (Romero-Romero *et al.*, 2017).

Mercury in the Indian River Lagoon: Research found that total mercury levels increased across trophic levels, with plankton showing mercury concentrations three times higher than those in pristine regions. A trophic magnification slope of 0.23 was calculated, suggesting that reducing mercury in plankton could lead to significant declines in mercury levels in higher trophic organisms (Beckett, 2015).

Methylmercury in Marine Plankton: A model showed that methylmercury biomagnification is more intense in larger zooplankton, influenced by factors such as food assimilation efficiency (Wu *et al.*, 2020). While biomagnification threatens top predators, including humans, it also raises concerns about ecosystem health and food safety. Understanding these processes is vital for effective environmental management and conservation.

7.0 Bioconcentration

Bioconcentration refers to the process where an organism absorbs and retains a substance directly through respiration, either from water in aquatic environments or from air in terrestrial ecosystems. This occurs without ingestion through food, as the substance is taken in solely from the surrounding medium (Alexander, 1999).

7.1 Bioconcentration factor

The bioconcentration factor (BCF) measures the extent to which a chemical accumulates in an organism relative to its environment, acting as an indicator of its potential for bioaccumulation (Petoumenou *et al.*, 2015). Determining BCF experimentally is often impractical, so Quantitative Structure-Activity Relationship (QSAR) models are used to estimate BCF values based on the molecular structures of chemicals (Garg and Smith, 2014; Bertato *et al.*, 2022). Research has demonstrated that BCF can vary considerably depending on a chemical's properties, such as hydrophobicity, with non-linear relationships observed for compounds that are highly hydrophobic (Garg and Smith, 2014). The bioconcentration factor (BCF) is used to assess the accumulation of heavy metals in organisms. Studies on mangrove trees have shown that BCF values highlight substantial metal uptake in these plants (Takarina and Pin, 2017).

7.2 Mechanism of bioconcentration

Bioconcentration involves various mechanisms, primarily observed in aquatic organisms such as fish, where chemicals are absorbed through their gill membranes. This process is greatly affected by the chemical characteristics of the substance, including its hydrophobic and lipophilic properties. Lipophilic substances, such as polycyclic aromatic hydrocarbons (PAHs), are prone to accumulate in the fatty tissues of organisms due to their ease of crossing biological membranes (Bertato *et al.*, 2022).

7.3 Chemical Properties of Pollutants

The extent to which a substance bioconcentrates is greatly affected by its chemical properties. Chemicals that are lipophilic (fat-attracting), like many persistent organic pollutants (POPs), are likely to accumulate in the fatty tissues of organisms due to their strong attraction to lipid-rich environments. This tendency is commonly quantified by the octanol-water partition coefficient (Kow), which measures the distribution of a chemical between organic and aqueous phases (Mackay and Fraser, 2000).

Transport Mechanisms: The process by which pollutants are transported during bioconcentration involves intricate interactions between the chemical characteristics of the pollutants, their biological uptake, and environmental conditions. Gaining insight into these mechanisms is essential for evaluating the ecological effects of persistent organic pollutants (POPs) in aquatic environments.

Hydrophobicity and Lipid Accumulation: Phytoplankton accumulate persistent organic pollutants (POPs) through lipid partitioning, a process influenced by the octanol/water partition coefficient (Kow). The relationship between bioconcentration factors (BCF) and Kow is non-linear, with dissolved organic carbon (DOC) impacting the rate of pollutant uptake (Seto and Handoh, 2009).

Transport via Biovectors: Migratory organisms, such as fish and birds, can concentrate contaminants and transport them across different ecosystems. This movement underscores the significance of biological pathways in the distribution of pollutants (Kallenborn and Blais, 2015).

Although biological transport mechanisms play a crucial role, they can sometimes be outweighed by physical transport processes, highlighting the need for a balanced perspective when analyzing pollutant dynamics within ecosystems.

8.0 Bioavailability

Bioavailability refers to the portion of a contaminant that organisms can absorb, whereas bioconcentration is the buildup of these contaminants in an organism's tissues from the surrounding environment (Shah and Kumar, 2022).

Factors Influencing Bioavailability

Environmental Conditions: The bioavailability of hydrophobic organic contaminants is influenced by soil characteristics and the desorption behavior of these substances, which is essential for evaluating their potential risks (Umeh *et al.*, 2020).

Chemical Properties: Research on perfluoroalkyl substances shows that their bioconcentration factors are strongly affected by their hydrophobicity and ability to permeate membranes, emphasizing the critical role of chemical structure in determining bioavailability (Chen *et al.*, 2016).

8.1 Key Physiological Factors Affecting Bioconcentration

Lipophilicity: The bioconcentration factor (BCF) is heavily influenced by the lipophilicity of compounds, with descriptors like XLogP commonly used in predictive models (Pramanik and Roy, 2014). Generally, higher lipophilicity is associated with increased BCF values.

Molecular Size: Studies indicate that the maximum cross-sectional diameter of molecules significantly impacts bioconcentration. A threshold of around 1.5 nm has been identified, beyond which the uptake mechanism may transition from passive diffusion to facilitated diffusion or active transport (Dimitrov *et al.*, 2002).

Chemical Structure: The presence of specific functional groups, such as halogens or hydrogen bonding sites, can either promote or hinder bioconcentration. Nonlinear relationships between BCF and molecular descriptors illustrate the complexity of these interactions (Wang *et al.*, 2014).

Ionization and Solubility: A chemical's ionization state, influenced by its pKa(The acidity of a specific molecule), along with its water solubility, are also key factors in determining BCF. New models have improved prediction accuracy by incorporating these factors (Dimitrov *et al.*, 2002).

8.2 Examples of Bioconcentration

Bioconcentration in aquatic environments is a growing concern, especially regarding pharmaceuticals and other contaminants. Numerous studies have shown the uptake of these substances by aquatic organisms, emphasizing the potential risks to ecosystems and food webs. Below are notable examples of bioconcentration observed in various aquatic species.

Pharmaceuticals in Aquatic Flora: Research conducted in the Yarrowee/ Leigh/Barwon River system found that aquatic plants like *Phragmites australis* and *Vallisneria australis* accumulated pharmaceuticals such as carbamazepine and venlafaxine, with concentrations in plant tissues reaching 31 $ng.g^{-1}$ (Harriage *et al.*, 2024).

Marine Organisms and Ionic Liquids: In marine mussels (*Mytilus trossulus*), long-chain imidazolium ionic liquids exhibited significant bioconcentration, with a bioconcentration factor (BCF) of 21,901 L/kg for one compound, indicating a high potential for bioaccumulation (Maculewicz *et al.*, 2023).

Non-Steroidal Anti-Inflammatory Drugs (NSAIDs): Research on marine bivalves revealed that NSAIDs like diclofenac and ibuprofen accumulated in tissues, leading to negative effects on immune responses (Almeida *et al.*, 2020).

These findings highlight the need to monitor bioconcentration in aquatic environments to assess the ecological risks posed by emerging contaminants. However, the variation in bioconcentration factors among species and compounds points to the need for further research to better understand the mechanisms and consequences of bioconcentration.

9.0 Detoxification in Aquatic Organisms

Detoxification refers to the body's natural process of eliminating harmful substances, primarily carried out by organs such as the liver, kidneys, and skin (Black, 2020).

9.1 Detoxification Mechanisms

Metal Sequestration: Aquatic life uses metallothioneins and similar peptides to capture non-essential metals, thereby preventing them from attaching to sensitive parts of cells (Campbell and Hare, 2009).

Microbial Detoxification: Certain microbes, such as Pseudomonas sp., can break down harmful substances like metalaxyl, highlighting the role of microbial communities in cleaning up aquatic environments (Massoud *et al.*, 2008).

Cyanotoxin Management: Organisms such as mussels and fish are capable of accumulating and neutralizing microcystins, which are dangerous cyanotoxins, thus helping to alleviate the impacts of eutrophication (Martins and Vasconcelos, 2009).

9.2 Importance of Detoxification

Ecosystem Health: Efficient detoxification processes are vital for preserving biodiversity and ecosystem stability, as they enable organisms to thrive in polluted habitats (Guerra-Rivas and Gomez-Gutierrez, 2016).

Public Health: Detoxification is essential for minimizing toxin build-up in edible aquatic species, thereby ensuring safer food sources and protecting human health (Martins and Vasconcelos, 2009).

However, chronic exposure to pollutants can exceed the capacity of these detoxification mechanisms, potentially causing significant harm to aquatic life and the overall health of ecosystems (Campbell and Hare, 2009).

9.3 Biotransformation process

Detoxification in aquatic organisms mainly involves biotransformation, which includes Phase I and Phase II reactions. These processes are essential for reducing the impact of environmental pollutants.

9.3.1 Phase I Reaction

Phase I reactions typically involve the addition or exposure of functional groups on xenobiotics, facilitated by enzymes such as cytochrome P450s. Research on salmonids, for instance, has demonstrated notable Phase I activity across various organs, showcasing their robust ability for initial chemical modifications (Franco *et al.*, 2024). Similarly, aquatic invertebrates have been observed transforming diclofenac into multiple oxidation products, underscoring the role of Phase I in altering the chemical structure of contaminants (Fu *et al.*, 2020).

9.3.2 Phase ii reaction

Phase II reactions involve conjugation processes, where the compounds modified during Phase I are attached to endogenous molecules. This step enhances the solubility of these compounds and facilitates their excretion, which is vital for detoxification. Such conjugation processes have been observed in the metabolism of various substances by aquatic organisms (Chen, 2024). The efficiency of Phase II enzymes, however, can vary significantly across different species and environmental conditions, impacting the overall effectiveness of detoxification (Zhang *et al.*, 2022).

Although biotransformation is a key defense mechanism, the formation of more toxic metabolites during these processes, as seen with diclofenac, raises concerns about potential risks associated with biotransformation products in aquatic ecosystems (Fu *et al.*, 2020).

9.4 Excretion of toxicants (kidneys, gills, etc.)

Detoxification in aquatic organisms primarily involves the elimination of toxic substances through various organs, such as the gills, kidneys, and intestines. This process is essential for maintaining internal balance and reducing the impact of harmful substances.

9.4.1 Mechanisms of Excretion

Detoxification in aquatic organisms involves multiple excretion mechanisms to remove toxic substances and maintain internal balance. Gills are crucial for expelling hydrophilic compounds, while the kidneys handle the elimination of small organic ions. Additionally, the gastrointestinal system, particularly the intestines, plays a significant role in processing and excreting xenobiotic compounds, with enterocyte proteins aiding in this detoxification. Together, these mechanisms ensure that toxins are effectively managed and excreted, helping aquatic organisms cope with environmental pollutants. Fish use their gastrointestinal tract to absorb and excrete xenobiotic compounds. They rely on enterocyte proteins within the gut for the detoxification of these substances (Luquet and Bieczynski, 2024).

9.4.2 Role of enzymes

CYP450 Enzymes in Aquatic Systems: CYP450 enzymes are crucial for the breakdown of antidiabetic drugs in aquatic environments, showing significant upregulation in areas affected by pollution (Wu *et al.*, 2024). In microalgae, these enzymes aid in the biotransformation of pharmaceuticals and personal care products (PPCPs), underscoring their potential role in bioremediation efforts (Li *et al.*, 2023).

CYP450 in Invertebrates: In chironomids, such as *Propsilocerus akamusi*, CYP450 genes are involved in detoxification processes, with their expression under pesticide stress being influenced by gut microbiota (Sun *et al.*, 2022). Research on *Daphnia magna* indicates that CYP450 biomarkers can effectively signal exposure to organophosphate pesticides, with specific enzymes like EROD being particularly useful (Melo de Almeida *et al.*, 2022).

CYP450 in Fish: In fish, CYP450 enzymes are distributed throughout various organs and are essential for metabolizing a broad spectrum of xenobiotics, playing a key role in detoxification (Chen, 2024). Despite their importance, the effectiveness of CYP450 enzymes can be affected by environmental factors and gut microbiota, reflecting the complex interactions within aquatic ecosystems.

9.5 Limitations of Detoxification Processes

Detoxification in aquatic organisms faces constraints due to various physiological and environmental factors, which limit their ability to manage toxic substances effectively. Recent research highlights several key limitations across different species and situations.

Threshold Responses to Metal Toxicity: Aquatic organisms often show a threshold response to non-essential metals, where their detoxification mechanisms can manage low levels of exposure but become overwhelmed at higher concentrations. This results in metal accumulation and toxicity, especially under conditions of chronic exposure (Campbell and Hare, 2009).

Herbivore Feeding Limitations: The detoxification limitation hypothesis suggests that herbivores' feeding rates are constrained by their capacity to detoxify secondary metabolites in their diet. This physiological limitation affects their overall fitness and feeding behavior (Sotka and Gantz, 2013).

Microbial Detoxification Capacity: Marine bacteria exhibit a low abundance of detoxification genes, which limits their ability to adapt to pollution. This lack of detoxification genes indicates a broader ecological vulnerability in marine environments (Bengtsson-Palme *et al.*, 2014).

Accumulation of Cyanotoxins: Cyanotoxins, such as microcystins, accumulate in aquatic organisms, with varying detoxification mechanisms among species. This accumulation poses risks not only to aquatic life but also to human health through the food chain (Martins and Vasconcelos, 2009).

Conversely, some microorganisms have shown effective detoxification abilities, suggesting potential strategies for bioremediation in aquatic systems (Massoud *et al.*, 2008). Nonetheless, the overall limitations in detoxification mechanisms across various organisms present a significant challenge for managing aquatic ecosystems.

10.0 Case Studies of Toxicants in Aquatic Systems

Toxic chemicals in aquatic ecosystems pose significant dangers to both ecosystems and human health, as highlighted by various studies examining the intricate interactions between pollutants and aquatic life.

Cadmium and Ketoprofen in Zebrafish: Research involving zebrafish exposed to cadmium (Cd) and ketoprofen (KPF) found that elevated levels of these chemicals led to decreased survival rates and developmental defects. Prolonged exposure caused the accumulation of toxins in specific organs and disrupted metabolic functions, underscoring the need for more research into the combined effects of these pollutants (Madesh *et al.*, 2024). This study highlights the importance of understanding the impact of long-term exposure to multiple toxins on aquatic organisms.

Effluent-Dominated Systems: In the Lubbock Canyon Lake System, high-resolution mass spectrometry was used to identify toxicity signatures in complex water samples. This advanced technique provided valuable insights into how pollutants impact different biological pathways, demonstrating the benefits of sophisticated analytical methods for evaluating toxicity (Kumar *et al.*, 2021). This approach illustrates the role of technological innovations in deepening our understanding of pollutant effects.

Ecological Risk Assessment in Italian Streams: An ecological risk assessment of the Adda River combined data on environmental concentrations with ecotoxicological information to identify priority chemical mixtures and assess their risks to aquatic life, particularly algae. This study emphasizes the importance of thorough risk assessments in freshwater systems to address and mitigate potential threats from chemical contaminants (Finizio *et al.*, 2022). It highlights the necessity for detailed evaluations to manage and reduce risks to aquatic ecosystems.

Polycyclic Aromatic Hydrocarbons (PAHs): Polycyclic Aromatic Hydrocarbons (PAHs) are prevalent pollutants that accumulate in aquatic food chains, particularly affecting fish populations in areas such as the Pearl River Estuary. As PAH levels increase with trophic levels, they lead to reproductive issues in species like *Danio rerio* and *Megalobrama hoffmanni*. This accumulation highlights the urgent need for effective strategies to manage and reduce PAH levels to safeguard aquatic ecosystems and maintain ecological balance (Mai *et al.*, 2024).

Psychotropic Drugs: Psychotropic drugs remain in aquatic environments due to inadequate wastewater treatment, resulting in harmful effects even at low concentrations. These substances disrupt reproductive and developmental processes in organisms such as zebrafish and freshwater mussels. The persistent presence of these drugs underscores the need for enhanced wastewater treatment technologies to protect aquatic health (Zhu *et al.*, 2023a).

Persistent Organic Pollutants (POPs): Persistent Organic Pollutants (POPs) often originate from landfills and accumulate in aquatic systems, posing long-term health risks to both aquatic life and humans. The review highlights the critical need for stringent policies to reduce POPs emissions and mitigate their environmental and health impacts. Addressing these challenges is essential for preventing further ecological and health problems (Ochs *et al.*, 2024).

Thiochemicals: A case study on thiochemicals showed that updated integrated testing approaches can improve toxicity predictions, revealing that the actual toxicity is lower than previously thought. This finding supports regulatory compliance and emphasizes the need for ongoing refinement of testing methods to ensure accurate toxicity assessments and effective chemical risk management (Nendza and Ahlers, 2022).

Emerging Pollutants: Emerging pollutants, including microplastics and pharmaceuticals, pose significant threats to aquatic ecosystems. Techniques such as phycoremediation using microalgae offer promising solutions for addressing these contaminants. These innovative methods provide potential strategies for mitigating the effects of emerging pollutants and protecting aquatic environments. Continued research and development are crucial for managing and reducing these pollutants' impact (Thanigaivel *et al.*, 2023).

11.0 Managing and Mitigating Toxicants in Aquatic Systems

Effective management of toxicants in aquatic systems is crucial for safeguarding both environmental and human health. Key strategies include pollution prevention, regulatory measures, and advanced monitoring systems.

11.1 Pollution Prevention Strategies

- **Bioremediation and Phytoremediation:** Biological methods, such as using microalgae, are effective for removing pollutants like pesticides and heavy metals from water. These techniques have demonstrated significant potential in improving water quality (AbuQamar *et al.*, 2024; Thanigaivel *et al.*, 2023).
- **Integrated Approaches:** Combining various methods, including chemical precipitation and adsorption, can effectively address potentially toxic elements (PTEs) in industrial wastewater (Singh *et al.*, 2022).

11.2 Regulation of Toxicants

- **International Treaties and National Regulations:** Coordinated regulatory frameworks, such as the establishment of unified chemical registries and effect-based approaches in Europe, highlight the importance of cohesive strategies for managing new and emerging pollutants (Sanganyado, 2022).
- **Analytical Considerations:** Developing robust detection and analytical methods is critical for monitoring hazardous substances and ensuring adherence to regulatory standards (Reyes-Calderón *et al.*, 2022).

11.3 Monitoring and Assessment

- **Innovative Monitoring Techniques:** Advances in both chemical and biological analysis techniques have improved the ability to identify emerging contaminants, enabling timely and effective interventions (Sanganyado, 2022).

11.4 Bioremediation

- Bioremediation involves the use of microorganisms to break down toxic pollutants, effectively cleaning contaminants from soil and water. Key techniques include biosparging, bioventing, and bioaugmentation (Sharma *et al.*, 2022). The use of genetically engineered microbes can further enhance bioremediation by targeting specific pollutants such as heavy metals and organic compounds (Sharma *et al.*, 2022).

11.5 Phytoremediation

- Phytoremediation utilizes plants to absorb, detoxify, and stabilize heavy metals and organic pollutants in contaminated areas. Techniques include phytoextraction, where plants take up contaminants, and rhizofiltration, where plant roots remove pollutants from water (Rai *et al.*, 2021). Integrating plant-microbe interactions can significantly boost the effectiveness of phytoremediation by utilizing natural detoxification processes (Álvarez *et al.*, 2021).

11.6 Sediment Dredging

- Sediment dredging is a physical method used to remove contaminated sediments from water bodies. This technique helps reduce toxic exposure to both aquatic life and humans by eliminating the source of contamination.

When applied effectively, these strategies play a crucial role in lowering the presence of toxicants in the environment and lessening their effects on ecosystems and human health.

Conclusion

Toxicants of chemical, biological, or physical nature present significant threats to both aquatic ecosystems and human health. Pollutants like pesticides, heavy metals, microplastics, and pharmaceuticals enter water bodies through various routes, leading to extensive ecological harm and health risks. The complex processes of bioaccumulation, biomagnification, bioconcentration, bioavailability and detoxification further exacerbate these impacts, underscoring the importance of thoroughly understanding their effects on aquatic life and human health. The rise of new pollutants introduces additional challenges, requiring ongoing research and advanced detection technologies to manage their interactions with existing contaminants. Effective strategies for managing and mitigating these pollutants are essential, including reducing their sources, improving waste treatment, and enforcing strong regulations. Public education is also crucial for reducing chemical use and promoting proper disposal methods. To tackle the issues posed by toxicants, a comprehensive approach that integrates scientific research, technological advancements, and proactive policy-making is necessary. Enhancing our knowledge of toxicant behavior and their effects on aquatic systems will help protect biodiversity, ensure the sustainability of water resources, and safeguard public health. Future efforts should focus on expanding our understanding, advancing monitoring techniques, and developing collaborative solutions to effectively address and mitigate the impacts of toxicants.

References

AbuQamar, S. F., El-Saadony, M. T., Alkafaas, S. S., Elsalahaty, M. I., Elkafas, S. S., Mathew, B. T., ... & El-Tarabily, K. A. (2024). Ecological impacts and management strategies of pesticide pollution on aquatic life and human beings. Marine Pollution Bulletin, 206, 116613.

Alexander, D. E. (1999). Bioaccumulation, bioconcentration, biomagnification. In Environmental geology. Encyclopedia of Earth Science. Springer. https://doi.org/10.1007/1-4020-4494-1_31

Almeida, Â., Solé, M., Soares, A. M., & Freitas, R. (2020). Anti-inflammatory drugs in the marine environment: Bioconcentration, metabolism and sub-lethal effects in marine bivalves. Environmental Pollution, 263, 114442.

Álvarez, S. P., Ardisana, E. F. H., Morales, S. G., & Monarez, A. P. (2021). Mitigation of Hazardous Contaminants: A Phyto-Microbiome Approach. Rhizobiont in Bioremediation of Hazardous Waste, 363-389.

Arthur, C., Baker, J. E., & Bamford, H. A. (2009). Proceedings of the International Research Workshop on the Occurrence, Effects, and Fate of Microplastic Marine Debris, September 9-11, 2008, University of Washington Tacoma, Tacoma, WA, USA.

Baldisserotto, B. (2021). General introduction to toxicology of aquatic animals. In Aquaculture Toxicology (pp. 17-24). Academic Press.

Beckett, K. (2015). Biomagnification of Mercury in the Indian River Lagoon, Florida.

Bengtsson-Palme, J., Alm Rosenblad, M., Molin, M., & Blomberg, A. (2014). Metagenomics reveals that detoxification systems are underrepresented in marine bacterial communities. BMC genomics, 15, 1-17.

Bertato, L., Chirico, N., & Papa, E. (2022). Predicting the bioconcentration factor in fish from molecular structures. Toxics, 10(10), 581.

Black, J. L. (2020). Detoxification and Biotransformation. Integrative and Functional Medical Nutrition Therapy: Principles and Practices, 205-212.

Campbell, P. G., & Hare, L. (2009). Metal detoxification in freshwater animals. Roles of metallothioneins.

Carpenter, S. R., Kitchell, J. F., & Hodgson, J. R. (1985). Cascading trophic interactions and lake productivity. BioScience, 35(10), 634-639.

Chen, C. H. (2024). Phase II Detoxification Enzymes. In Activation and Detoxification Enzymes: Functions and Implications (pp. 59-70). Cham: Springer Nature Switzerland.

Chen, F., Gong, Z., & Kelly, B. C. (2016). Bioavailability and bioconcentration potential of perfluoroalkyl-phosphinic and-phosphonic acids in zebrafish (Danio rerio): Comparison to perfluorocarboxylates and perfluorosulfonates. Science of the Total Environment, 568, 33-41.

Cope, W. G., Leidy, R. B., & Hodgson, E. (2004). Classes of toxicants: use classes. A textbook of modern toxicology, 49-74.

Del Carmen Gómez-Regalado, M., Martín, J., Santos, J. L., Aparicio, I., Alonso, E., & Zafra-Gómez, A. (2023). Bioaccumulation/bioconcentration of pharmaceutical active compounds in aquatic organisms: Assessment and factors database. Science of the Total Environment, 861, 160638.

Díaz-Delgado, E., Girolametti, F., Annibaldi, A., Trueman, C. N., & Willis, T. J. (2024). Mercury bioaccumulation and its relationship with trophic biomarkers in a Mediterranean elasmobranch mesopredator. Marine Pollution Bulletin, 201, 116218.

Dimitrov, S., Dimitrova, N. C., Walker, J. D., Veith, G. D., & Mekenyan, O. G. (2002). Predicting bioconcentration factors of highly hydrophobic chemicals. Effects of molecular size. Pure and Applied Chemistry, 74(10), 1823-1830.

EEA (2019). Urban waste water treatment for 21st century challenges. Accessed on 08-09-2024. https://www.eea.europa.eu/publications/urban-waste-water-treatment-for

Erzinger, G. S., Strauch, S. M., Fröhlich, M., Machado, C. K., & del Ciampo, L. (2021). Pharmaceutical Pollutants in Aquatic Ecosystems. Anthropogenic Pollution of Aquatic Ecosystems, 229-243. Doi: 10.1007/978-3-030-75602-4_11

Finizio, A., Grenni, P., Petrangeli, A. B., Caracciolo, A. B., Santoro, S., & Di Guardo, A. (2022). Use of large datasets of measured environmental concentrations for the ecological risk assessment of chemical mixtures in Italian streams: A case study. Science of the Total Environment, 806, 150614.

Fleeger, J. W., Carman, K. R., & Nisbet, R. M. (2003). Indirect effects of contaminants in aquatic ecosystems. Science of the total environment, 317(1-3), 207-233.

Franco, M. E., Schönenberger, R., Hollender, J., & Schirmer, K. (2024). Organ-specific biotransformation in salmonids: Insight into intrinsic enzyme activity and biotransformation of three micropollutants. Science of The Total Environment, 925, 171769.

Fu, Q., Fedrizzi, D., Kosfeld, V., Schlechtriem, C., Ganz, V., Derrer, S., ... & Hollender, J. (2020). Biotransformation changes bioaccumulation and toxicity of diclofenac in aquatic organisms. Environmental science & technology, 54(7), 4400-4408.

Garg, R., & Smith, C. J. (2014). Predicting the bioconcentration factor of highly hydrophobic organic chemicals. Food and chemical toxicology, 69, 252-259.

Gray, J. S. (2002). Biomagnification in marine systems: the perspective of an ecologist. Marine pollution bulletin, 45(1-12), 46-52.

Guerra-Rivas, G., & Gomez-Gutierrez, C. M. (2016). Glycoproteins and detoxification in the marine environment. In Marine Glycobiology (pp. 181-188). CRC Press.

Harriage, S., Harasheh, A., Schultz, N., & Long, B. (2024). Bioconcentration of pharmaceuticals by aquatic flora in an Australian river system. Science of The Total Environment, 946, 174361.

Howarth, R. W., Sharpley, A., & Walker, D. (2002). Sources of nutrient pollution to coastal waters in the United States: Implications for achieving coastal water quality goals. Estuaries, 25, 656-676.

Huertos, L. M. (2020). Ecology and management of inland waters: A Californian perspective with global applications. Elsevier: San Diego, CA, USA; ISBN 978-0-12-814266-0.

Hunt, D., Branson, M., Putnam, E., & Pershouse, M. (2020). Toxicoinformatics today. In Information Resources in Toxicology (pp. 51-62). Academic Press.

Ikwuemesi, J., Egesi, C., Irozuru, O., & Ezeamaka, S. N. (2023). Bioaccumulation of Heavy Metals in Gills and Muscles of Selected Fish Species from Owerrinta Section of Imo River. Journal of Aquatic Sciences, 38(1).

Jaingam, W. (2018). Biomagnification process of total mercury in the aquatic ecosystem of an enclosed Isahaya Bay, Kyushu, western Japan (Doctoral dissertation).

Jeong, H., Byeon, E., Kim, D. H., Maszczyk, P., & Lee, J. S. (2023). Heavy metals and metalloid in aquatic invertebrates: a review of single/mixed forms, combination with other pollutants, and environmental factors. Marine Pollution Bulletin, 191, 114959.

Kallenborn, R., & Blais, J. M. (2015). Tracking contaminant transport from biovectors. Environmental Contaminants: Using natural archives to track sources and long-term trends of pollution, 461-498.

Kieliszek, M., Błażejak, S., Gientka, I., & Bzducha-Wróbel, A. (2015). Accumulation and metabolism of selenium by yeast cells. Applied microbiology and biotechnology, 99, 5373-5382.

Kumar, N., Zhao, H. N., Awoyemi, O., Kolodziej, E. P., & Crago, J. (2021). Toxicity testing of effluent-dominated stream using predictive molecular-level toxicity signatures based on high-resolution mass spectrometry: a case study of the Lubbock Canyon Lake System. Environmental Science & Technology, 55(5), 3070-3080.

Li, S., Chu, Y., Ren, N., & Ho, S. H. (2023). Cytochrome P450 enzyme-based biotransformation of pharmaceuticals and personal care products (PPCPs) by microalgae in the aquatic environment. Chemical Engineering Journal, 476, 146557.

Likens, G. E. (1985). An experimental approach for the study of ecosystems. Journal of ecology, 73(2).

Liu, Y., & Huang, Q. (2020). Time-delayed models for the effects of toxicants on populations in contaminated aquatic ecosystems. Mathematics in Applied Sciences and Engineering, 1-15.

Lu, T., Zhang, Q., Zhang, Z., Hu, B., Chen, J., Chen, J., & Qian, H. (2021). Pollutant toxicology with respect to microalgae and cyanobacteria. Journal of Environmental Sciences, 99, 175-186. doi: 10.1016/J.JES.2020.06.033

Luquet, C. M., & Bieczynski, F. (2023). Absorption and excretion: Detoxification of wastes and toxins in the fish gastrointestinal system.

Mackay, D., & Fraser, A. (2000). Bioaccumulation of persistent organic chemicals: mechanisms and models. Environmental pollution, 110(3), 375-391.

Maculewicz, J., Dołżonek, J., Sharma, L., Białk-Bielińska, A., Stepnowski, P., & Pazdro, K. (2023). Bioconcentration of imidazolium ionic liquids: In vivo evaluation in marine mussels Mytilus trossulus. Science of the Total Environment, 858, 159388.

Madesh, S., Sudhakaran, G., Ramamurthy, K., Kathiravan, M. K., Almutairi, M. H., Almutairi, B. O., ... & Arockiaraj, J. (2024). Cadmium and ketoprofen accumulation influences aquatic ecosystem demonstrated using in-vivo zebrafish model. Drug and Chemical Toxicology, 1-16.

Mai, Y., Wang, Y., Geng, T., Peng, S., Lai, Z., Wang, X., & Li, H. (2024). A systematic toxicologic study of polycyclic aromatic hydrocarbons on aquatic organisms via food-web bioaccumulation. Science of The Total Environment, 929, 172362.

Martins, J. C., & Vasconcelos, V. M. (2009). Microcystin dynamics in aquatic organisms. Journal of Toxicology and Environmental Health, Part B, 12(1), 65-82.

Massoud, A. H., Derbalah, A. S., & Belal, E. S. B. (2008). Microbial detoxification of metalaxyl in aquatic system. Journal of Environmental Sciences, 20(3), 262-267.

Melo de Almeida, E., Tisserand, F., Faria, M., & Chèvre, N. (2022). Efficiency of several cytochrome P450 biomarkers in highlighting the exposure of Daphnia magna to an organophosphate pesticide. Toxics, 10(8), 482.

Meybeck, M. (1982). Carbon, nitrogen, and phosphorus transport by world rivers. American journal of Science, 282(4), 401-450.

Milan, Hait., Nand, Kumar, Kashyap., Arvind, Kumar, Bhardwaj. (2024). Emerging Pollutants from Urbanization and Industrialization in the Global South. 39-87. doi: 10.1007/978-981-97-1658-6_2

Moore, D., Lord, J. C., & Smith, S. M. (2000). Pathogens. In Alternatives to pesticides in stored-product IPM (pp. 193-227). Boston, MA: Springer US.

Nendza, M., & Ahlers, J. (2022). Aquatic toxicity integrated testing and assessment strategies (ITS) for difficult substances: case study with thiochemicals. Environmental Sciences Europe, 34(1), 17.

Ochs, C., Garrison, K., Saxena, P., Romme, K., & Sarkar, A. (2024). Contamination of aquatic ecosystems by persistent organic pollutants (POPs) originating from landfills in Canada and the United States: A rapid scoping review. Science of The Total Environment, 171490.

Petoumenou, M. I., Pizzo, F., Cester, J., Fernández, A., & Benfenati, E. (2015). Comparison between bioconcentration factor (BCF) data provided by industry to the European

Chemicals Agency (ECHA) and data derived from QSAR models. Environmental Research, 142, 529-534.

Pramanik, S., & Roy, K. (2014). Modeling bioconcentration factor (BCF) using mechanistically interpretable descriptors computed from open source tool "PaDEL-Descriptor". Environmental Science and Pollution Research, 21, 2955-2965.

Radomyski, A., Lei, K., Giubilato, E., Critto, A., Lin, C., & Marcomini, A. (2018). Bioaccumulation of trace metals in aquatic food web. A case study, Liaodong Bay, NE China. Marine pollution bulletin, 137, 555-565.

Rai, G. K., Bhat, B. A., Mushtaq, M., Tariq, L., Rai, P. K., Basu, U., ... & Bhat, J. A. (2021). Insights into decontamination of soils by phytoremediation: A detailed account on heavy metal toxicity and mitigation strategies. Physiologia Plantarum, 173(1), 287-304.

Rajak, P., Ganguly, A., Nanda, S., Mandi, M., Ghanty, S., Das, K., ... & Sarkar, S. (2024). Toxic contaminants and their impacts on aquatic ecology and habitats. In Spatial Modeling of Environmental Pollution and Ecological Risk (pp. 255-273). Woodhead Publishing. Doi: 10.1016/b978-0-323-95282-8.00040-7

Reyes-Calderón, A., Pérez-Uribe, S., Ramos-Delgado, A. G., Ramalingam, S., Oza, G., Parra-Saldívar, R., ... & Sharma, A. (2022). Analytical and regulatory considerations to mitigate highly hazardous toxins from environmental matrices. Journal of Hazardous Materials, 423, 127031.

Rezania, S., Taib, S. M., Din, M. F. M., Dahalan, F. A., & Kamyab, H. (2016). Comprehensive review on phytotechnology: heavy metals removal by diverse aquatic plants species from wastewater. Journal of hazardous materials, 318, 587-599.

Romero-Romero, S., Herrero, L., Fernandez, M., Gomara, B., & Acuña, J. L. (2017). Biomagnification of persistent organic pollutants in a deep-sea, temperate food web. Science of the Total Environment, 605, 589-597.

Saadoun, I. M. (2015). Impact of oil spills on marine life. Emerging pollutants in the environment-current and further implications, 10, 60455.

Saha, R., & Dutta, S. M. (2024). Pesticides' Mode of Action on Aquatic Life.

Sanganyado, E. (2022). Policies and regulations for the emerging pollutants in freshwater ecosystems: Challenges and opportunities. Emerging freshwater pollutants, 361-372.

Seto, M., & Handoh, I. C. (2009). Mathematical explanation for the non-linear hydrophobicity-dependent bioconcentration processes of persistent organic pollutants in phytoplankton. Chemosphere, 77(5), 679-686.

Shah, S., & Kumar, A. (2022). Bioavailability, Bioconcentration, and Biomagnification of Pollutants. In Environmental Toxicology and Ecosystem (pp. 179-214). CRC Press.

Shanmuganathan, R., Kadri, M. S., Mathimani, T., Le, Q. H., & Pugazhendhi, A. (2023). Recent innovations and challenges in the eradication of emerging contaminants from aquatic systems. Chemosphere, 332, 138812.

Sharma, P., Parakh, S. K., Singh, S. P., Parra-Saldívar, R., Kim, S. H., Varjani, S., & Tong, Y. W. (2022). A critical review on microbes-based treatment strategies for mitigation of toxic pollutants. Science of The Total Environment, 834, 155444.

Singh, A., Chauhan, S., Varjani, S., Pandey, A., & Bhargava, P. C. (2022). Integrated approaches to mitigate threats from emerging potentially toxic elements: A way forward for sustainable environmental management. Environmental Research, 209, 112844.

Sotka, E. E., & Gantz, J. (2013). Preliminary evidence that the feeding rates of generalist marine herbivores are limited by detoxification rates. Chemoecology, 23, 233-240.

Sun, Z., Liu, Y., Xu, H., & Yan, C. (2022). Genome-wide identification of P450 genes in Chironomid Propsilocerus akamusi reveals candidate genes involved in gut microbiota-mediated detoxification of Chlorpyrifos. Insects, 13(9), 765.

Takarina, N. D., & Pin, T. G. (2017). Bioconcentration factor (BCF) and translocation factor (TF) of heavy metals in mangrove trees of Blanakan fish farm. Makara Journal of Science, 77-81.

Teuten, E. L., Saquing, J. M., Knappe, D. R., Barlaz, M. A., Jonsson, S., Björn, A., ... & Takada, H. (2009). Transport and release of chemicals from plastics to the environment and to wildlife. Philosophical transactions of the royal society B: biological sciences, 364(1526), 2027-2045.

Thanigaivel, S., Vinayagam, S., Gnanasekaran, L., Suresh, R., Soto-Moscoso, M., & Chen, W. H. (2023). Environmental fate of aquatic pollutants and their mitigation by phycoremediation for the clean and sustainable environment: A review. Environmental Research, 117460.

Tong, Y., Huang, Z., Janssen, A. B., Wishart, M., He, W., Wang, X., & Zhao, Y. (2022). Influence of social and environmental drivers on nutrient concentrations and ratios in lakes: A comparison between China and Europe. Water Research, 227, 119347. https://doi.org/10.1016/j.watres.2022.119347

Umeh, A. C., Naidu, R., Owojori, O. J., & Semple, K. T. (2020). Bioavailability and bioaccessibility of hydrophobic organic contaminants in soil and associated desorption-based measurements. Bioavailability of Organic Chemicals in Soil and Sediment, 293-350.

Wang, W. X. (2016). Bioaccumulation and biomonitoring. In Marine ecotoxicology (pp. 99-119). Academic Press.

Wang, Y., Wen, Y., Li, J. J., He, J., Qin, W. C., Su, L. M., & Zhao, Y. H. (2014). Investigation on the relationship between bioconcentration factor and distribution coefficient based on class-based compounds: The factors that affect bioconcentration. Environmental toxicology and pharmacology, 38(2), 388-396.

Wu, L., Luo, H., Xu, J., Yu, L., Xiong, J., Liu, Y., & Zou, X. (2024). Vital role of CYP450 in the biodegradation of antidiabetic drugs in the aerobic activated sludge system and the mechanisms. Journal of Hazardous Materials, 469, 134056.

Wu, P., Zakem, E. J., Dutkiewicz, S., & Zhang, Y. (2020). Biomagnification of methylmercury in a marine plankton ecosystem. Environmental Science & Technology, 54(9), 5446-5455.

Yadav, S., Chauhan, A. K., Kumar, S., & Kataria, N. (2022). Advanced membrane technology for the removal of pesticides from water and wastewater. In Pesticides remediation technologies from water and wastewater (pp. 143-156). Elsevier. https://doi.org/10.1016/B978-0-323-90893-1.00007-6.

Zhang, W., Miao, A. J., Wang, N. X., Li, C., Sha, J., Jia, J., & Ok, Y. S. (2022). Arsenic bioaccumulation and biotransformation in aquatic organisms. Environment international, 163, 107221.

Zhang, X., Xiong, W., Wu, Q., Nian, K., Pan, X., Crump, D., ... & Zhang, R. (2023). Bioaccumulation, trophic transfer, and biotransformation of polychlorinated diphenyl ethers in a simulated aquatic food chain. Environmental Science & Technology, 57(14), 5751-5760.

Zhu, M., Chen, J., Peijnenburg, W. J., Xie, H., Wang, Z., & Zhang, S. (2023). Controlling factors and toxicokinetic modeling of antibiotics bioaccumulation in aquatic organisms: A review. Critical Reviews in Environmental Science and Technology, 53(15), 1431-1451.

Zhu, X., Luo, T., Wang, D., Zhao, Y., Jin, Y., & Yang, G. (2023a). The occurrence of typical psychotropic drugs in the aquatic environments and their potential toxicity to aquatic organisms-A review. Science of The Total Environment, 165732.

10

Emerging Contaminant in Aquatic System

Nazeefah Hassan[1], Monisa Malik[1*], Ifra Tabassum[1] Adnan Abubakr[1] and Shabana Arjamand[2]

[1]*Division of Aquatic Environmental Management, Faculty of Fisheries Rangil SKUAST-Kashmir*

[2]*AAAM, Govt. Degree College Bemina Srinagar*

**Corresponding Author Email: monisamalik@gmail.com*

Abstract

Emerging contaminants (ECs) are a class of newly discovered contaminants that have been identified as major water pollutants that have negative impacts on human health in recent years. These micropollutants are said to bioaccumulate in people, other aquatic food web creatures, and macro invertebrates. Conventional treatment methods are unable to eliminate them. Only advancements in analytical instrumentation made it possible to identify these pollutants, even at minuscule amounts. As new compounds are discovered over time, concerns are raised about their origins, fate, transportation, transformations, and effects on aquatic habitats. The purpose of this chapter is to provide an overview of the contaminants that are categorized as "emerging," together with information on their sources, entry points into the environment, and associated threats to environment and possible treatments to somehow eliminate them.

***Keywords**: Emerging contaminants, Micropollutants and Aquatic System.*

Introduction

Unchecked growth in a number of human endeavors, such as industrial, transportation, agricultural, and urbanization, has affected the world's growth metrics (Gavrilescu *et al.,* 2015). Pollution has increased due to rising consumer demands and living standards (Giudice *et al.,* 2021). Examining surface and groundwater confirmed the existence of emerging contaminants, but the observed concentration in the surface water is lesser than that at effluent wastewater treatment facility (Arman *et al.*, 2021). In addition, chemicals, microelements, industrial effluents, spills, and soil contamination caused by the spread of pesticides and sludge, as well as the dumping of hazardous waste, are discovered in water bodies (Galindo-Miranda *et al.,* 2019). Antibiotics and endocrine disruptors are two examples of toxins with low concentrations but major ecological effects that are spreading rapidly as environmental needs increase (Cheng *et al.*, 2021). A class of chemical pollutants called as emerging pollutants or contaminants (EPs) poses potential threats to human health and the ecological balance of the environment. These extremely complex chemical compounds are most commonly found in water (Cheng *et al.,* 2021). The United States Geological Survey defines ECs as substances of synthetic or natural origin, or microorganisms that are not widely monitored in the environment but may have negative consequences on the ecosystem or human health (Kumar *et al.,* 2022). It is critical to highlight that the vast majority of emerging contaminants are not new or recently introduced pollutants into the environment. However, the vast majority of newly found contaminants are already well-known pollutants with a novel adverse impact or Mode of Action (MOA). As a result, the term "emerging" refers to both the contamination and the growing worry about it as a problem. Contaminants are considered "emerging" when they have a new source, a different path to people, or innovative treatment options (Gogoi *et al.,* 2018).

Many different classes of chemicals, including pharmaceuticals and hormones, illegal drugs, pesticides, personal care products, artificial sweeteners, perfluorinated compounds, disinfection byproducts, UV filters, and other industrial chemicals, have been found in surface and groundwater according to recent studies. These chemicals have been detected in the range of ng/L to µg/L. (Kasprzyk-Hordern *et al.,* 2008; Lapworth *et al.,* 2012; Petrovic *et al.,* 2004; Richardson and Kimura, 2016; Zuccato and Castiglioni, 2009). Typically, these contaminants tend to bioaccumulate. Among them Disruptive endocrine chemicals (EDCs) are a specific class of EC that have drawn a lot of attention recently. EDCs are defined by the Endocrine Society as follows: "any exogenous (non-natural) chemical or combination of chemicals that obstructs the function of hormones in any way. (Kumar *et al.,* 2022) Numerous investigations have been conducted to ascertain the toxicity of a relatively small number of ECs to ecosystems, with data provided by the OECD (Boxal, 2012) and US EPA (US EPA, 2015a). According to their research, human and veterinary medications can harm fish and crayfish's hormones (Daughton and Ternes, 1999); interfere with insects' physiology and development (Floate *et al.,* 2005, Sommer and Bibby, 2002); alter fish biochemistry (Lavelle *et al.,* 2004); as well as jeopardise the invertebrate community's ability to develop (Pascoe *et al.,* 2003).The main issue with ECs is that available data for most of these contaminants is scarce and relatively limited, and respective detection methods and instrumentation either do not yet exist or are in the early stages of development. This is the reason why the detection of these micro-contaminants in the environment was only made possible by the steady advancement of analytical detecting and measuring techniques. To put it another way, the degree of risk that many ECs pose to the environment and public health remains unclear, and both effective technologies and monitoring strategies to reduce contamination are still being developed. Since the great majority of ECs are not currently regulated, it is necessary to continuously monitor and report any potential presence of ECs in water supplies and effluent discharges (Stefanakis *et al.,* 2016). As a result of the chemical diversity of these compounds, their typically low concentrations (parts per billion or part per trillion) and the complexity of their matrices, the analysis of emerging contaminants is a true analytical challenge (Petrović *et al.,* 2004).It is clear that an integrated research and approach to these contaminant groups should be multidisciplinary, involving disciplines such as chemistry, biology, and engineering (Stefanakis *et al.,* 2016).An important environmental issue is raised by the ineffective removal. As a result, there is increasing interest in this subject from a regulatory and research standpoint (Thomaidis *et al.,* 2012).

Classification of Emerging Contaminants

Pharmaceuticals, personal care products (PCPs), Endocrine Disrupting Chemicals (EDCs), steroids, hormones, surfactants and surfactant metabolites, flame retardants, pesticides, industrial additives, nanomaterials, and gasoline additives are among the compounds classified as ECs. Typically, emerging contaminants that have been in surface water for a long time but whose presence was only recently discovered are pharmaceuticals and PCPs.

Pharmaceuticals & PPCPs

Personal care and pharmaceutical products (PPCPs) often include a polar functional moiety encased in a non-polar core, though their chemical characteristics can vary greatly. Environmentally persistent products could result from their transition. (Clarke and Smith, 2011; Fent *et al.,* 2006). All across the world, medicines and their metabolites are released into the aquatic environment (Beek *et al.,* 2016). PPCPs and their metabolites, which were initially found in wastewater, are frequently referred to as "effluent-derived" contaminants (Thomaidis *et al.,* 2013*).* The detection of these substances at trace levels was enabled by recent advancements in analytical techniques and improvements in analytical instrumentation (Steafkins *et al.,* 2016). The biotic and abiotic changes in pharmaceutical, both anaerobic and aerobic conditions, as well as contact with sediments and soil, bank filtration, wastewater treatment, and sewers themselves can produce chemicals and hormones. In addition, ultraviolet light in surface waters and oxidative treatment procedures like ozonation, advanced oxidation, and chlorine disinfection cause transformation (Richardson *et al.,* 2016). Pharmaceuticals are classified according to their purpose and biological activity, such as antibiotics, analgesics, anti-neoplastic, disinfectants, anti-flammatory compounds, X-ray contrast media, and so on. Another classification can be made based on the mode of action, such as anti-metabolites, alkyl-ating agents, and so on. Pharmaceuticals have a wide range of biological and physicochemical properties due to the large number of compounds. The typical concentrations of these compounds range from micrograms per litre (g/L) to nanograms per liter (ng/L), which explains why they are also known as micropollutants (Steafkinis et al,). The propensity of many pharmaceuticals and PPCPs to evade treatment completely or to only partially degrade during treatment through sewage overflows increases their load in receiving waters, many of which are sources of drinking and recreational water (Clarke and Smith, 2011; Hernando *et al.,* 2011; Oulton *et al.,* 2010; Reif *et al.,* 2011; Wells *et al.,* 2010).

Personal Care Products (PCPs) Detected in Waters Include

1. **DEET:** The most common active component found in insect repellents is N, N-diethyl-meta-toluamide.
2. **Parabens:** p-hydroxybenzoic acid esters, which have been used for decades as bacteriostatic and fungistatic agents in cosmetics, drugs, and foods,
3. **Triclosan:** a bacteriocide as well as antifungal agent found in household items such as toothpaste and soap.
4. **Polycyclic musk's:** PCPs, which include tonalide along with galaxolide, are employed as fragrances in washing and cleaning agents.
5. **UV filters/sunscreen:** Some of such substances are benzophenones and methoxycinnamates.
6. **Caffine and nicotine:** substances associated with the contemporary culture and way of living.

Hormones and Steroids

Natural endogenous steroids such as sex hormones (androgens such as androstenedione and testosterone, oestrogens such as oestrone, oestriol, and progesterone), phytoestrogens, faecal indicators, and plant sterols are all examples of ECs that are excreted from the human body. Synthetic androgens include oxandrolone, nandrolone, and synthetic oestrogens (xenoestrogens) used as contraception, such as diethylstilbestrol. As free active steroids, natural and synthetic steroids and hormones can be found in wastewater influents and effluents. The majority of these compounds are classified as EDCs. The four most frequently detected compounds in groundwater are etrone, E2, NP, and bisphenol A (Lapworth *et al.,* 2012). Environmental biota has suffered as a result of estrogenic activity in WWTP effluents (Fan *et al.,* 2011).

Flame Retardants

By obstructing the combustion of polymers, these compounds are used in furnishing foam, textiles, and plastics to lessen their flammability. They might be compounds that have been brominated or halogenated. Polybrominated diphenyl ethers (PBDEs) are classified as environmentally dangerous compounds (EDCs) due to their bioaccumulative nature (Rahman, Langford, Scrimshaw, & Lester, 2001).

Chlorinated straight-chain hydrocarbons make up the technical mixtures of polychlorinated alkanes (PCAs), also known as chlorinated paraffins (CPs), a class of industrial chemicals. Waste effluent that has been treated may contain traces. Further investigation is required to evaluate the risk that the industrial use of this chemical group poses to human health and the environment, as indicated by the concentrations of PCAs in sewage sludge, toxicity data, and evidence of accumulation in human and environmental biota. (Thomaidis *et al.,*2012).

Artificial Sweeteners

In recent times, it has been determined that sucrose and other artificial sweeteners are persistent emerging pollutants. (Richardson and Ternes, 2011). Sucralose recognised as one of the new artificial sweeteners. Although it may seem strange to list this substance as an emerging contaminant, it is now widely present in environmental waters and has a very long half-life (up to several years). Research has expanded to include other artificial sweeteners, such as acesulfame, saccharin, cyclamate, and aspartame. Several research groups have reported measurements of sucralose in the environment (including river water, groundwater, and coastal waters). (Scheurer *et al.,* 2009; Richarson, 2012). Acesulfame and sucrose were thought to be the perfect markers for identifying household wastewater in environmental waters, especially groundwater, due to their resistance to transformation. (Buerge *et al.,* 2009; Oppenheimer *et al.,* 2011).

Pesticides

Pesticides can contaminate drinking water due to carelessness, such as reverse syphoning, spraying on fields and golf courses, and a significant unintended spill (Srivastav 2020). The soil has easily absorbed pesticide metabolites (Nowak *et al.,* 2020). The variability in the runoff specimens to various pesticide applications in sparsely populated urban areas; pesticide metabolites were found in ground and surface water (Reemtsma *et al.,* 2013). Numerous metabolites discovered in water matrices indicate that pesticide use is every day in urban areas (Gallé *et al.,* 2020, Reemtsma e*t al.,* 2013). According to a study by Belenguer *et al.* (2014), the presence and concentration of pesticides such as prochloraz, clofenvinphos, pyriproxyfen, imazalil, and dichlofenthion have been linked to the region's heavy agricultural activity (Belenguer *et al.,* 2014). In recent times, substances like atrazine, which were deemed to be highly hazardous to the environment, have been phased out of use; however, the substances that were substituted (like diuron) may also have unfavorable effects. Metaldehyde is currently one of the pesticides of special interest because it has occasionally been found to be present in excess of the EU's pesticide limit in drinking water (Stefanakis *et al.,* 2016).

EPs from Perfluorinated Compounds (PFCs)

A class of man-made chemicals known as perfluorochemicals (PFCs) is used to make products resistant to water, heat, oil, stains, and grease. PFCs are ubiquitous and long-lasting in the environment. These compounds are unique in that they have one of the strongest known chemical bonds (C-F) and are both lipophobic and hydrophobic. Although the exact toxicological reaction in humans and the environment to such exposure is unknown, it may include endocrine disruption among other things. (Clarke and Smith, 2011; Martin *et al.,* 2010). Perfluoroalkyl sulphonates (PFASs), with the primary representative compound being perfluorooctane sulphonate (PFOS), and perfluoroalkyl carboxylates (PFACs), with the primary representative compound being perfluorooctanoic acid (PFOA), are the two most frequently measured and detected PFC groups in environmental matrices. (Thomaidis *et al*., 2012).

According to Sznajder-Katarzyńska *et al.* (2019), PFOA has been found in processing and potable water due to emissions from industrial sites, WWTPs, and non-point sources. Because of the contaminated groundwater plume, it may also be detected in drinking water wells. Similar to this, PFOA from adjacent industrial sites may infiltrate the air, land on the soil's surface, and then seep into the groundwater (Xu *et al.,* 2021). Furthermore, it has been reported that potable water sources approximately 20 miles away from the target site were affected when an industrial facility released PFOAs into the environment without first treating them (Pitter *et al.,* 2020, Steenland *et al.,* 2020).

Microplastics

The growing global contamination of rivers, lakes, and oceans over the past few years has drawn a lot of attention to microplastics, or MPs. MPs, which are particles of less than 5 mm in size, may be found in the environment as a result of larger plastic materials degrading in the environment or as a result of their direct usage in products. The Microbead-Free Waters Act, a recent legislation, forbids the use of microbeads in toothpaste and cosmetics produced or sold in the United States. (www.congress.gov/bill/114th-congress/house-bill/1321/text).

The most common type of MPs are fibres, which are easily released from clothing during washing, according to recent studies (particularly from polyester fleece Clothing) (Richardson *et al.,* 2017).

Primary microplastics are present in a variety of household products, such as exfoliators, toothpaste, and cleaning supplies, and are specifically made to be either millimetric or sub-millimetric in size. Primary microplastics include pre-production plastic pellets, which are used as feedstock to make plastic products. These tiny pellets are commonly lost during handling, as a result of spills during transit, or at manufacturing sites. (McVeigh *et al.,* 2021). Additionally, microplastics can enter the (marine) environment by being directly released into residential and commercial wastewater systems (Andrady *et al.,* 2011). Microplastics are widely distributed throughout the environment and can be found in food, water, and soil, among other spheres (pironti *et al.,* 2021).

As a result, scientists have lately started using a new category for these called plastcides. (Da Costa *et al.,* 2020). These additives, which are typically non-polymeric and low molecular weight, may cause biochemical reactions and effects. It Should be mentioned that residual monomers, which can also migrate like other adsorbed organic pollutants and often show a significant level of toxicity, may also be present to a lesser extent as a result of incomplete polymerization reactions. (Kokalj *et al.,* 2021).

According to certain research, the release of additives rather than the ingestion of the particles themselves is primarily responsible for the ecotoxicity of microplastics in water that has been observed (Beiras *et al.,* 2021), and additive release may affect plastic ecotoxicity more than previously believed (Zimmermann *et al.,* 2021).

Sources of Emerging Contaminants

Sources of ECs differ in terms of quantity and kind. According to Lapworth *et al.* (2012), they can be divided into point sources and diffuse pollution sources.

Point sources include drug flushing, bathing, household use, and excretion in human and animal urine and feces. At the outflow of wastewater treatment plants in urban, industrial, and agricultural areas, treated municipal and industrial wastewater is the main point source of pollution (Glassmeyer *et al.,*

2005; Pal *et al.,* 2010; Lapwoth *et al.,* 2012). Because of their high solubility, low volatility, and, in many cases, low degradability, ECs are able to flow through various treatment stages.

The sources of ECs are different for each of these treatments. Conventional biological wastewater treatment plants (such as activated sludge systems) and combined sewage-stormwater overflow treatment facilities have been shown to be insufficient in their ability to treat most of these contaminants, so ECs are still present in these treatment facilities' effluent, albeit at low concentrations (Asano, Burton, Leverenz, Tsuchihashi, & Tchobanoglous, 2007; Stefanakis and *et al.,* 2014).

While diffuse pollution sources of ECs can extend over wide geographic areas, it is more challenging to identify them, making it challenging to define the precise sources (Lapworth *et al.,* 2012). Stormwater runoff, terrestrial runoff from roads, urban areas, highways, and agricultural land are examples of sources falling under this category. Diffuse sources typically release smaller loads of pollutants into the surrounding environment than point sources do (Lapworth *et al.,* 2012).

Hospitals, the pharmaceutical industry, and municipal wastewater effluent are common places to find pharmaceuticals such as antidepressants, anticancer drugs, non-steroidal anti-inflammatory drugs, anti-retroviral drugs, and antimicrobial agents (Sanganyado *et al.,* 2017). On the other hand, veterinary medications and pesticides are often found in large quantities in agricultural runoff and occasionally in urban stormwater (Pereira *et al.,* 2016). Landfill leachate is frequently associated with the discharge of industrial chemicals such as flame retardants, plasticizers, and rare earth elements (Gwenzi *et al.,* 2018). Consequently, the source frequently affects the contaminant profile in the runoff and effluent. According to Du Lio *et al.* (2018), the Network of Reference Laboratories, Research Centers, and Associated Organizations for Monitoring of Emerging Environmental Substances (NORMAN) in Europe has identified over 700 new pollutants in aquatic systems.

Potential Impacts of Emerging Contaminants

Effects on Human Health

These days, people are exposed to an increasing number of organic pollutants (OPs), such as perchlorate, artificial muscles, triclosan, bisphenol-A, organophosphate flame retardants, phthalate substitutes, and polycyclic siloxanes. These pollutants have the potential to harm human health (Adeel *et al.* 2017). Bisphenol is one of the plasticizers that is most frequently used worldwide. It affects many different endocrine diseases in different systems, including the thyroid, salivary glands, men's genetic system, and many others. Women are more likely to develop breast cancer as a result of the hormonal effects of bisphenol, which also acts as an anti-androgen and has a negative impact on men's virility (Kawa *et al.*, 2021, Patel, 2017).

Workplace exposure of pesticide is associated with broad genotoxic effects, neurotoxicity, cancer-causing agents, and microbial induction. Brominated flame retardants have the potential to develop neuroprotective, endocrine, pulmonary, impermeable, and cancerous effects. According to Modesto *et al.* (2015), pentabromodiphenyl ether, hexabromodiphenyl ether, and decatabromodiphenyl ether are carcinogenic substances that affect thyroid function, neuronal function, brain development, fertility, and general growth. Furthermore, tetrabromodiphenyl ether can result in DNA damage, hormone imbalances, endocrine disruption, and mitochondrial damage. Moreover, because of possible genetic toxic effects, such as the implantation of abnormalities in female anatomy, the breast, and the liver, brominated flame retardants have been linked to an increase in interrupted immune cells in relation to thyroid hormones, androgens, progestin, and estrogen formation (Pereira *et al.* 2015).

Exposure to elevated levels of phthalates has been linked to miscarriages and difficult pregnancies. Hormonal side effects of chemical pesticides include ovarian tumours, thyroid problems, shrinkage of eggshells, and behavioural abnormalities in men. An ingredient found in many fragrance products, musk xylol, has been shown in a rat bacterial genome to be highly carcinogenic. It can also cause brain damage (Lei *et al.* 2015).

Effects of EP on the Environment

Establishing a clear causal relationship between the visibility of contaminants and their impact on human health is a challenging task. However, because many organisms exist at varying levels of biological complexity and because the assessment of field data is complicated by multiple external factors, it is even more difficult to do so at the environmental level. Surfactants are harmful to organisms. There are some substances that have a high potential to harm the environment. According to Pereira *et al.* (2015), bisphenol-A is neurotoxic and impairs plant growth and cell division. Unfavourable side effects have been associated with alkylphenol exposure in experimental mice during the spermatogenesis pathway.

It has been demonstrated that alkylphenols have a variety of estrogenic effects on marine life, including a decline in male sexual competence. The toxic-like tendency of furans, polychlorinated naphthalene, and brominated dioxins can occasionally impair the hormone system, immunity, and reproductive growth in animals, as well as cause aberrant congenital disabilities. Perchlorate mainly impacts the thyroid and brain development at lower concentrations; it has been observed that rodent pups born to infected mothers were especially vulnerable to negative effects (Shahid *et al.* 2021).

Perfluorochemicals can either promote or hinder the development of proteins that transfer oxidative metabolism via the outer membrane of aquatic species. Animals exposed to phthalates during pregnancy developed severe cognitive toxic effects, primarily in the male reproductive organ. Phthalates can induce thyroid, renal, and liver disease in old livestock. In research lab animal tests, polybrominated diphenyl ethers have been linked to a number of negative health outcomes, including thyroxine disruption, persistent thinking and memory issues, behavioral changes, partial deafness, late onset of adolescence, reduced sperm quality, and foetal deformities. Specific carrier proteins in marine invertebrates cause nitro-muscular and polycyclic musk chemicals to block the mnemiopsis cell membrane resistance detection systems for a relatively long time. Triclosan is recognized to be toxic to certain marine creatures. Biodiversity is frequently contaminated by the breakdown products of triclosan. Fish contain methyl triclosan, a breakdown product of triclosan that is primarily detected in fat cells (Rathi *et al.* 2021). An improved omics-related approach to classifying the harmful effects of a complex mixture of pollutants from wastewater treatment may be able to supplement the wide range of biological tests used to assess the ecotoxicological environmental effects of developing contaminants (Noguera-Oviedo & Aga 2016, R. K. Mishra *et al.,* 2023).

Treatment of Emerging Contaminants

Major issues in the reuse of water resources are the existence of trace organic contaminants in wastewaters, the way they behave during wastewater treatment, and how drinking water is produced.Sewage effluents are one of the main sources of various emerging contaminants and their treatment-resistant metabolites because wastewater-treatment plants (WWTPs) have been found to be relatively low in their elimination of various classes of emerging contaminants, such as pharmaceuticals, polar pesticides, and surfactant degradants (Petrovic *et al.,* 2003).

A variety of technologies have been used to eliminate ECs during the last several decades, including physical, chemical, and biological methods (Zhang *et al.,* 2008; Grover *et al.,* 2011). A blend of two or more treatment technologies, like membrane treatment with micro-filtration and reverse osmosis (Rodriguez *et al.,* 2015), simultaneous adoption of membrane ultra-filtration, and activated Carbon adsorption. Ultrasound irradiation (Secondes *et al.,* 2014), or a combination of two or more treatment technologies such as enzymatic catalysis and electrocoagulation, is presumed to be more effective in eliminating ECs as Well as the utilization of ultraviolet light and hydrogen peroxide during Advanced oxidation before membrane bio-filtration (Keen *et al.,* 2012; Keen *et al.,* 2016; Lester *et al.,* 2014].

Physio-chemical Treatment

Traditional water treatment frequently uses physical–chemical procedures to eliminate pathogens, manage taste and odour problems, and lessen turbidity. Additionally, these procedures could reduce the number of micropollutants in the final drinking water. The removal is usually insufficient, though (Benner *et al.,* 2013).

Coagulation and Flocculation

Over time, molecules in colloidal particles agglomerate and settle due to a chemical process called coagulation. Using aluminum sulfate (A12SO4)3 in combination with coagulation-flocculation combined with sand filtration, Huerta-Fontela *et al.* found that medications like warfarin, betaxolol, and hydrochlorothiazide could be effectively removed (with an 80% removal efficiency).Hospital wastewater was found to have high levels of eliminated Musk compounds (personal care products), specifically tonalide, galaxolide, and celestolide, with considerable removal rates of 83%, 79%, and 78%, respectively (Suarez *et al.,* 2009).It has been demonstrated that chemical treatment methods including coagulation, flocculation, and lime softening are unsuccessful at eliminating EDCs and PPCPs. Metal salt coagulants, such as aluminum sulfate and ferric sulfate, were not able to remove the chemicals tested, carbadox, sulfadimethoxine, and trimethoprim (Adams *et al.,* 2002).

Membrane Technology

The removal of new pollutants was not the original purpose of traditional wastewater treatment (WWT) Due to their general inefficiency, the industries will require a variety of state-of-the-art treatment technologies, such as activated carbon adsorption, membrane separation, enhanced oxidation, and ozonation (Mishra *et al.,* 2023). Memnrane technology is one of the methods used for EC removal from wastewater. Different materials are used to create membranes, and these materials give rise to distinct filtering features (such as pore size, surface charge, and hydrophobicity) that dictate the kind of contaminants that can be retained (Schäfer *et al.,* 2011; Simmons *et al.,* 2011).

Membrane filtration falls into four categories: reverse osmosis (RO), forward osmosis (FO), nanofiltration (NF), microfiltration (MF), and ultrafiltration (UF). Because microfiltration can be done at atmospheric pressure, it is frequently used. Despite its many benefits, MF is ineffective for removing ECs and is unable to remove contaminants smaller than 1 μm, or dissolved solids (Zhou *et al.,* 2010; *Lu et al.,* 2010). A wide range of newly discovered contaminants have been eliminated through ultrafiltration because their pore sizes are smaller than MF's (between 0.001 and 0.1 μm) (Acero *et al.,* 2010; Cissé *et al.,* 2011; Gao *et al.,* 2011; Liden *et al.,* 2016]. The type of membrane and the type of contamination have a direct impact on the removal efficiency (Heo *et al.,* 2012). For instance, UF membranes made of polysulfone and polyvinylidene were used to test the removal of bisphenol A from water. While the latter was able to remove up to 98% of the load in the experimental influent, the former was only able to achieve 75% removal (Heo *et al.,* 2012; Melo *et al.,* 2013). Because nanofiltration has tiny pores (between 10 and 100 Å), it can be used to remove ECs (Han *et al.,* 2013; Vatanpour *et al.,* 2011). NF membrane processes run with a minimal feed rate. Water force (Liden *et al.,* 2016; Figoli *et al.,* 2010; Setiawan *et al.,* 2011). Which implies a big benefit if operating costs are taken into account. It has been shown that NF is more effective than UF at removing some ECs. For instance, the efficiency of removing caffeine using UF ranged from 2-21%, while the efficiency of removing caffeine using NF was reported to be 46-84%, and the same trend was found for ketorolac tromethamine (Acero *et al.,* 2010).

By filtering the biomass through a membrane, membrane bioreactor technology can replace both physical separation and biodegradation processes. Because of this, significantly more high-quality water is produced than with conventional treatment, negating the need for a second tertiary procedure (Ma *et al.,* 2018; Jalilnejad *et al.,* 2020).

Adsorption-oriented Processes for the Removal of Emerging Contaminants

One important method for removing contaminants from wastewater or sources of drinking water is the use of commercial activated carbons for liquid-solid adsorption (Crini, 2010).

Despite not being the most recent technology available for eliminating pollutants, adsorption is unquestionably the most significant globally due to the challenges associated with replacing and discarding the adsorbents. (Ngueagni *et al.,* 2020, Nithya *et al.,* 2020). In fact, this industrial method of eliminating contaminants not only produces water of excellent quality, but it is also a process that is both economically viable and simple in terms of technology. (Crini, 2010).Sorbents include zeolites, mineral matter, metal-organic frameworks, organic carbon frameworks, activated carbon, hydrochar, biochar, carbon nanotubes, graphene, carbon composite materials, polymer, and mesoporous clay nanocomposites (Mishra *et al.,* 2023).Diverse nanocomposites were employed as adsorbents to eliminate novel pollutants from the discharge of wastewater(Lima *et al.,* 2019) Commercial activated carbons, whether in granular or powder form, are well-known to be effective adsorbents for treating a variety of contaminants, such as pesticides and medications, for decolorizing water, for adsorbing organic matter to lower chemical and biochemical oxygen demands, and for treating compounds that give off particular tastes or odours. (Crini and Badot 2007, 2010; Couto *et al.,* 2015; Ferreira *et al.,* 2015; Crini *et al.,* 2019; Khan *et al.,* 2020).

When Snyder *et al.* examined the removal effectiveness of powdered activated carbon (PAC) for 66 pharmaceuticals and personal care products (PPCPs) at a concentration of 5 mg/liter and a contact time of 5 hours, they discovered that only nine of them were removed with less than 50%.

Powdered activated carbon (PAC) or granulated activated carbon (GAC) are the two most regularly employed adsorbents to remove biological and chemical contaminants in raw drinking water and industrial wastewater. Their use is advantageous because of their highly sophisticated chemistry, porosity, and surface area (Nazifa *et al.,* 2018). Activated carbon powder has been used to eliminate emerging contaminants like ofloxacin, clofibric acid, diclofenac, gallic acid, ibuprofen, fluoroquinolonic acids, caffeine, and salicylic acid (Lofrano, 2012).

The best adsorbent seems to be powdered activated carbon, particularly for materials containing refractory organic, non-biodegradable compounds (Zeid *et al.,* 1995). According to Schafer *et al.*, powdered activated carbon has a 90% chance of removing EDC.

Biological Treatments

There are two categories of biological treatment technologies: anaerobic and aerobic processes. Among the efficient aerobic treatments are a sequencing batch reactor, membrane bioreactors, and active sludge. Environmentally friendly anaerobic sludge reactors and anaerobic film reactors are two examples of anaerobic technologies (Kumar *et al.,* 2022).

For each of the 55 PPCPs under study, the application of activated sludge treatment produced an average removal efficiency of more than 85% (Kasprzyk-Hordern *et al.,* 2009; Zupanc *et al.,* 2013). The capacity of biological processes to eliminate micropollutants through biodegradation has been studied, as is becoming more widely acknowledged (Lin *et al.,* 2021; Tran *et al.,* 2021). Pharmaceuticals, personal care items, and some environmental contaminants (EDCs) were eliminated using anaerobic, anoxic, and aerobic membrane bioreactors (Nguyen *et al.,* 2009). The nitrification and denitrification processes have low removal efficiencies for a wide range of ECs. Still, to improve removal efficiency, they could be coupled with MBR and other procedures (Oller *et al.,* 2020; Ahmed *et al.,* 2017; Afroz *et al.,* 2021; Ahmed *et al.,* 2021).

It was discovered that the fate of the target compounds in wastewater treatment processes was the determining factor in an activated sludge study by Urase and Kikuta in the removal of three estrogens, two endocrine disruptors, and pharmaceuticals.

They reported that at neutral pH, medications (especially acidic medications like ibuprofen, diclofenac, and clofibric acid) showed up as ions in the water phase and did not get removed or adsorbed in the activated sludge. The transition of substances from the water phase to the sludge phase was the removal process' limiting step. To guarantee that materials were transferred from the water phase into the sludge phase by adsorption rather than biodegradation, the ideal removal conditions for the activated sludge were acidic. It has been documented that anaerobic procedures can remove both synthetic and natural estrogens as well as nonylphenol derivatives in mesophilic and thermophilic environments, with removal rates ranging from 60% to 90% of ECs (Paterakis *et al.,* 2012). The researchers point out that one of the primary challenges in using biological processes to remove ECs is the absence of precise analytical techniques capable of locating and measuring these substances in such a complicated matrix. Further research on the development of extraction techniques for the isolation and quantification of ECs in activated sludge and/or other biological process by-products is made possible by this knowledge gap. Furthermore, a growing field of opportunity is the pursuit of precise analytical technologies for the identification of ECs in complex matrices (Rodriguez *et al.,* 2017).

Advanced Oxidation Process

The ability of advanced oxidation processes (AOPs) to remove pollutants more effectively than traditional water treatment methods has sparked interest in AOPs in recent years. The primary feature of AOPs, the production of hydroxyl radicals (oxidation potential, 2.8 V), has been linked to the high removal rates. The processes that are part of the AOPs group may use various materials and have distinct work environments in addition to distinct pathways for producing free radicals. (Rodriguez *et al.,* 2017).

According to research by Ikehata *et al.,* (2006); Ikehata *et al.,* (2008), free radicals, especially hydroxyl radicals, are formed during advanced oxidation processes (AOP) and help transform pollutants into less dangerous and more biodegradable compounds.

Photocatalysts are semiconductor metal oxides that are perfect for photocatalysis because they have a narrow energy band gap. Among many other reasons, titanium is one of the most researched heterogeneous photocatalysts due to its photostability, inert nature, and affordability (Gaya *et al.,* 2008). As an alternative, pesticides like aldrin, diazinon, and malathion as well as some antibiotics like amoxicillin, ampicillin, and chloxacclin can be removed with a high removal efficiency (99–100%) using photocatalysis in the presence of hydrogen peroxide (Prieto *et al.,* 2013).

Constructed Wetlands

According to Kaur *et al.* (2020), constructed wetlands are artificial ecosystems that depend on plant activity for water remediation. They can be classified based on various operational parameters, including the kind of few horizontal versus vertical and free water versus sub-surface—the hydraulic regime, and the species of plants used for Phyto depuration. These systems, like natural wetlands, make use of a variety of physical, chemical, and biological processes that take set at the interface between the wastewater being treated and the root system. These processes encompass volatilization, uptake, and degradation, as well as adsorption on soil and sediment. (Gorito *et al.,* 2017). Constructed wetlands are an inexpensive wastewater treatment option that can reduce waterborne pollutants, particularly those identified in personal care and pharmaceutical products (Al-Baldawi *et al.,* 2021). In a variety of artificial wetlands, the efficacy of anti-inflammatory medications, cholesterol regulators, anti-epileptic medications, fragrance products, surfactants, and estrogens are studied. It has been shown that the oxygen capacity and the interactions between adsorbents are critical for the maximum extraction of numerous Eps among the factors influencing how well wetland's function (Mpatani *et al.,* 2021).

The combined action of plant roots and the microbiome they are associated with banishes contaminants from artificial wetlands. Plant roots serve as an environmental niche that is ideal for microorganisms. These microorganisms are especially attracted to the roots through their discharge, and they can aid the plants in surviving unfavourable circumstances such as pollution (Rolli *et al.,* 2021). Remarkably, microorganisms can directly contribute to the breakdown of organic contaminants. Single or mixed plant species, primarily from the genera Typha, Phragmites, Iris, and Juncus, can be used to design constructed wetlands. The concentration of pathogens and organic matter in the effluent were successfully diminished by a constructed wetland system that was realized by co-cultivating the Halophytes species *Tamarix parvifora*, *Juncus acutus*, *Sarcocornia perrenis*, and *Limoniastrum monopetalum*

(Fountoulakis *et al.,* 2017). In both spiked experiments with 36 multi-class pollutants and non-spiked fresh-water aquaculture efuents containing atrazine, isoproturon, perfuorooctanesulfonic acid, clarithromycin, erythromycin, fuoxetine, norfuoxetine, and 2-ethylhexyl-4-methoxycinnamate, high removal efficiencies were seen for all considered micropollutants, with the exception of 2-ethylhexyl-4-methoxycinnamate (Gorito *et al.* 2018). Recent years have seen an increase in interest in the use of constructed wetlands for the removal of emerging organic contaminants. Yet, it has never been evident what factors mostly affect the separation process (Ding *et al.,* 2017).

Conclusion

Emerging contaminants are man-made toxic compound having adverse effects on living organisms when they are in contact with them. These are composed of different types; pharmaceuticals, PPCPs, food additives, pesticides and many other. Recently focused among them are endocrine disruptive chemicals because these are the contaminants directly causing alteration to endocrine system. Our WTPs are are not able to completely remove them, leading to their pathway in waterbodies causing disastrous changes in aquatic life new techniques have been tried for their elimination like Membrane technology, Adsorption, constructed wetlands etc. These technologies have both advantages as well as disadvantages. Scientists should now focus on new technologies that will be environmentally friendly as well as cost effective.

References

Acero, J. L., Benitez, F. J., Teva, F., & Leal, A. I. (2010). Retention of emerging micropollutants from UP water and a municipal secondary effluent by ultrafiltration and nanofiltration. Chemical Engineering Journal, 163(3), 264-272.

Adams, C., Wang, Y., Loftin, K., & Meyer, M. (2002). Removal of antibiotics from surface and distilled water in conventional water treatment processes. Journal of environmental engineering, 128(3), 253-260.

Adeel, M., Song, X., Wang, Y., Francis, D., & Yang, Y. (2017). Environmental impact of estrogens on human, animal and plant life: A critical review. Environment international, 99, 107-119.

Afroz, A. S., Romano, D., Inglese, F., & Stefanini, C. (2021). Towards bio-hybrid energy harvesting in the real-world: pushing the boundaries of technologies and strategies using bio-electrochemical and bio-mechanical processes. Applied Sciences, 11(5), 2220.

Ahmed, M. B., Zhou, J. L., Ngo, H. H., Guo, W., Thomaidis, N. S., & Xu, J. (2017). Progress in the biological and chemical treatment technologies for emerging contaminant removal from wastewater: a critical review. Journal of hazardous materials, 323, 274-298.

Ahmed, S. F., Mofijur, M., Nuzhat, S., Chowdhury, A. T., Rafa, N., Uddin, M. A., ... & Show, P. L. (2021). Recent developments in physical, biological,chemical, and hybrid treatment techniques for removing emerging contaminants from wastewater. Journal of hazardous materials, 416, 125912.

Al-Baldawi, I. A., Mohammed, A. A., Mutar, Z. H., Abdullah, S. R. S., Jasim, S. S., & Almansoory, A. F. (2021). Application of phytotechnology in alleviating pharmaceuticals and personal care products (PPCPs) in wastewater: Source, impacts, treatment, mechanisms, fate, and SWOT analysis. Journal of Cleaner Production, 319, 128584.

Andrady, A. L. (2011). Microplastics in the marine environment. Marine pollution bulletin, 62(8), 1596-1605.

Arman, N. Z., Salmiati, S., Aris, A., Salim, M. R., Nazifa, T. H., Muhamad, M. S., & Marpongahtun, M. (2021). A review on emerging pollutants in the water environment: Existences, health effects and treatment processes. Water, 13(22), 3258.

Asano, T., Burton, F. L., Leverenz, H. L., Tsuchihashi, R., & Tchobanoglous, G. (2007). Water reuse: Issues, technologies, and applications. New York: McGraw-Hill.

Aus der Beek, T., Weber, F. A., Bergmann, A., Hickmann, S., Ebert, I., Hein, A., & Küster, A. (2016). Pharmaceuticals in the environment—Global occurrences and perspectives. Environmental toxicology and chemistry, 35(4), 823-835.

Beiras, R., Verdejo, E., Campoy-Lopez, P., & Vidal-Linan, L. (2021). Aquatic toxicity of chemically defined microplastics can be explained by functional additives. Journal of Hazardous Materials, 406, 124338.

Belenguer, V., Martinez-Capel, F., Masiá, A., & Picó, Y. (2014). Patterns of presence and concentration of pesticides in fish and waters of the Júcar River (Eastern Spain). Journal of hazardous materials, 265, 271-279.

Benner, J., Helbling, D. E., Kohler, H. P. E., Wittebol, J., Kaiser, E., Prasse, C., ... & Boon, N. (2013). Is biological treatment a viable alternative for micropollutant removal in drinking water treatment processes?. Water research, 47(16), 5955-5976

Buerge, I. J., Buser, H. R., Kahle, M., Muller, M. D., & Poiger, T. (2009). Ubiquitous occurrence of the artificial sweetener acesulfame in the aquatic environment: an ideal chemical marker of domestic wastewater in groundwater. Environmental science & technology, 43(12), 4381-4385

Cheng, N., Wang, B., Wu, P., Lee, X., Xing, Y., Chen, M., & Gao, B. (2021). Adsorption of emerging contaminants from water and wastewater by modified biochar: A review. Environmental Pollution, 273, 116448.

Cissé, M., Vaillant, F., Pallet, D., & Dornier, M. (2011). Selecting ultrafiltration and nanofiltration membranes to concentrate anthocyanins from roselle extract (Hibiscus sabdariffa L.). Food Research International, 44(9), 2607-2614.

Clarke, B. O., & Smith, S. R. (2011). Review of 'emerging'organic contaminants in biosolids and assessment of international research priorities for the agricultural use of biosolids. Environment international, 37(1), 226-247.

Couto Jr, O. M., Matos, I., da Fonseca, I. M., Arroyo, P. A., da Silva, E. A., & de Barros, M. A. S. D. (2015). Effect of solution pH and influence of water hardness on caffeine adsorption onto activated carbons. The Canadian Journal of Chemical Engineering, 93(1), 68-77.

Crini, G. (2010). Wastewater treatment by sorption. Sorption Process and Pollution-Conventional and non-conventional sorbents for pollutant removal from wastewaters, 39-78.

Crini, G., Lichtfouse, E., Wilson, L. D., & Morin-Crini, N. (2019). Conventional and non-conventional adsorbents for wastewater treatment. Environmental Chemistry Letters, 17, 195-213.

Crini, G., Montiel, A. J., & Badot, P. M. (2007). Traitement et épuration des eaux industrielles polluées: Procédés membranaires, bioadsorption et oxydation chimique. Presses Univ. Franche-Comté.

Da Costa, J. P., Rocha-Santos, T., & Duarte, A. C. (2020). The environmental impacts of plastics and micro-plastics use, waste and pollution: EU and national measures.

Daughton, C. G., & Ternes, T. A. (1999). Pharmaceuticals and personal care products in the environment: agents of subtle change?. Environmental health perspectives, 107(suppl 6), 907-938.

Ding, H., Ding, Y., & Bai, S. (2017, June). Emerging organic contaminant removal in constructed wetlands. In 2017 6th International Conference on Energy and Environmental Protection (ICEEP 2017) (pp. 451-454). Atlantis Press.

Dulio, V., van Bavel, B., Brorström-Lundén, E., Harmsen, J., Hollender, J., Schlabach, M., ... & Koschorreck, J. (2018). Emerging pollutants in the EU: 10 years of NORMAN in support of environmental policies and regulations. Environmental Sciences Europe, 30(1), 1-13.

Fan, Z., Wu, S., Chang, H., & Hu, J. (2011). Behaviors of glucocorticoids, androgens and progestogens in a municipal sewage treatment plant: comparison to estrogens. Environmental Science & Technology, 45(7), 2725-2733.

Ferreira, R. C., De Lima, H. H. C., Cândido, A. A., Junior, O. C., Arroyo, P. A., De Carvalho, K. Q., ... & Barros, M. A. S. D. (2015). Adsorption of paracetamol using activated carbon of dende and babassu coconut mesocarp. International Journal of Biotechnology and Bioengineering, 9(7), 717-722.

Figoli, A., Cassano, A., Criscuoli, A., Mozumder, M. S. I., Uddin, M. T., Islam, M. A., & Drioli, E. (2010). Influence of operating parameters on the arsenic removal by nanofiltration. Water research, 44(1), 97-104.

Floate, K. D., Wardhaugh, K. G., Boxall, A. B., & Sherratt, T. N. (2005). Fecal residues of veterinary parasiticides: nontarget effects in the pasture environment. Annu. Rev. Entomol., 50, 153-179.

Fountoulakis, M. S., Daskalakis, G., Papadaki, A., Kalogerakis, N., & Manios, T. (2017). Use of halophytes in pilot-scale horizontal flow constructed wetland treating domestic wastewater. Environmental Science and Pollution Research, 24, 16682-16689.

Galindo-Miranda, J. M., Guízar-González, C., Becerril-Bravo, E. J., Moeller-Chávez, G., León-Becerril, E., & Vallejo-Rodríguez, R. (2019). Occurrence of emerging contaminants in environmental surface waters and their analytical methodology–a review. Water Supply, 19(7), 1871-1884.

Gallé, T., Bayerle, M., Pittois, D., & Huck, V. (2020). Allocating biocide sources and flow paths to surface waters using passive samplers and flood wave chemographs. Water Research, 173, 115533.

Gao, W., Liang, H., Ma, J., Han, M., Chen, Z. L., Han, Z. S., & Li, G. B. (2011). Membrane fouling control in ultrafiltration technology for drinking water production: A review. Desalination, 272(1-3), 1-8.

Gavrilescu, M., Demnerová, K., Aamand, J., Agathos, S., & Fava, F. (2015). Emerging pollutants in the environment: present and future challenges in biomonitoring, ecological risks and bioremediation. New biotechnology, 32(1), 147-156.

Gaya, U. I., & Abdullah, A. H. (2008). Heterogeneous photocatalytic degradation of organic contaminants over titanium dioxide: a review of fundamentals, progress and problems. Journal of photochemistry and photobiology C: Photochemistry reviews, 9(1), 1-12.

Giudice, L. C., Llamas□Clark, E. F., DeNicola, N., Pandipati, S., Zlatnik, M. G., Decena, D. C. D., ... & FIGO Committee on Climate Change and Toxic Environmental Exposures. (2021). Climate change, women's health, and the role of obstetricians and gynecologists in leadership. International Journal of Gynecology & Obstetrics, 155(3), 345-356.

Glassmeyer, S. T., Furlong, E. T., Kolpin, D. W., Batt, A. L., Benson, R., Boone, J. S., ... & Wilson, V. S. (2017). Nationwide reconnaissance of contaminants of emerging concern in source and treated drinking waters of the United States. Science of the Total Environment, 581, 909-922.

Glassmeyer, S. T., Furlong, E. T., Kolpin, D. W., Cahill, J. D., Zaugg, S. D., Werner, S. L., ... & Kryak, D. D. (2005). Transport of chemical and microbial compounds from known wastewater discharges: potential for use as indicators of human fecal contamination. Environmental science & technology, 39(14), 5157-5169.

Gogoi, A., Mazumder, P., Tyagi, V. K., Chaminda, G. T., An, A. K., & Kumar, M. (2018). Occurrence and fate of emerging contaminants in water environment: a review. Groundwater for Sustainable Development, 6, 169-180.

Gorito, A. M., Ribeiro, A. R., Almeida, C. M. R., & Silva, A. M. (2017). A review on the application of constructed wetlands for the removal of priority substances and contaminants of emerging concern listed in recently launched EU legislation. Environmental Pollution, 227, 428-443.

Gorito, A. M., Ribeiro, A. R., Gomes, C. R., Almeida, C. M. R., & Silva, A. M. (2018). Constructed wetland microcosms for the removal of organic micropollutants from freshwater aquaculture effluents. Science of the total environment, 644, 1171-1180.

Grover, D. P., Zhou, J. L., Frickers, P. E., & Readman, J. W. (2011). Improved removal of estrogenic and pharmaceutical compounds in sewage effluent by full scale granular activated carbon: Impact on receiving river water. Journal of Hazardous Materials, 185(2-3), 1005-1011.

Gwenzi, W., Mangori, L., Danha, C., Chaukura, N., Dunjana, N., & Sanganyado, E. (2018). Sources, behaviour, and environmental and human health risks of high-technology rare earth elements as emerging contaminants. Science of the Total Environment, 636, 299-313.

Han, X., Liang, C. F., Li, T. Q., Wang, K., Huang, H. G., & Yang, X. E. (2013). Simultaneous removal of cadmium and sulfamethoxazole from aqueous solution by rice straw biochar. Journal of Zhejiang University Science B, 14, 640-649.

Heo, J., Flora, J. R., Her, N., Park, Y. G., Cho, J., Son, A., & Yoon, Y. (2012). Removal of bisphenol A and 17β-estradiol in single walled carbon nanotubes–ultrafiltration (SWNTs–UF) membrane systems. Separation and Purification Technology, 90, 39-52.

Huerta-Fontela, M., Galceran, M. T., & Ventura, F. (2011). Occurrence and removal of pharmaceuticals and hormones through drinking water treatment. Water research, 45(3), 1432-1442.

Ikehata, K., Gamal El-Din, M., & Snyder, S. A. (2008). Ozonation and advanced oxidation treatment of emerging organic pollutants in water and wastewater. Ozone: Science and Engineering, 30(1), 21-26.

Ikehata, K., Jodeiri Naghashkar, N., & Gamal El-Din, M. (2006). Degradation of aqueous pharmaceuticals by ozonation and advanced oxidation processes: a review. Ozone: Science and Engineering, 28(6), 353-414.

Jalilnejad, E., Sadeghpour, P., & Ghasemzadeh, K. (2020). Advances in membrane bioreactor technology. In current trends and future developments on (bio-) membranes (pp. 1-29). Elsevier.

Kasprzyk-Hordern, B., Dinsdale, R. M., & Guwy, A. J. (2008). The occurrence of pharmaceuticals, personal care products, endocrine disruptors and illicit drugs in surface water in South Wales, UK. Water research, 42(13), 3498-3518.

Kasprzyk-Hordern, B., Dinsdale, R. M., & Guwy, A. J. (2009). The removal of pharmaceuticals, personal care products, endocrine disruptors and illicit drugs during wastewater treatment and its impact on the quality of receiving waters. Water research, 43(2), 363-380.

Kaur, R., Talan, A., Tiwari, B., Pilli, S., Sellamuthu, B., & Tyagi, R. D. (2020). Constructed wetlands for the removal of organic micro-pollutants. In Current Developments in Biotechnology and Bioengineering (pp. 87-140). Elsevier.

Kawa, I. A., Fatima, Q., Mir, S. A., Jeelani, H., Manzoor, S., & Rashid, F. (2021). Endocrine disrupting chemical Bisphenol A and its potential effects on female health. Diabetes & Metabolic Syndrome: Clinical Research & Reviews, 15(3), 803-811.

Keen, O. S., Baik, S., Linden, K. G., Aga, D. S., & Love, N. G. (2012). Enhanced biodegradation of carbamazepine after UV/H2O2 advanced oxidation. Environmental science & technology, 46(11), 6222-6227.

Keen, O. S., Love, N. G., Aga, D. S., & Linden, K. G. (2016). Biodegradability of iopromide products after UV/H2O2 advanced oxidation. Chemosphere, 144, 989-994.

Khan, N. A., Khan, S. U., Ahmed, S., Farooqi, I. H., Yousefi, M., Mohammadi, A. A., & Changani, F. (2020). Recent trends in disposal and treatment technologies of emerging-pollutants-A critical review. TrAC Trends in Analytical Chemistry, 122, 115744

Kokalj, A. J., Hartmann, N. B., Drobne, D., Potthoff, A., & Kühnel, D. (2021). Quality of nanoplastics and microplastics ecotoxicity studies: refining quality criteria for nanomaterial studies. Journal of hazardous materials, 415, 125751.

Kumar, R., Qureshi, M., Vishwakarma, D. K., Al-Ansari, N., Kuriqi, A., Elbeltagi, A., & Saraswat, A. (2022). A review on emerging water contaminants and the application of sustainable removal technologies. Case Studies in Chemical and Environmental Engineering, 6, 100219.

Lapworth, D. J., Baran, N., Stuart, M. E., & Ward, R. S. (2012). Emerging organic contaminants in groundwater: a review of sources, fate and occurrence. Environmental pollution, 163, 287-303.

Lavelle, P., Charpentier, F., Villenave, C., Rossi, J. P., Derouard, L., Pashanasi, B., ... & Bernier, N. (2004). Effects of earthworms on soil organic matter and nutrient dynamics at a landscape scale over decades. Earthworm ecology, 2, 145-160.

Lei, M., Zhang, L., Lei, J., Zong, L., Li, J., Wu, Z., & Wang, Z. (2015). Overview of emerging contaminants and associated human health effects. BioMed research international, 2015.

Lester, Y., Ferrer, I., Thurman, E. M., & Linden, K. G. (2014). Demonstrating sucralose as a monitor of full-scale UV/AOP treatment of trace organic compounds. Journal of hazardous materials, 280, 104-110.

Lidén, A., & Persson, K. M. (2016). Comparison between ultrafiltration and nanofiltration hollow-fiber membranes for removal of natural organic matter: a pilot study. Journal of Water Supply: Research and Technology—AQUA, 65(1), 43-53.

Lima, D. R., Gomes, A. A., Lima, E. C., Umpierres, C. S., Thue, P. S., Panzenhagen, J. C., ... & de Alencar, W. S. (2019). Evaluation of efficiency and selectivity in the sorption process assisted by chemometric approaches: removal of emerging contaminants from water. Spectrochimica Acta Part A: Molecular and Biomolecular Spectroscopy, 218, 366-373.

Lin, C., Cheruiyot, N. K., Hoang, H. G., Le, T. H., Tran, H. T., & Bui, X. T. (2021). Benzophenone biodegradation and characterization of malodorous gas emissions during co-composting of food waste with sawdust and mature compost. Environmental Technology & Innovation, 21, 101351.

Lofrano, G. (Ed.). (2012). Emerging compounds removal from wastewater: natural and solar based treatments. Springer Science & Business Media.

Lu, N. C., & Liu, J. C. (2010). Removal of phosphate and fluoride from wastewater by a hybrid precipitation–microfiltration process. Separation and Purification Technology, 74(3), 329-335.

Ma, J., Dai, R., Chen, M., Khan, S. J., & Wang, Z. (2018). Applications of membrane bioreactors for water reclamation: micropollutant removal, mechanisms and perspectives. Bioresource technology, 269, 532-543.

McVeigh, K. (2021). Nurdles: The Worst Toxic Waste You've Probably Never Heard Of: Billions of these tiny plastic pellets are floating in the ocean, causing as much damage as oil spills, yet they are still not classified as hazardous. The Guardian.

Melo-Guimarães, A., Torner-Morales, F. J., Durán-Álvarez, J. C., & Jiménez-Cisneros, B. E. (2013). Removal and fate of emerging contaminants combining biological, flocculation and membrane treatments. Water science and technology, 67(4), 877-885.

Mishra, R. K., Mentha, S. S., Misra, Y., & Dwivedi, N. (2023). Emerging pollutants of severe environmental concern in water and wastewater: A comprehensive review on current developments and future research. Water-Energy Nexus.

Modesto, T., Tiemeier, H., Peeters, R. P., Jaddoe, V. W., Hofman, A., Verhulst, F. C., & Ghassabian, A. (2015). Maternal mild thyroid hormone insufficiency in early pregnancy and attention-deficit/hyperactivity disorder symptoms in children. JAMA pediatrics, 169(9), 838-845.

Mpatani, F. M., Aryee, A. A., Kani, A. N., Han, R., Li, Z., Dovi, E., & Qu, L. (2021). A review of treatment techniques applied for selective removal of emerging pollutant-trimethoprim from aqueous systems. Journal of Cleaner Production, 308, 127359.

Nazifa, T. H., Habba, N., Salmiati, Aris, A., & Hadibarata, T. (2018). Adsorption of Procion Red MX-5B and Crystal Violet Dyes from Aqueous Solution onto Corncob Activated Carbon. Journal of the Chinese Chemical Society, 65(2), 259-270.

Ngueagni, P. T., Woumfo, E. D., Kumar, P. S., Siéwé, M., Vieillard, J., Brun, N., & Nkuigue, P. F. (2020). Adsorption of Cu (II) ions by modified horn core: Effect of temperature on adsorbent preparation and extended application in river water. Journal of molecular liquids, 298, 112023.

Nguyen, T. B., Bui, X. T., Dang, B. T., Tra, V. T., Tran, H. T., Tran, L. L., & Ngo, H. H. (2020). Anaerobic membrane bioreactors for industrial wastewater treatment. In Current Developments in Biotechnology and Bioengineering (pp. 167-196). Elsevier.

Nithya, K., Sathish, A., & Kumar, P. S. (2020). Packed bed column optimization and modeling studies for removal of chromium ions using chemically modified Lantana camara adsorbent. Journal of Water Process Engineering, 33, 101069.

Noguera-Oviedo, K., & Aga, D. S. (2016). Lessons learned from more than two decades of research on emerging contaminants in the environment. Journal of hazardous materials, 316, 242-251.

Nowak, K. M., Miltner, A., Poll, C., Kandeler, E., Streck, T., & Pagel, H. (2020). Plant litter enhances degradation of the herbicide MCPA and increases formation of biogenic non-extractable residues in soil. Environment international, 142, 105867.

Oller, I., Malato, S., & Sánchez-Pérez, J. (2011). Combination of advanced oxidation processes and biological treatments for wastewater decontamination—a review. Science of the total environment, 409(20), 4141-4166.

Oppenheimer J., Eaton A., Badruzzaman M., Haghani A. W., Jacangelo J. G., (2011), Occurrence and Suitability of sucralose as an indicator compound of wastewater loading to surface waters in rbanized Regions, Water Res., 45, 4019-4027.

Pal, A., Gin, K. Y. H., Lin, A. Y. C., & Reinhard, M. (2010). Impacts of emerging organic contaminants on freshwater resources: review of recent occurrences, sources, fate and effects. Science of the total environment, 408(24), 6062-6069.

Pascoe, D., Karntanut, W., & Müller, C. T. (2003). Do pharmaceuticals affect freshwater invertebrates? A study with the cnidarian Hydra vulgaris. Chemosphere, 51(6), 521-528.

Patel, S. (2017). Disruption of aromatase homeostasis as the cause of a multiplicity of ailments: A comprehensive review. The Journal of steroid biochemistry and molecular biology, 168, 19-25.

Paterakis, N., Chiu, T. Y., Koh, Y. K. K., Lester, J. N., McAdam, E. J., Scrimshaw, M. D., ... & Cartmell, E. (2012). The effectiveness of anaerobic digestion in removing estrogens and nonylphenol ethoxylates. Journal of hazardous materials, 199, 88-95.

Pereira, L. C., de Souza, A. O., Bernardes, M. F. F., Pazin, M., Tasso, M. J., Pereira, P. H., & Dorta, D. J. (2015). A perspective on the potential risks of emerging contaminants to human and environmental health. Environmental Science and Pollution Research, 22, 13800-13823.

Pereira, V.J., da Cunha, J.P.A.R., de Morais, T.P., de Oliveira, J.P.R., de Morais, J.B., 2016. Physical-chemical properties of pesticides: concepts, applications, and interactions with the environment. Biosci. J. 32, 627–641. https://doi.org/10.14393/BJ-v32n3a2016-31533.

Petrovic, M., Eljarrat, E., Lopez de Alda, M. J., & Barceló, D. (2004). Endocrine disrupting compounds and other emerging contaminants in the environment: a survey on new monitoring strategies and occurrence data. Analytical and Bioanalytical Chemistry, 378, 549-562.

Petrović, M., Gonzalez, S., & Barceló, D. (2003). Analysis and removal of emerging contaminants in wastewater and drinking water. TrAC Trends in Analytical Chemistry, 22(10), 685-696.

Pironti, C., Ricciardi, M., Motta, O., Miele, Y., Proto, A., & Montano, L. (2021). Microplastics in the environment: intake through the food web, human exposure and toxicological effects. Toxics, 9(9), 224.

Pitter, G., Da Re, F., Canova, C., Barbieri, G., Zare Jeddi, M., Daprà, F., ... & Russo, F. (2020). Serum levels of perfluoroalkyl substances (PFAS) in adolescents and young adults exposed to contaminated drinking water in the Veneto region, Italy: a cross-sectional study based on a health surveillance program. Environmental health perspectives, 128(2), 027007.

Prieto-Rodríguez, L., Oller, I., Klamerth, N., Agüera, A., Rodríguez, E. M., & Malato, S. (2013). Application of solar AOPs and ozonation for elimination of micropollutants in municipal wastewater treatment plant effluents. Water research, 47(4), 1521-1528.

Rahman, F., Langford, K. H., Scrimshaw, M. D., & Lester, J. N. (2001). Polybrominated diphenyl ether (PBDE) flame retardants. Science of the Total Environment, 275(1-3), 1-17.

Rathi, B. S., Kumar, P. S., & Show, P. L. (2021). A review on effective removal of emerging contaminants from aquatic systems: Current trends and scope for further research. Journal of hazardous materials, 409, 124413.

Reemtsma, T., Alder, L., & Banasiak, U. (2013). Emerging pesticide metabolites in groundwater and surface water as determined by the application of a multimethod for 150 pesticide metabolites. Water research, 47(15), 5535-5545.

Richardson S.D., (2012), Environmental Mass Spectrometry: Emerging Contaminants and Current Issues, Anal. Chem., 84, 747–778.

Richardson S.D., Ternes T.A., (2011) Water Analysis: Emerging Contaminants and Current Issues, Anal. Chem., 83, 4614–4648.

Richardson, S. D., & Kimura, S. Y. (2016). Water analysis: emerging contaminants and current issues. Analytical chemistry, 88(1), 546-582

Richardson, S. D., & Ternes, T. A. (2017). Water Analysis: Emerging Contaminants and Current Issues. Analytical Chemistry, 90(1), 398-428.

Rodriguez-Mozaz, S., Ricart, M., Köck-Schulmeyer, M., Guasch, H., Bonnineau, C., Proia, L., ... & Barceló, D. (2015). Pharmaceuticals and pesticides in reclaimed water: efficiency assessment of a microfiltration–reverse osmosis (MF–RO) pilot plant. Journal of hazardous materials, 282, 165-173.

Rodriguez-Narvaez, O. M., Peralta-Hernandez, J. M., Goonetilleke, A., & Bandala, E. R. (2017). Treatment technologies for emerging contaminants in water: A review. Chemical Engineering Journal, 323, 361-380.

Rolli, E., Vergani, L., Ghitti, E., Patania, G., Mapelli, F., & Borin, S. (2021). 'Cry-for-help'in contaminated soil: a dialogue among plants and soil microbiome to survive in hostile conditions. Environmental Microbiology, 23(10), 5690-5703.

Sanganyado, E., Lu, Z., Fu, Q., Schlenk, D., & Gan, J. (2017). Chiral pharmaceuticals: A review on their environmental occurrence and fate processes. Water Research, 124, 527-542.

Schäfer, A. I., Akanyeti, I., & Semião, A. J. (2011). Micropollutant sorption to membrane polymers: a review of mechanisms for estrogens. Advances in colloid and interface science, 164(1-2), 100-117.

Schäfer, A. I., Nghiem, L. D., & Waite, T. D. (2003). Removal of the natural hormone estrone from aqueous solutions using nanofiltration and reverse osmosis. Environmental science & technology, 37(1), 182-188.

Scheurer M., Brauch H.-J., Lange F. T., (2009) Analysis and occurrence of seven artificial sweeteners in German waste water and surface water and in soil aquifer treatment (SAT), Anal. Bioanal. Chem., 394, 1585–1594.

Secondes, M. F. N., Naddeo, V., Belgiorno, V., & Ballesteros Jr, F. (2014). Removal of emerging contaminants by simultaneous application of membrane ultrafiltration, activated carbon adsorption, and ultrasound irradiation. Journal of hazardous materials, 264, 342-349.

Setiawan, L., Wang, R., Li, K., & Fane, A. G. (2011). Fabrication of novel poly (amide–imide) forward osmosis hollow fiber membranes with a positively charged nanofiltration-like selective layer. Journal of membrane science, 369(1-2), 196-205.

Shahid, M. K., Kashif, A., Fuwad, A., & Choi, Y. (2021). Current advances in treatment technologies for removal of emerging contaminants from water–A critical review. Coordination Chemistry Reviews, 442, 213993.

Simmons, F. J., Kuo, D. H. W., & Xagoraraki, I. (2011). Removal of human enteric viruses by a full-scale membrane bioreactor during municipal wastewater processing. Water research, 45(9), 2739-2750.

Snyder, S. A., Wert, E. C., Rexing, D. J., Zegers, R. E., & Drury, D. D. (2006). Ozone oxidation of endocrine disruptors and pharmaceuticals in surface water and wastewater. Ozone: Science and Engineering, 28(6), 445-460.

Sommer, C., & Bibby, B. M. (2002). The influence of veterinary medicines on the decomposition of dung organic matter in soil. European Journal of Soil Biology, 38(2), 155-159.

Srivastav, A. L. (2020). Chemical fertilizers and pesticides: role in groundwater contamination. In Agrochemicals detection, treatment and remediation (pp. 143-159). Butterworth-Heinemann.

Steenland, K., Fletcher, T., Stein, C. R., Bartell, S. M., Darrow, L., Lopez-Espinosa, M. J., ... & Savitz, D. A. (2020). Evolution of evidence on PFOA and health following the assessments of the C8 Science Panel. Environment International, 145, 106125.

Stefanakis, A. I., & Becker, J. A. (2016). A review of emerging contaminants in water: Classification, sources, and potential risks. Impact of water pollution on human health and environmental sustainability, 55-80.

Stefanakis, A. I., Akratos, C. A., & Tsihrintzis, V. A. (2014). Vertical Flow Constructed Wetlands: Eco-engineering systems for wastewater and sludge treatment. Amsterdam, The Netherlands: Elsevier Publishing

Suarez, S., Lema, J. M., & Omil, F. (2009). Pre-treatment of hospital wastewater by coagulation–flocculation and flotation. Bioresource technology, 100(7), 2138-2146.

Sznajder-Katarzyńska, K., Surma, M., & Cieślik, I. (2019). A review of perfluoroalkyl acids (PFAAs) in terms of sources, applications, human exposure, dietary intake, toxicity, legal regulation, and methods of determination. Journal of Chemistry, 2019.

Thomaidis, N. S., Asimakopoulos, A. G., & Bletsou, A. A. (2012). Emerging contaminants: a tutorial mini-review. Global NEST Journal, 14(1), 72-79.

Tran, H. T., Lin, C., Bui, X. T., Itayama, T., Dang, B. T., Cheruiyot, N. K., & Vu, C. T. (2021). Bacterial community progression during food waste composting containing high dioctyl terephthalate (DOTP) concentration. Chemosphere, 265, 129064.

Urase, T., & Kikuta, T. (2005). Separate estimation of adsorption and degradation of pharmaceutical substances and estrogens in the activated sludge process. Water research, 39(7), 1289-1300.

Vatanpour, V., Madaeni, S. S., Moradian, R., Zinadini, S., & Astinchap, B. (2011). Fabrication and characterization of novel antifouling nanofiltration membrane prepared from oxidized multiwalled carbon nanotube/polyethersulfone nanocomposite. Journal of membrane science, 375(1-2), 284-294.

Xu, B., Liu, S., Zhou, J. L., Zheng, C., Weifeng, J., Chen, B., & Qiu, W. (2021). PFAS and their substitutes in groundwater: Occurrence, transformation and remediation. Journal of Hazardous Materials, 412, 125159.

Zeid, N. A., Nakhla, G., Farooq, S., & Osei-Twum, E. (1995). Activated carbon adsorption in oxidizing environments. Water Research, 29(2), 653-660.

Zhang, Y., & Zhou, J. L. (2008). Occurrence and removal of endocrine disrupting chemicals in wastewater. Chemosphere, 73(5), 848-853.

Zhou, J. E., Chang, Q., Wang, Y., Wang, J., & Meng, G. (2010). Separation of stable oil–water emulsion by the hydrophilic nano-sized ZrO2 modified Al2O3 microfiltration membrane. Separation and Purification Technology, 75(3), 243-248.

Zimmermann, L., Bartosova, Z., Braun, K., Oehlmann, J., Völker, C., & Wagner, M. (2021). Plastic products leach chemicals that induce *in vitro* toxicity under realistic use conditions. Environmental science & technology, 55(17), 11814-11823.

Zuccato, E., & Castiglioni, S. (2009). Illicit drugs in the environment. Philosophical Transactions of the Royal Society A: Mathematical, Physical and Engineering Sciences, 367(1904), 3965-3978.

Zupanc, M., Kosjek, T., Petkovšek, M., Dular, M., Kompare, B., Širok, B., & Heath, E. (2013). Removal of pharmaceuticals from wastewater by biological processes, hydrodynamic cavitation and UV treatment. Ultrasonics sonochemistry, 20(4), 1104-1112.

11

Heavy Metal Pollution: A Hidden Threat to Aquatic Biodiversity

Gora Shiva Prasad[1], Gopikrishna[2], Shilpa Koleya[1], Suman Karmakar[3], Sulav Mandal[1] and S.K. Rout[1*]

[1]*Department of Aquatic Environment Management, Faculty of Fishery Sciences, West Bengal University of Animal and Fishery Sciences, Kol- 700094*

[2]*Aquaculture Production and Environment Division, ICAR-Central Institute of Freshwater Aquaculture, Kausalyaganga, Bhubaneswar-751002, Orissa, India*

[3]*Division of Fisheries Science, The Neotia University, Sarisha, Diamond Harbour West Bengal- 743368, India*

Abstract

Aquatic ecosystems are seriously threatened by heavy metal pollution from urban, agricultural, and industrial processes. Aquatic animals can accumulate toxic metals like mercury, lead, cadmium, arsenic and chromium, which can have detrimental physiological, ecological, and toxicological impacts. Oxidative stress, organ damage, poor reproduction, and disruption of ecosystem functioning, including loss of biodiversity and changes in food web dynamics, can all result from exposure to these contaminants. Human health is also at risk from heavy metals moving via aquatic food networks. The effects of heavy metal pollution on aquatic life are examined in this chapter, along with mitigation strategies.

Keywords: *Heavy metals, pollution, toxicity, aquatic ecosystems and mitigation strategies*

Introduction

Water is one of the most valuable natural resources. The quality of water is of vital concern for mankind since it is directly link with human welfare (Kumar, 2004). Global industrialization has advanced and made life easier, but uncontrolled disposal of waste accidental and process spillage, mining and smelting of metalliferous ores, sewage sludge application causing pollutants to get released into the air, soil, and water (Abubakar *et al.*, 2024). Over the past few decades, there has been a growing focus on the connection between heavy metal conformation and its impact on aquatic organism because according to the US Environmental Protection Agency (EPA) heavy metals are among the most prevalent pollutants in aquatic environments and pose dangerous effect to the environment and human health (Sonone *et al.*, 2020). It is widely accepted that anthropogenic activities significantly increases the overall aquatic burden of toxic metals through the possibility of both point source and non-point source pollution. Point sources typically results from direct mining and ore processing and non point source contamination usually arises from agricultural, industrial, and urban effluents that reach the coast by way of waterways, surface runoff, and precipitation (Gupta & Singh, 2011). Heavy metal pollution is a serious environmental issue that affects the health of aquatic ecosystems, terrestrial ecosystems, and human populations because they have the ability to harm cellular structures, impede metabolic processes, and interfere with the activity of enzymes (Bhan & Sarkar, 2005). Heavy metals can enter the bodies of aquatic animals by contact to water, sediment, or food sources. Heavy metals, including lead (Pb), mercury (Hg), cadmium (Cd), arsenic (As), copper (Cu), nickel (Ni) and chromium (Cr) are persistent pollutants that have the ability to bioaccumulate and biomagnify in food webs (Abubakar *et al.*, 2024).

Therefore it is important to monitor and control the level of these heavy metals for its long residence time within food chains and potential risk of human exposure (Giarratano & Amin, 2010).

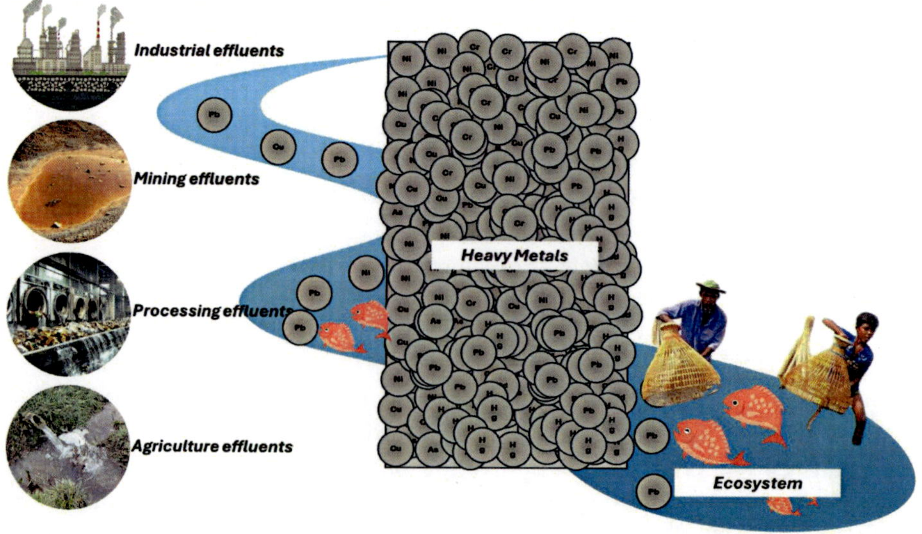

Pictorial representation of the potential sources of Heavy metal pollution

This paper reviews the impact of various heavy metals especially on aquatic biota, it's mode of action and solution to get rid of such heavy metal toxicity conditions.

Chromium (Cr)

As an element, Cr is very stable but is not usually found pure in nature. The principal ore of Cr is Chromite. Eisler (1985); Bakshi (2016).The metal may be present in divalent (Cr^{+2}), trivalent (Cr^{+3}), and hexavalent (Cr^{+6}) forms but Cr^{+3} and Cr^{+6} being the most predominant and stable forms (Velma et al, 2009). In the ecosystem, Cr is usually found in the trivalent form and this form is reported as an essential element in mammals as it takes an e-ective role in glucose, lipid, and protein metabolism (Steven et al, 1976). The toxicity of Cr^{+3} is lower than Cr^{+6} (Ahmed et al, 2013)due to poor membrane permeability, non-corrosiveness and less tendency to biomagnify in the food chain.After entering the cell, the Cr^{+6} readily reduces to Cr^{+3} and complexes with intracellular macromolecules even with genetic materials (Farag et al, 2006; Li et al, 2010).

Cr^{+6} has been reported to be different toxicologically from other heavy metals (Doudoroff& Katz, 1953) as it can readily penetrate the gill membrane by the process of passive diffusion mediated by the pH of the system. Cr enters the body of the fishes mainly through the gills (Velma et al, 2009). Its concentration arises through bio-magnification at each trophic level and accumulates maximum in carnivores (Obasohan, 2007). The general pattern of distribution of Cr^{+6} in fishes is Gills > Liver > Skin > Muscles (Avenant& Marx, 2000). The toxicity of Crincreases with its concentration as well as temperature but decreases with increasing salinity and sulfate concentration (Wang & Shi., 2001).It has been reported that the range of pH has tremendous influence in determining the bioavailability of metal to the fish. Gills contain more amount of Cr at pH 6.5 than other internal organs whereas, the reverse is evident at pH 7.8 (Van Der Putte et al, 1981). The primary storage and detoxification site for chromium is said to be liver in the experimental condition (Eisler. R, 1985). It affects the physiological, behavioral, histological, biochemical, genetic, and immunological conditions of the experimental organism. Prolonged exposure to Cr^{+6} is shown to induce adaptability in fish.

In the case of acute exposure at 50% lethal concentration, fishes have been found to lose their body balance with restlessness, lowered breathing rate, and a higher rate of mucus secretion. Cr exerts its effect on the epithelial cells of the intestine and can also modify the rate of glucose transport (Stokes, 1965). The osmoregulatory function of various fish fauna is affected by the influence of Cr^{+3} (Subashini, 2005). Cr^{+6} also retard the action of ATPase in gills, kidney, and intestine of coastal teleost (Tan F, 2008). Lethal effects have been reported to range from 17 to 118 mg/1 (and higher) for fish with "safe" levels ranging between 0.3 and 1.0 mg/1 whereas lethal levels for invertebrates and algae are generally in the range from 0.032 to 6.4 mg/l.

Copper (Cu)

Cu is an essential trace nutrient that is required in small amounts by humans, other mammals, fish, and shellfish for carbohydrate metabolism and the functioning of more than 30 enzymes. Copper is moderately soluble in water and binds easily to sediments and organic matter.Cupric ion (Cu^{+2}) is found to be most toxic form of Cu. The concentration of organic Cu in organisms is tightly regulated (Pena, 1999; Camakaris, 1999). Despite its weak effect on terrestrial vertebrates, it is considered as an extremely toxic cation, as it can affect the physiological status of invertebrates and protists at nanomolar concentrations (Viarengo, 1989).

The effects of this metal on aquatic organisms can be directly or indirectly lethal. Gills become frayed and lose their ability to regulate the transport of some important salt used for the normal functioning of the cardiovascular and nervous systems such as sodium chloride and potassium chloride into and out of fish. This may lead to death as a result of the disruption of the salt balance between a Cu exposed fish and the surrounding water. It forms complex with dissolved organic carbon (DOC), therefore, be less bioavailable and provides some protection on the gills. In acidic, poor buffer and soft water, the toxicity of Cu in the fish gill is seen to be higher. The calcium ion provides protection and reduces its toxicity in living cells.Canada has outlined Water Quality Standards for the protection of aquatic life and is stricter for soft water than for hard water (Bradl, 2005; Nriagu, 1980; Wright and Welbourn, 2002). Research on marine mussels has demonstrated the inhibition of in-vitro Ca-ATPase activity of gill cell plasma membranes at micromolarconcentrations (Viarengo, 1993).

The direct contact of fish olfactory tissues with the surrounding water facilitates Cu uptake and adversely affects olfaction in fish. They also reduced sperm and egg production in many species of fish and aquatic organisms as well as early hatching, smaller fry and increased incidence of abnormalities, and reduced survival in the fry (Taub, 2004). Different reports have shown that Cu may affect the rate of oxygen consumption in various aquatic organisms (Viarengo, 1989). Fish and crustaceans are 10 to 100 times more sensitive to the toxic effects of Cu than mammals, whereas algae especially blue-green algae are 1,000 times more sensitive to its toxicity(Forstner and Wittman., 1979; Hodson., 1979 and Wright and Welbourn., 2002). Since Cu is used as an algaecide to control nuisance algae in lakes and rivers, it is not surprising that it causes decreased algal growth when inadvertently discharged to a waterbody. The algal biomass present in an aquatic ecosystem will affect the amount of food availability for aquatic animals including zooplankton, insects, shellfish, fish, and aquatic mammals.

Nickel (Ni)

Ni is considered as a transitional metal, essential for the fish and other aquatic invertebrates. Ni is a significant component of enzymes and plays a critical role in different nutrient cycles (Ragsdale, 1998).In the aquatic ecosystem, it acts as a regulator of the various physiological functions of fishes and freshwater invertebrates(Muyseen *et al.*, 2004). But when it comes to the negative impact on aquatic biota, it is always well recognized for itsdeleterious effects. Excess Ni uptake creates complications in gas exchange (Pane etal., 2003a, 2003b), inhibition of osmoregulation(Leonard *et al.*, 2009, 2014b)and the promotion of oxidative stress (Kubrak et al, 2013,2014). In freshwater biota, Ni gets accumulated mainly in the kidney followed by skin, skeleton, scales, gonads, and muscles (Ptashynski and Klaverkump 2002;Sreedevi *et al.*, 1992).In theopen ocean and coastal areas, the Ni concentration level ranges from 0.2-2 μg/l (Boyden, 1975 WHO, 1991) and 250 μg/l (Boyden,1975;Well *et al.*, 2000;Ezekwe and Edoghotu, 2015; Knauer, 1997; Eisler,1998) respectively. Industrial activity is considered as the main cause of Ni pollution in coastal waters.

There are few data on nickel accumulation in marine invertebrates. It was found that the largest amount of Ni accumulation in the body of Green Shore Crabs is in its carapace and at the very next gill is the second greatest amount of Ni accumulation center (Blewett *et al.*, 2015).Thus, respiration and ion regulation gets interrupted due to its accumulation in anterior and posterior gill respectively.A study on coral found that the most sensitive coral species *Acroporaaspera* with NOEC of < 280 μg Ni/l. In *A. digitifera*, fertilization success was checked by 10% at a concentration of 1580-2480 μg Ni/l. Coral has severe fertilization problems even at a concentration of >920 μg Ni/l(Gissi *et al.*, 2017).

Lead (Pb)

Leads are found in trace amounts on Earth. Various anthropogenic activities like mining, manufacturing batteries, paints, and cement, etc. are responsible for high levels of Pb in the environment (Kim and Kang, 2014). Pb can form a stable complex with oxygen and sulphur in a protein which increases the affinity of Pb to the given protein (Verstraeten *et al.*, 2008). Excess accumulation leads to hypercalcemia due to the inhibition of basolateral transportation of ions in the gill epithelium. Since Pb has a higher affinity towards Ca^{2+} ATPase, Na^{+}/

Ca^{2+} exchanger and Na^+/K^+ ATPase will disturb the ion exchange mechanism and the electrochemical gradient of the ionocytes. The toxic effects of lead are caused bythe bioaccumulation process (Eroglu *et al.*, 2015). This toxicity will affect the physiological, biochemical, and behavioral activity of aquatic organisms. This will also disrupt the function of the central nervous system, peripheral nervous system, membrane structures, even the liver and kidney functions(Hsu and Guo, 2002). Researchers found that the accumulation of Pb is high in the gills and kidney of fishes among all the heavy metals.

Research on juvenile Rockfishes(***Sebsteschlegilii***) found thatthe accumulation of Pb is mainly in the liver, spleen, intestine, and gill. The same is in the case of Starry Flounder (***Platichthysstellatus)*** (Huwang *et al.*, 2016), which undergoes oxidative stress. It is due to the production of ROS like hydrogen peroxide, hydroxide, superoxide radical(Eroglu *et al.*, 2015) etc.Excess accumulation of Pb in the coral causes reduction in fertilization success, larval survivability and skeletal development (Reichelt – Brusheet and Harrison, 2005).Some coral species like ***Poriteslutea***, ***Goniastrearetiformis***, ***Galaxeafascicularis***, ***Stylophorapistillata***, ***Pavonadecussata***, ***Montiporafoliosa***, ***Acroporaaspera*** and ***Pocilloporadamicornis*** was found to have high levels of Pbconcentration.

Mercury (Hg)

Hg is a conservative type of neurotoxic pollutant and is biologically converted into the most dangerous organic methyl mercury (Bhattacharyya *et al.*, 2010). Hg has been found in various natural ecosystems and the bioaccumulation in the aquatic organisms varies. The maximum accumulation of Hg in the river Gomti, Lucknow was found in *Mastacembelusarmatus* followed by ***Clariusbatrachus***, ***Mystuscavasius*** (Agarwal *et al.*, 2007) etc. Among IMC, it is found that Catla fingerlings are more sensitive to Hg and showed low LC50' value (Kumar and Gupta, 2006).

Histopathological examination including electron and light microscopy has been used as a valuable tool to assess the histological modifications in fishes induced by Hg pollutants (de Oliveira Ribeiro*et al.*, 2002). The toxicity of Hg on fishes has been studied by several researchers and it was found that the liver, kidney, and brain are more sensitive towards Hg pollution (Bhardwaj and Choudhary, 2015). Ghosh and Mandal (2012) observed expanded renal tubules, separation of epithelial lining from the tubular cell, shrinkage of the glomerulus, necrosis, and pyknotic nuclei in mesenchymal tissue in the kidney of *Labeo rohita* by exposure of $HgCl_2$. They also affect enzymatic activity as they are sensitive to metal pollutants. Hg disrupts the Na^+/K^+-ATPase system in gill and finally results in an imbalance of ion-regulatory activity (Poopal *et al.*, 2013).

Not only in finfishes but the presence of Hg have also been documented in some shellfish species such as Dog cockle (***Glycymerisglycymeris***), Manila clam (***Ruditapesphilippinarum***), Atlantic jackknife clam (***Ensisdirectus***), Blue mussel (***Mytilusedulis***), Pacific oyster (***Crassostreagigas***), Great scallop (***Pectenmaximus***), Common cockle (***Cardiumedule***), Hard clam (***Mercenariamercenaria***) (Szkoda *et al.*, 2015). Most of these species occupy the intertidal zone as sedentary benthic fauna.

Cadmium (Cd)

The free ion Cd^{2+} is usually considered the most bioavailable form. The ionic form of cadmium (Cd^{2+}) is combined in normal with ionic forms of oxygen (cadmium oxide, CdO_2), chlorine (cadmium chloride, $CdCl_2$) or with sulfur (cadmium sulfate, $CdSO_4$). Majority of Cd content in fish or other seafood is highly absorbable in $CdCl_2$ form. It is taken up across gills through transport systems initially designed for Ca (Verbost*et al.* 1989; Thomann*et al.* 1997; Craig *et al.* 1999). Cd occurs in the aquatic organisms and marine environment only in trace concentrations. The water salinity can affect Cd speciation. It negatively affects several organs in humans: kidney, lung, bones, placenta, brain and the central nervous system (Castro-González and Méndez-Armenta, 2008). Cadmium concentrations are also at levels in tissues below those known to affect the growth, reproduction and survival of marine organisms (Canli& Furness, 1995; Jennings & Rainbow, 1979; Olla, B.L. *et al.*, 1988). Other observed damages include reproductive toxicity, hepatic, hematological and immunological effects (ATSDR, 2008). Cadmium preferentially accumulates in internal organs such as the liver and kidney and but not in muscles, where the concentrations are usually very low (Storelli*et al.,* 2005). A major route of metal accumulation in marine organisms is via diet. Eg. Chub and horse mackerel feed on fish, crustaceans, and cephalopods. There is evidence for Cd bioaccumulation, particularly in freshwater organisms.

Arsenic (As)

Both Cd and As were the subject of intense scientific investigation and attracted public attention and concern particularly with respect to bio-transference and bio-magnification. Total arsenic concentrations are similar to those found in other marine ecosystems and thus unlikely to cause adverse effects to aquatic organisms. In the case of arsenic, both inorganic compounds, as arsenates (As^{5+}) or as arsenites (As^{3+}), and organic compounds, such as arsenobetaine and arsenocholine are of importance.

Inorganic Arsenic is considered carcinogenic and is related mainly to lung, kidney, bladder, and skin disorders. The influence of species and feeding habits on As accumulation is not clear in marine mammals (Storelli and Marcotrigiano, 2004). There is a significant relationship observed by Cain *et al.* (1992), in the molting of crayfish with the adsorption of As to the carapace. The molt events can lead to substantial decontamination of As. However, organisms feeding on crustaceans and algae appear to retain higher As concentrations than piscivorous species (Storelli *et al.*, 2005). The highest levels are usually found in the muscle tissues (Falco *et al.*, 2006; Visnjic-Jeftic *et al.*, 2010).

Table: Effect of metal pollution in some aquatic animals

Species	Metal	Concentration	Effect	Reference
Clar-iasgariepineus	Pb	0.1-0.5 mg/l	Irregular head, notochord defects, finfold effect, yolk sac edema	Osman *et al.* (2007)
Prochiloduslin-eatus	Pb	95mg/l	Epithelial lifting, lamellar aneurism, hyperplasia in gill	Martinez *et al.* (2004)
Channapunc-tatus	Lead acetate	9.43mg/l	Proliferation of white blood lymphocyte, inhibited the release of antimicrobial substances	Paul *et al.* (2014)
Clariasbatra-chus	Lead nitrate	5, 10, 15 mg/l	Decreased hepatic 5'-D enzyme activity	Chaurasia et al (1996)
Fish (not specified)	Cd	-	Necrosis in pancreatic cell, engorgement of blood vessel, vacuolar degeneration of hepatocytes, oxidative stress	Kumar and Singh (2010)
Rhamdiaquelen	Cd	0.1, 1 and 100 µg/l	Increase hepatic lesion index, reduced LPO activity and catalase activity in kidney, decresed liver glutathione S-transferaseactivity	Pereira *et al.* (2016)
Zebrafish (AB strain)	Cd	5mg/l	Decreased survival, impaired locomotary activity	Xia *et al.* (2020)
Tilapia sparr-manii	Cr	0.0098 mg/l (acute)	Reduced RBC, WBC count	Khangarot *et al.* (1999)
Labeorohita	Cr	39.4 mg/l (acute)	Decreased protein, lipid and glycogen content.	Vutukuru, (2003)
Oreochromis Mossambicus	Cr(IV)	7.5µg/fish (chronic)	Splenic weight reduction, decreased antibody production and lymphocyte count	Arunkumar et al (2000)
Daniorerio	Cu	0.068-0.244 mg/l	Lateral line deformities in embryonic stage	Sfakianakis *et al.* (2015)

Channapunc-tatus	Cu	10 ppb	Increased total iron-binding capacity and serum ceruloplasmin level, reduced glutathione level	Parvez *et al.* (2003)
Prochiloduslin-eatus	Cu	-	Increased hexokinaseand pyruvate kinase; decreased phosphofructokinase, ; lactate dehydrogenase and G6PDH enzyme in liver	Carvalho and Fernandes, (2008)
Labeorohita	As	1 g/100 ml	Decreased the level of alkaline phosphate, phosphatase, glutamate-pyruvate transaminase in liver and muscles.	Humtsoe *et al.* (2007)
Oncorhyn-chusmykiss	As	9.64 mg/l	Anemia, liver degeneration, gallbladder inflammation	Kumari *et al.* (2017)
Labeorohita	As	-	Increased Alanine aminotransferase in serum, hepatic damage, distress	Vutukuru *et al.* (2007)
Oreochro-misniloticus	Zn	2 mg/l	Pycnosis of liver nuclei, increased kuffer cells in liver	Abdel *et al.* (2011)
		4 mg/l	Pycnosis of liver nuclei, dilated sinusoids and congested blood vessel	
Channapunc-tatus	Zn	-	enlargement of renal tubules, necrosis and pycnotic nuclei in the mesencymal tissue	Gupta and Srivastava (2004)

Mitigation Strategies

The recalcitrance and persistence nature of heavy metals in the environment make heavy metal remediation a top priority. The traditional treatment technologies for extracting these hazardous heavy metals are ineffective and generate a significant amount of toxic chemical sludge which includes ion exchangers, reverse osmosis, chemical precipitation, chemical oxidation or reduction, electro dialysis, ultra filtration etc. Remediation of heavy metal ions from aqueous solutions using agricultural waste materials is found to be a potential adsorbent (Sud *et al.*, 2007). Agricultural wastes are composed of cellulose, hemicellulose, lignin and many other components. Phenolic, carbonyl, amido, amino, structural polysaccharides, esters, sulphydryl carboxyl groups are the functional groups found in biomass which have affinity towards metal complexation include both selective as well as non-selective affinity. Several agricultural waste such as orange peels, maize cobs, soyabean hulls, hazelnut shells, jack fruit etc. has been reported to have a great redemediation capacity of Cr from aqueous solution (Kurniawan et al, 2006). A successful

Pb removal upto 98% has been documented by Johns *et al.* (1998) using rice straw, sugarcane bagasse, soyabean hulls, walnut shells and peanut shells. Many other researchers (Mohan and Pittman, 2007; Rajeshwarisivaraj and subburam, 2002; Garg *et al.* 2007; Ajmal *et al.* 2005; Min *et al.* 2004) has been revealed the metal remediation ability of many agricultural wastes which also help to reduce the possible pollutions from wastes.

Maleki *et al.* (2019) used the magnetic property of metals and showed 92%, 96%, 92%, 99% and 99% removal efficiency for sorption of Hg, Cd, Ni, Pb and Cu respectively using magnetic geo-polymer base on bentonite clay. Despite of developing modern technologies for solving metal pollution, phytoremediation still play an important role in natural water bodies especially in wetlands. Various types of aquatic plants have shown significant removal capacity of metals from the water. Weeds like water fern, cattail, poplar trees, duckweed, water hyacinth, water lettuce, salt marsh bulrush, zebra rush etc. proved themselves as good phytoremediator of metals from the wetlands (Rai, 2008) and contributes towards good health of the aquatic ecosystem.

Conclusion

The accumulation of Cd, As, Cr, Cu, Hg, Ni, Pb etc. to hazardous levels in the aquatic ecosystem has become a problem of increasing concern. Some sedentary aquatic animals such as bivalves, gastropods, etc. have been used as a bio-indicator to study heavy metals pollution both from sediment and water (Gupta and Singh, 2011). A well-structured study of possible hazard and carcinogenic risks of heavy metals on fisheries and aquaculture have yet to be accomplished. Industrial contamination is the major route of higher concentration of heavy metal in aquatic organisms. Disruptive effects of pollutants on the fish gonads and gamete development, hormonal and pathological activities are observed in past findings. Minamata disease is an example of such a disaster in Japan. Organic Hg particularly methyl mercury discharged from the Chisso Company built-in 1902 (produced lime nitrate, ammonium sulfate fertilizer, acetaldehyde using mercury as a catalyst) contaminated the aquatic life such as fish and shellfish which were being caught and consumed by the people as a major component of the diet and became poisoned that led to uncountable mortality and severe damage to the embryo in the human uterus in the year 1956 along the sea of Shiranui (Harada, 1995, 1982). The second Minamata disease epidemic happened in Niigata (Japan) along the Agano river basin in January 1965 (Hachiya, 2006).With the growing global appetite for fish consumption, the heavy metal concentration must not exceed the permissible limits for the intake of aquatic foods. Proper waste management technologies, strict legislation, proper monitoring, research on the effect of these metals on the different organisms in detail and precautionary approaches, etc. are needed to protect the environment from such pollution.

Reference

Abdel-Warith, A. A., Younis, E. M., Al-Asgah, N. A., &Wahbi, O. M. (2011).Effect of zinc toxicity on liver histology of Nile tilapia, Oreochromisniloticus. Scientific Research and Essays, 6(17), 3760-3769.

Agarwal, R., Kumar, R., &Behari, J. R. (2007).Mercury and lead content in fish species from the river Gomti, Lucknow, India, as biomarkers of contamination. Bulletin of Environmental contamination and Toxicology, 78(2), 118-122.

Ahmed, M. K., Kundu, G. K., Al-Mamun, M. H., Sarkar, S. K., Akter, M. S., & Khan, M. S. (2013). Chromium (VI) induced acute toxicity and genotoxicity in freshwater stinging catfish, Heteropneustesfossilis. Ecotoxicology and environmental safety, 92, 64-70.

Ajmal, M., Rao, R.A.K., Khan, M.A., 2005.Adsorption of Cu from aqueous solution on Brassica cumpestris (mustard oil cake). J. Hazard. Mater. B122, 177–183

Arunkumar RI, Rajashekaran P, Michael RD. Differential effect of chromium compounds on the immune response of the African mouth breeder Oreochromismossambicus (Peter). Fish and Shellfish Immunology.2000; 10:667- 676.

ATSDR, (2008). Draft Toxicological Profile for Cadmium. US Department of Health and Humans Services, Public Health Human Services, Centers for Diseases Control, Atlanta.

Avenant-Oldewage, A., & Marx, H. M. (2000).Bioaccumulation of chromium, copper and iron in the organs and tissues of Clariasgariepinus in the Olifants River, Kruger National Park. Water SA, 26(4), 569-582.

Bakshi, A. (2016). Analysis of anthropogenic disturbances and impact of pollution on fish fauna of River Churni with special reference to chromium pollution. pp188.

Bhardwaj, S., &Choudhary, L. (2015). Effect of heavy metal pollution on survival and protein changes in liver and intestine of fresh water fish Tilapia mossambica. Journal of Advanced Laboratory Research in Biology, 6(1), 40-42.

Bhattacharyya, S., Chaudhuri, P., Dutta, S., &Santra, S. C. (2010). Assessment of total mercury level in fish collected from East Calcutta Wetlands and Titagarh sewage fed aquaculture in West Bengal, India. Bulletin of environmental contamination and toxicology, 84(5), 618-622.

Blewett, T. A., Glover, C. N., Fehsenfeld, S., Lawrence, M. J., Niyogi, S., Goss, G. G., & Wood, C. M. (2015). Making sense of nickel accumulation and sub-lethal toxic effects in saline waters: Fate and effects of nickel in the green crab, Carcinusmaenas. Aquatic Toxicology, 164, 23-33.

Boyden, C. (1975). Distribution of some trace metals in Poole Harbour, Dorset. Marine Pollution Bulletin, 6(12), 180-187.

Cain, D. J., Luoma, S. N., Carter, J. L., & Fend, S. V. (1992). Aquatic insects as bioindicators of trace element contamination in cobble-bottom rivers and streams. Canadian Journal of Fisheries and Aquatic Sciences, 49(10), 2141-2154.

Camakaris, J., Voskoboinik, I., & Mercer, J. F. (1999).Molecular mechanisms of copper homeostasis. Biochemical and biophysical research communications, 261(2), 225-232.

Canli, M., & Furness, R. W. (1995).Mercury and cadmium uptake from seawater and from food by the Norway lobster Nephropsnorvegicus. Environmental Toxicology and Chemistry: An International Journal, 14(5), 819-828.

Carvalho, C. d. S., &Fernandes, M. N. (2008).Effect of copper on liver key enzymes of anaerobic glucose metabolism from freshwater tropical fish Prochiloduslineatus. Comparative Biochemistry and Physiology Part A: Molecular & Integrative Physiology, 151(3), 437-442.

Castro-González, M. I., & Méndez-Armenta, M. (2008). Heavy metals: Implications associated to fish consumption. Environmental toxicology and pharmacology, 26(3), 263-271.

Chaurasia, S. S., Gupta, P., Kar, A., &Maiti, P. K. (1996). Lead induced thyroid dysfunction and lipid peroxidation in the fish Clariasbatrachus with special reference to hepatic type I-5′-monodeiodinase activity. Bulletin of environmental contamination and toxicology, 56(4), 649-654.

Craig, A., Hare, L., &Tessier, A. (1999). Experimental evidence for cadmium uptake via calcium channels in the aquatic insect Chironomusstaegeri. Aquatic Toxicology, 44(4), 255-262.

de Oliveira Ribeiro, C. A., Belger, L., Pelletier, E., &Rouleau, C. (2002). Histopathological evidence of inorganic mercury and methyl mercury toxicity in the arctic charr (Salvelinusalpinus). Environmental Research, 90(3), 217-225.

Doudoroff, P., & Katz, M. (1953). Critical review of literature on the toxicity of industrial wastes and their components to fish: II. the metals, as salts. Sewage and Industrial Wastes, 802- 839.

Eisler, R. (1998). Nickel hazards to fish, wildlife, and invertebrates: a synoptic review. 1998–0001. Biological Science Report, United States Geological Survey, Biological Resources Division, Washington, DC.

Eroglu, A., Dogan, Z., Kanak, E. G., Atli, G. Ü.., &Canli, M. (2015).Effects of heavy metals (Cd, Cu, Cr, Pb, Zn) on fish glutathione metabolism. Environmental Science and Pollution Research, 22(5), 3229-3237.

Ezekwe, C. I., &Edoghotu, M. I. (2015).Water quality and environmental health indicators in the Andoni River estuary, Eastern Niger Delta of Nigeria. Environmental Earth Sciences, 74(7), 6123-6136.

Falcó, G., Llobet, J. M., Bocio, A., & Domingo, J. L. (2006).Daily intake of arsenic, cadmium, mercury, and lead by consumption of edible marine species. Journal of Agricultural and Food Chemistry, 54(16), 6106-6112.

Farag, A. M., May, T., Marty, G. D., Easton, M., Harper, D. D., Little, E. E., & Cleveland, L. (2006). The effect of chronic chromium exposure on the health of Chinook salmon (Oncorhynchustshawytscha). Aquatic toxicology, 76(3-4), 246-257.

Förstner, U., &Wittmann, G. T. (2012). Metal pollution in the aquatic environment.Springer Science & Business Media.

Garg, U.K., Kaur, M.P., Garg, V.K., Sud, D., 2007.Removal of hexavalent Cr from aqueous solutions by agricultural waste biomass.

Ghosh, D., &Mandal, D. K. (2012).Histopathological effects and bioaccumulation of mercury in the kidney of an indian major carp, Labeorohita (Hamilton). Bulletin of environmental contamination and toxicology, 89(3), 479-483.

Gissi, F., Stauber, J., Reichelt-Brushett, A., Harrison, P. L., &Jolley, D. F. (2017).Inhibition in fertilisation of coral gametes following exposure to nickel and copper. Ecotoxicology and environmental safety, 145, 32-41.

Gupta, P., &Srivastava, N. (2006). Effects of sub-lethal concentrations of zinc on histological changes and bioaccumulation of zinc by kidney of fish Channapunctatus(Bloch). Journal of Environmental Biology, 27(2), 211-215.

Gupta, S. K., & Singh, J. (2011). Evaluation of mollusc as sensitive indicator of heavy metal pollution in aquatic system: a review. The IIOAB Journal, 2(1), 49-57.

Hachiya, N. (2006). The history and the present of Minamata disease. Japan Medical Association Journal, 49, 112-118.

Harada, M. (1982).Minamata disease.In Adverse effects of foods (pp. 135-148).Springer, Boston, MA.

Harada, M. (1995).Minamata disease: methylmercury poisoning in Japan caused by environmental pollution. Critical reviews in toxicology, 25(1), 1-24.

Hodson, P. V., Borgmann, U., & Shear, H. (1979).Toxicity of copper to aquatic biota. Copper in the Environment, 2, 308-372.

Hsu, P. C., &Guo, Y. L. (2002).Antioxidant nutrients and lead toxicity. Toxicology, 180(1), 33-44.

Humtsoe, N., Davoodi, R., Kulkarni, B. G., &Chavan, B. (2007). Effect of arsenic on the enzymes of the rohu carp Labeorohita (Hamilton, 1822). Raffles Bull Zool, 14, 17-19.

Hwang, I. K., Kim, K. W., Kim, J. H., & Kang, J. C. (2016).Toxic effects and depuration after the dietary lead (II) exposure on the bioaccumulation and hematological parameters in starry flounder (Platichthysstellatus). Environmental toxicology and pharmacology, 45, 328-333.

Jennings, J. R., & Rainbow, P. S. (1979).Accumulation of cadmium by Artemiasalina. Marine Biology, 51(1), 47-53.

Johns, M.M., Marshall, W.E., Toles, C.A., 1998. Agricultural byproducts as granular activated carbons for adsorbing dissolved metals and organics. J. Chem. Technol. Biotechnol. 71, 131–140.

Khangarot BS, Rathore RS, Tripathi DM. Effects of chromium on humoral and cell-mediated immune responses and host resistance to disease in a Freshwater catfish, Saccobranchusfossilis (Bloch). Ecotoxicology and Environmental Safety.1999; 43:11-20.

Kim, J. H., & Kang, J. C. (2014). The selenium accumulation and its effect on growth, and haematological parameters in red sea bream, Pagrus major, exposed to waterborne selenium. Ecotoxicology and environmental safety, 104, 96-102.

Knauer, G. A. (1977). Immediate industrial effects on sediment metals in a clean coastal environment. Marine Pollution Bulletin, 8(11), 249-254.

Kubrak, O. I., Husak, V. V., Rovenko, B. M., Poigner, H., Kriews, M., Abele, D., &Lushchak, V. I. (2013). Antioxidant system efficiently protects goldfish gills from Ni2+-induced oxidative stress. Chemosphere, 90(3), 971-976.

Kubrak, O. I., Poigner, H., Husak, V. V., Rovenko, B. M., Meyer, S., Abele, D., &Lushchak, V. I. (2014). Goldfish brain and heart are well protected from Ni2+-induced oxidative stress. Comparative Biochemistry and Physiology Part C: Toxicology & Pharmacology, 162, 43-50.

Kumar, A., & Gupta, A. K. (2006). Acute toxicity of mercury to the fingerlings of Indian major carps(catla, rohu and mrigal) in relation to water hardness and temperature. Journal of Environmental Biology, 27(1), 89-92.

Kumar, P., & Singh, A. (2010). Cadmium toxicity in fish: An overview. GERF Bulletin of Biosciences, 1(1), 41-47.

Kumari, B., Kumar, V., Sinha, A. K., Ahsan, J., Ghosh, A. K., Wang, H., &DeBoeck, G. (2017). Toxicology of arsenic in fish and aquatic systems. Environmental chemistry letters, 15(1), 43-64.

Kurniawan, T.A., Chan, G.Y.S., Lo, W.H., Babel, S., 2006. Comparison of low-cost adsorbents for treating wastewater laden with heavy metals. Sci. Total Environ. 366, 409–426.

Leonard, E. M., Marentette, J. R., Balshine, S., & Wood, C. M. (2014).Critical body residues, Michaelis–Menten analysis of bioaccumulation, lethality and behaviour as endpoints of waterborne Ni toxicity in two teleosts. Ecotoxicology, 23(2), 147-162.

Leonard, E. M., Nadella, S. R., Bucking, C., & Wood, C. M. (2009).Characterization of dietary Ni uptake in the rainbow trout, Oncorhynchusmykiss. Aquatic toxicology, 93(4), 205-216.

Maleki, A., Hajizadeh, Z., Sharifi, V., & Emdadi, Z. (2019).A green, porous and eco-friendly magnetic geopolymer adsorbent for heavy metals removal from aqueous solutions. Journal of cleaner production, 215, 1233-1245.

Martinez, C. B. R., Nagae, M. Y., Zaia, C. T. B. V., &Zaia, D. A. M. (2004). Acute morphological and physiological effects of lead in the neotropical fish Prochiloduslineatus. Brazilian Journal of Biology, 64(4), 797-807.

Min, S.H., Han, J.S., Shin, E.W., Park, J.K., 2004. Improvement of cadmium ion removal by base treated juniper fiber. Water Res. 38, 1289–1295.

Mohan, D., Pittman, C.U., 2007. Arsenic removal from water/wastewater using adsorbents – a critical review. J. Hazard. Mater. 142, 1–53.

Muyssen, B. T., Brix, K. V., DeForest, D. K., & Janssen, C. R. (2004). Nickel essentiality and homeostasis in aquatic organisms. Environmental reviews, 12(2), 113-131.

Obasohan, E. E. (2007). Heavy metals concentrations in the offal, gill, muscle and liver of a freshwater mudfish (Parachannaobscura) from Ogba River, Benin city, Nigeria. African Journal of Biotechnology, 6(22).2620–2627.

Olla, B. L., Estelle, V. B., Braun, G., Swartz, R. C., &Studholme, A. L. (1988). Responses of polychaetes to cadmium-contaminated sediment: Comparison of uptake and behavior. Environmental Toxicology and Chemistry: An International Journal, 7(7), 587- 592.

Osman, A.G.M., Wuertz, S., Mekkawy, I.A., Exner, H.-J., Kirschbaum, F., 2007. Lead induced malformations in embryos of the African catfish Clariasgariepinus (Burchell, 1822). Environ. Toxicol. 22, 375–389.

Pane, E. F., Richards, J. G., & Wood, C. M. (2003a). Acute waterborne nickel toxicity in the rainbow trout (Oncorhynchusmykiss) occurs by a respiratory rather than ionoregulatory mechanism. Aquatic toxicology, 63(1), 65-82.

Pane, E. F., Smith, C., McGeer, J. C., & Wood, C. M. (2003b).Mechanisms of acute and chronic waterborne nickel toxicity in the freshwater cladoceran, Daphnia magna. Environmental science & technology, 37(19), 4382-4389.

Parvez, S., Sayeed, I., Pandey, S., Ahmad, A., Bin-Hafeez, B., Haque, R., Ahmed. I., &Raisuddin, S. (2003). Modulatory effect of copper on nonenzymatic antioxidants in freshwater fish Channapunctatus (Bloch.). Biological trace element research, 93(1), 237-248.

Paul, N., Chakraborty, S., &Sengupta, M. (2014). Lead toxicity on non-specific immune mechanisms of freshwater fish Channapunctatus. Aquatic Toxicology, 152, 105-112.

Penna, M. M., Lee, J., & Thiele, D. J. (1999). A delicate balance: homeostatic control of copper uptake and distribution. The Journal of nutrition, 129(7), 1251-1260.

Pereira, L. S., Ribas, J. L. C., Vicari, T., Silva, S. B., Stival, J., Baldan, A. P., ...& de Assis, H. S. (2016). Effects of ecologically relevant concentrations of cadmium in a freshwater fish. Ecotoxicology and environmental safety, 130, 29-36.

Poopal, R. K., Ramesh, M., & Dinesh, B. (2013).Short-term mercury exposure on Na+/K+-ATPase activity and ionoregulation in gill and brain of an Indian major carp, Cirrhinus mrigala. Journal of Trace Elements in Medicine and Biology, 27(1), 70-75.

Ptashynski, M. D., &Klaverkamp, J. F. (2002). Accumulation and distribution of dietary nickel in lake whitefish (Coregonusclupeaformis). Aquatic toxicology, 58(3-4), 249-264.

Ragsdale, S. W. (1998). Nickel biochemistry. Current opinion in chemical biology, 2(2), 208-215.

Rai, P. K. (2008). Heavy metal pollution in aquatic ecosystems and its phytoremediation using wetland plants: an ecosustainable approach. International journal of phytoremediation, 10(2), 133-160.

Rajeshwarisivaraj, Subburam, V., 2002. Activated parthenium carbon as an adsorbent for the removal of dyes and heavy metal ions from aqueous solution.Biores. Technol. 85, 205–206.

Reichelt-Brushett, A. J., & Harrison, P. L. (2005). The effect of selected trace metals on the fertilization success of several scleractinian coral species. Coral Reefs, 24(4), 524-534.

Sfakianakis, D. G., Renieri, E., Kentouri, M., &Tsatsakis, A. M. (2015).Effect of heavy metals on fish larvae deformities: a review. Environmental research, 137, 246-255.

Sreedevi, P., Suresh, A., Sivaramakrishna, B., Prabhavathi, B., &Radhakrishnaiah, K. (1992). Bioaccumulation of nickel in the organs of the freshwater fish, Cyprinuscarpio, and the freshwater mussel, Lamellidensmarginalis, under lethal and sublethal nickel stress. Chemosphere, 24(1), 29-36.

Steven, J. D., Davies, L. J., Stanley, E. K., Abbott, R. A., Ihnat, M., Bidstrup, L., &Jaworski, J. F.(1976). Effects of chromium in the Canadian environment. Nat. Res. Coun. Canada, NRCC, (15017), 168.

Stokes, R. M., & Fromm, P. O. (1965).Effect of chromate on glucose transport by the gut of rainbow trout. Physiological Zoology, 38(3), 202-205.

Storelli, M. M., &Marcotrigiano, G. O. (2004).Interspecific variation in total arsenic body concentrations in elasmobranch fish from the Mediterranean Sea. Marine pollution bulletin, 48(11-12), 1145-1149.

Storelli, M. M., Giacominelli-Stuffler, R., Storelli, A., &Marcotrigiano, G. O. (2005). Accumulation of mercury, cadmium, lead and arsenic in swordfish and bluefin tuna from the Mediterranean Sea: a comparative study. Marine pollution bulletin, 50(9), 1004-1007.

Subashini, P., Manavalaramanujam, R., Ramesh, M., &Geetha, N. (2005). Changes in selected biomarkers in freshwater teleost fish, Cyprinuscarpio var. communis exposed to sublethal concentrations of chromium sulphate toxicity. Journal of environmental science & engineering, 47(1), 65-68.

Sud, D., Mahajan, G., &Kaur, M. P. (2008).Agricultural waste material as potential adsorbent for sequestering heavy metal ions from aqueous solutions–A review. Bioresource technology, 99(14), 6017-6027.

Szkoda, J., Durkalec, M., Nawrocka, A., &Michalski, M. (2015).Mercury concentration in bivalve molluscs. Bulletin of the Veterinary Institute in Pulawy, 59(3), 357-360.

Tan, F., Wang, M., Wang, W., & Lu, Y. (2008). Comparative evaluation of the cytotoxicity sensitivity of six fish cell lines to four heavy metals in vitro. Toxicology in vitro, 22(1), 164-170.

Taub Frieda, B. (2004). Fish 430 lectures (Biological Impacts of Pollutants on Aquatic Organisms). University of Washington College of Ocean and Fishery Sciences, Seattle.

Thomann, R. V., Shkreli, F., & Harrison, S. (1997). A pharmacokinetic model of cadmium in rainbow trout. Environmental Toxicology and Chemistry, 16(11), 2268-2274.

Van der Putte, I., Lubbers, J., &Kolar, Z. (1981).Effect of pH on uptake, tissue distribution and retention of hexavalent chromium in rainbow trout (Salmogairdneri). Aquatic Toxicology, 1(1), 3-18.

Velma, V., Vutukuru, S. S., &Tchounwou, P. B. (2009). Ecotoxicology of hexavalent chromium in freshwater fish: a critical review. Reviews on environmental health, 24(2), 129-145.

Verbost, P. M., Van Rooij, J., Flik, G., Lock, R. A. C., &Bonga, S. W. (1989). The movement of cadmium through freshwater trout branchial epithelium and its interference with calcium transport. Journal of experimental biology, 145(1), 185-197.

Verstraeten, S. V., Aimo, L., &Oteiza, P. I. (2008).Aluminium and lead: molecular mechanisms of brain toxicity. Archives of toxicology, 82(11), 789-802.

Viarengo, A. (1989). Heavy metals in marine invertebrates: mechanisms of regulation and toxicity at the cellular level. Rev. Aquat. Sci, 1(2), 295-317.

Viarengo, A., Mancinelli, G., Pertica, M., Fabbri, R., &Orunesu, M. (1993). Effects of heavy metals on the Ca2+-ATPase activity present in gill cell plasma-membrane of mussels

(Mytilusgalloprovincialis Lam.). Comparative Biochemistry and Physiology Part C: Pharmacology, Toxicology and Endocrinology, 106(3), 655-660.

Visnjic-Jeftic, Z., Jaric, I., Jovanovic, L., Skoric, S., Smederevac-Lalic, M., Nikcevic, M., &Lenhardt, M. (2010). Heavy metal and trace element accumulation in muscle, liver and gills of the Pontic shad (AlosaimmaculataBennet 1835) from the Danube River (Serbia). Microchemical Journal, 95(2), 341-344.

Vutukuru SS. Chromium induced alterations in some biochemical profiles of the Indian major carp, Labeorohita (Hamilton). Bulletin of Environmental and Contamination Toxicology. 2003; 70(1):118-23.

Vutukuru, S. S., Prabhath, N. A., Raghavender, M., &Yerramilli, A. (2007).Effect of arsenic and chromium on the serum amino-transferases activity in Indian major carp, Labeorohita. International journal of environmental research and public health, 4(3), 224-227.

Wang, S., & Shi, X. (2001).Molecular mechanisms of metal toxicity and carcinogenesis. Molecular and cellular biochemistry, 222(1-2), 3-9.

Wells, M. L., Smith, G. J., &Bruland, K. W. (2000).The distribution of colloidal and particulate bioactive metals in Narragansett Bay, RI. Marine Chemistry, 71(1-2), 143-163.

World Health Organization. (1991). Environmental health criteria: 108. Nickel. World Health Organization, Geneva.

Wright, D. A., &Welbourn, P. (2002). Environmental toxicology (Vol. 11).Cambridge University Press.

Xia, Y., Zhu, J., Xu, Y., Zhang, H., Zou, F., &Meng, X. (2020).Effects of ecologically relevant concentrations of cadmium on locomotor activity and microbiota in zebrafish. Chemosphere, 257, 127220.

12

An Ambitious Carbon Sequestration by Phytoplankton in the Marine Coastal Systems

Narshivudu Daggula[1], Ramachandra Naik A.T.[2] and Mukkeri Kranthirekha[3]

[1] National Fisheries Development Board, Ministry of Fisheries, Animal Husbandry and Dairying, Hyderabad–500 052, Telangana

[2] Department of Aquatic Environment Management, College of Fisheries, Karnataka Veterinary, Animal and Fisheries Sciences University, Mangaluru–575 002, Karnataka

[3] Central Institute of Fisheries Education, Mumbai

Abstract

Coastal and marine ecosystems are important for climate change mitigations. The coastal ecosystems of mangroves, seagrass meadows and tidal marshes mitigate climate change by sequestering carbon dioxide from the atmosphere and oceans at significantly higher rates, per unit area, than terrestrial forests. Phytoplankton are vital autotrophic organisms in marine and freshwater ecosystems, playing a central role in carbon sequestration through the biological carbon pump. By converting atmospheric CO_2 *into organic carbon via photosynthesis, phytoplankton sustain aquatic food webs and facilitate carbon transport through the food web to the ocean floor as particulate organic carbon. This process, influenced by microbial activity and the sulfur cycle, results in the long-term storage of carbon in deep-sea sediments, reducing atmospheric* CO_2 *levels. Notably, phytoplankton exhibit high carbon fixation rates and lipid productivity, particularly among the chlorophyceae group, enhancing their effectiveness as a natural carbon sink. Biological carbon fixation is an attractive strategy for climate mitigation, offering a net-zero carbon emission approach to* CO_2 *capture and storage. Assessing the efficiency of phytoplankton in carbon sequestration, through factors like biomass, growth rates, and oceanic conditions, is essential for understanding and potentially amplifying the ocean's role as a carbon sink, which is critical in addressing global climate change.*

Keywords: *Phytoplankton, Carbon sequestration, Climate change and Coastal waters.*

Introduction

The ocean, atmosphere, soil, and terrestrial forest ecosystems have been the largest natural carbon sinks, new research on the role of vegetated coastal ecosystems has highlighted their potential as highly efficient carbon sinks, and led to the scientific recognition of the term "Blue Carbon". It is designating carbon that is fixed via coastal ocean ecosystems, rather than traditional land ecosystems, like forests. Although the ocean's vegetated habitats cover less than 0.5% of the seabed, they are responsible for more than 50%, and potentially up to 70%, of all carbon storage in ocean sediments. Coastal ecosystems are the most productive on earth and home to a wealth of biodiversity. They provide with essential services such as coastal protection from storms and nursery grounds for fish. Their role in sequestering and storing 'blue' carbon from the atmosphere and oceans is also increasing and recognized by policymakers. Coastal blue carbon ecosystems are found along the coasts of every continent except Antarctica The different coastal wetlands types such as mangroves, seagrasses, tidal marshes and coral reefs commonly referred to as blue carbon ecosystems, provides numerous benefits and services that contribute to mitigate and adapt to the impacts of climate change. Mangroves, salt marshes and seagrasses make up the majority of the ocean's vegetated habitats but only equal 0.05% of the plant biomass on land. Despite their small footprint, they can store a comparable amount of carbon per year and are highly efficient carbon sinks. Seagrasses, mangroves and salt marshes can capture carbon dioxide (CO_2) from the atmosphere by sequestering the Carbon in their underlying sediments, in underground and below-ground biomass, and in dead biomass.

Importance of the Coastal Blue Carbon Ecosystem

When protected or restored, blue carbon ecosystems sequester and store carbon. When degraded or destroyed, these ecosystems emit the carbon they have stored for centuries into the atmosphere and oceans and become sources of greenhouse gases. Experts estimate that as much as 1.02 billion tons of carbon dioxide are being released annually from degraded coastal ecosystems, which is equivalent to 19% of emissions from tropical deforestation globally. Mangroves, tidal marshes and seagrasses are critical along the world's coasts, supporting coastal water quality, healthy fisheries, and coastal protection against floods and storms. For example, mangroves are estimated to be worth at least US$1.6 billion each year in ecosystem services that support coastal livelihoods and human populations around the world.

Coastal Biological Ecosystems as a Carbon Sequesters

Mangroves, seagrasses, saltmarshes, microalgae, coral reef, vegetated sediments, and benthic coastal ocean ecosystems covers 5.3 × 109 ha globally and play an important role in carbon sequestration (Alongi, 2012). Mangroves store five times more organic carbon than tropical upland forests (Donato *et al.*, 2011). Comparatively the mangrove forests play a crucial role in carbon cycling and budgets in the coastal landscape. Although coastal forests account for approximately 0.5% of total area of coastal ecosystems, mangroves contribute 10-15% of global carbon storage. A combination of high productivity and slow soil decomposition rates of mangroves significantly increases its ability to capture and store organic carbon in soils (Alongi, 2012).

Types of Blue Carbon Ecosystems

Mangroves: Mangroves are a type of tropical forest, found at the edge of land and sea and flooded regularly by tidal water. Mangroves are among the most carbon-rich forests in the tropics. It is estimated that the - average annual carbon sequestration rate for mangroves averages between 6 to 8 Mg CO_2 e/ha (tons of CO_2 equivalent per hectare). These rates are about two to four times greater than global rates observed in mature tropical forests. Mangroves provide at least US $1.6 billion each year in ecosystem services, which include: supporting fisheries by providing important spawning grounds for commercial fish species; filtering pollutants and contaminants from coastal waters and contributing to healthy coastal marine water quality; and protecting coastal development and communities against storms, floods and erosion. In the last 50 years, between 30-50% of mangroves have been lost globally and they continue to be lost at a rate of 2% each year. Major causes of destruction to mangrove ecosystems include deforestation for construction of aquaculture ponds and other forms of unsustainable coastal development. Experts estimate that emissions from the degradation of mangroves can be as high as 10% of total emissions from deforestation globally, even though mangroves account for only 0.7% of tropical forest area.

Tidal marshes: Tidal marshes are coastal wetlands with deep soils that are built through the accumulation of mineral sediment and organic material and then flooded with salty water brought in by the tides. Almost all of the carbon in tidal marsh ecosystems is found in the soil, which can be several meters deep. Similarly, to mangrove and seagrass habitats, marshes also serve as important carbon sinks, Marshes sequester C in underground biomass due to high rates of organic sedimentation and anaerobic- dominated decomposition. Salt marshes cover approximately 22,000 to 400,000 km^2 globally, with an estimated carbon burial rate of 210 g C m−2 yr−1.It is estimated that the average annual carbon sequestration rate for tidal marshes averages between 6 to 8 Mg CO_2 e/ha (Mg of CO_2 equivalent per hectare). These rates are about two to four times greater than those observed in mature tropical forests. Tidal marshes filter pollutants from land runoff and hence help maintain water quality in coastal areas. They provide critical habitat for many stages of the life cycle of important marine species, which is essential for healthy fisheries and coastal marine ecosystems. Tidal marshes also serve as a buffer to coastal communities, absorbing some energy from storms and floods and helping to prevent erosion. Tidal and freshwater marshes are being lost at a rate of 1-2% per year. Major threats to tidal marsh ecosystems include draining for coastal development, conversion to agriculture, and rising sea levels.

Seagrasses: Seagrasses are submerged flowering plants with deep roots that are found in meadows along the shore of every continent except Antarctica. Carbon accumulates in seagrasses over time which are anoxic and thus continually preserve organic carbon from decadal-millennial time scales and is stored almost entirely in the soils, which have been measured up to four meters deep. Carbon sequestration rates in seagrass meadows vary depending on the species, characteristics of the sediment, and depth of the habitats, but on average the carbon burial rate is approximately 138 g C m−2 yr−1. Although seagrasses account for less than 0.2% of the world's oceans, they sequester approximately 10% of the carbon buried in ocean sediment annually (27.4Tg of carbon per year). Per hectare, seagrasses can store up to twice as much carbon than terrestrial forests. The global seagrass ecosystem organic carbon pool could be as high as 19.9 billion metric tons. Seagrass meadows filter sediment and other nutrients from the water and are constantly building and securing sediment, which buffers coasts from erosion, storms and flooding. They are also important habitats for fisheries and flagship marine species, such as sea turtles and manatees. Seagrasses are among the world's most threatened ecosystems, with annual global loss of around 1.5% and accelerating in recent decades. Globally, about 29% of Earth's seagrass ecosystems have been lost. Major threats to seagrasses include degradation of water quality due to poor land use, such as deforestation and dredging.

Algae: Macro-algae and microalgae are being investigated as possible means of carbon sequestration. Because algae lack the complex lignin associated with terrestrial plants, the carbon in algae is released into the atmosphere more rapidly than carbon captured on land. Algae have been proposed as a short-term storage pool of carbon that can be used as a feedstock for the production of various biogenic fuels. Many cyanobacteria, microalgae, and microalgae species can utilize carbonate as a carbon source for photosynthesis. In South Korea, microalgae have been utilized as part of a climate change mitigation program. The country has established the Coastal CO_2 Removal Belt which is composed of artificial and natural ecosystems. The goal is to capture carbon using large areas of kelp forest.

Phytoplankton: The Jewel of the Marine World

Phytoplankton has evolved over many millions of years, interacting with other organisms to produce a great diversity of life forms (Chhapgar and Kothari, 2005). This group is the basis of the marine food web and a major driver in the cycling of elements. Being primary producers, they are involved in oceans' ecological and biogeochemical services. Diatoms and coccolithophores play a major role in the regulation of atmospheric carbon dioxide. Biological carbon pump means the carbon uptake by phytoplankton and its export as organic matter to the ocean interior which reduces the partial pressure of carbon dioxide in the upper waters and also the diffusive drawdown of atmospheric carbon dioxide.

Phytoplankton in Coastal Ecosystems

Phytoplankton is the one that makes largest contribution to the fundamental life processes through photosynthesis. A great majority of the conversion of carbon and nitrogen into a living matter is through phytoplankton. Thus every gram of fish or any other consumable organism from the environment is a final product of chain of intricate biological processes. For several pelagic fishes, plankton form a chief food therefore, the fluctuation of fisheries of a region can partly be attributed to the physical and chemical changes and the resultant changes in the productivity of plankton. Like land plants, phytoplankton has chlorophyll to capture sunlight and they use photosynthesis to turn it into chemical energy. They absorb carbon dioxide and release oxygen. All phytoplankton photosynthesize, but some get additional energy by consuming other organisms (Mee, 2012).

Life Inside the Phytoplankton

Phytoplankton growth depends on the availability of carbon dioxide, sunlight and nutrients. Phytoplankton, like land plants, require nutrients such as nitrate, phosphate, silicate and calcium at various levels depending on the species. Some phytoplankton can fix nitrogen and can grow in areas where nitrate concentrations are low. They also require trace amounts of iron which limits phytoplankton growth in large areas of the ocean because iron concentrations are very low. Other factors influence phytoplankton growth rates, including water temperature and salinity, water depth, wind and also predators that are grazing on them. Through photosynthesis, phytoplankton consumes carbon dioxide on a scale equivalent to forests and other land plants. Some of this carbon is carried to the deep ocean when phytoplankton dies and some are transferred to different layers of the ocean as phytoplankton are eaten by other creatures, which reproduce, generate waste and die. Phytoplankton is responsible for most of the transfer of carbon dioxide from the atmosphere to the ocean. Carbon dioxide is consumed during photosynthesis and the carbon is incorporated in the phytoplankton just as carbon is stored in the wood and leaves of a tree (Zhang *et al.,* 2017). Most of the carbon is returned to near-surface waters when phytoplankton is eaten or decomposed but some fall into the ocean grounds.

Composition of Phytoplankton

Phytoplankton is the one that makes the largest contribution to the fundamental life processes through photosynthesis. A great majority of the conversion of carbon and nitrogen into living matter is through phytoplankton. Thus, every gram of fish or any other consumable organism from the environment is the final product of a chain of intricate biological processes. Thus, phytoplankton assumes a threshold position in the estuarine food chain. For several pelagic fishes, plankton form a chief food therefore, the fluctuation of fisheries of a region can partly be attributed to the physical and chemical changes and the resultant changes in the productivity of plankton.

Phytoplankton constitutes mostly unicellular algae and the main components are diatoms, dinoflagellates, coccolithophores, silicoflagellates, green algae and blue green algae. They have a wide distribution in all habitats and play a major role in the food chain of aquatic ecosystems as a major source of organic carbon (Okeke *et al.,* 2022). The sensitivity and large variations in species composition are often a reflection of significant alterations in ambient conditions within the ecosystem (Adesalu and Nwankwo, 2010) slight fluctuation in the water quality may affect the population of phytoplankton and be related to nutrient availability (Irkin, 2019), based on the phytoplankton size structure, the total chlorophyll concentration is classified into three functional classes Mohan *et al.* (2021), phytoplankton contributes around 50% of the total carbon that gets sequestered in the form of biomass. The biomass productivity of phytoplankton is higher than for any conventional plant crop. It is reported as being up to 15 to 45 g/ m^2/day under optimum scaled-up conditions in an open environment for much more promising phytoplankton. Although the accumulated carbon percentage varies, in dried biomass, the majority of phytoplankton contains 20 % to 65 % carbon in dried conditions, although the dissolution rate of carbon dioxide at normal atmospheric pressure and temperature was investigated by (Mishra *et al.,* 2022).

The global algal net primary productivity is 111 to 117 Pg C per year, which includes both freshwater and ocean ecosystems. Parveen and Patidar (2022) phytoplankton are microscopic photosynthetic organisms that may be unicellular or multicellular and essentially contain Chlorophyll-*a*, they may be comprised of both prokaryotic and eukaryotic cells. They synthesize bioorganic macromolecules and reserve storage material in the form of chrysolaminarin (a polymer of carbohydrate that is especially found in diatoms), starch, and Triacylglycerols (Sivakumar *et al.,* 2022).

Biochemical Composition of Phytoplankton

A large number of studies on the chemical composition of plankton in different geographical zones have been made but most are focused on qualitative studies and do not include the total elemental or biochemical composition. (Libes, 2009), these nutrients enter the estuarine and coastal system in the form of organic and inorganic nitrogen (N) and phosphorus (P) through rivers and the atmosphere, jeopardizing the ecosystem (Galloway *et al.,* 2004). Measurements of carbon, nitrogen and phosphorous are numerous. These nutrients enter the estuarine and coastal system in the form of organic and inorganic nitrogen (N) and phosphorus (P) through rivers and the atmosphere, jeopardizing the ecosystem (Halsey and Jones, 2015), consequently, these ecosystems sustain high primary productivity and elevated heterotrophy which has been discussed in many previous works in the Arabian Sea and adjacent coastal waters of the west coast of India (Gupta *et al.,* 2016).

The main chemical components of phytoplankton are proteins, carbohydrates, lipids, phosphorus compounds and total chlorophylls (Basu and Mackey, 2018), the chemical composition of phytoplankton depends on nutrient availability and other environmental conditions (Paerl and Justic, 2013; Vasanthi *et al.,* 2014). Roy *et al.* (2017) the lack of adequate direct measurements on carbohydrate, protein and lipid overlapping the temporal coverage of the ocean-colour data have restricted rigorous validation of the satellite-derived estimates. Therefore, new in situ measurements of phytoplankton macromolecules across various oceanic conditions should be a priority for increasing the reliability and reducing the bias and uncertainties of the satellite-based estimates. The average ratio of elements in marine plankton has traditionally been thought to centre on the Redfield ratio at 106 moles of carbon (C): 16 moles of nitrogen (N): 1 mole of phosphorus (P) Kolmakova and Kolmakov (2019), however, the stoichiometry of elements within marine organic particles is variable between biogeographical provinces (Martiny *et al.,* 2013) and within the phytoplankton isolates, which suggests that the average oceanic C:N:P (Sharoni and Halevy 2020). Environmental factors like light, temperature and nutrients influence phytoplankton physiology and cellular elemental content, potentially molding relationships between phytoplankton phylogeny and cellular elemental stoichiometry (Toseland *et al.*, 2013; Garcia *et al.*, 2018).

Phytoplankton-mediated Carbon Sequestration

Carbon sequestration can be defined as the capture and secure storage of carbon from the atmosphere. The idea is to keep carbon emissions produced by anthropogenic activities from the atmosphere by capturing and diverting them to secure storage or to remove carbon from the atmosphere by various means and storing it, being a primary producer (autotrophs) phytoplankton interacting with other organisms to produce a great diversity of life forms. This group is the basis of the marine food web and it is a major driver in the cycling of elements. These primary producers involved oceans' ecological and biogeochemical services. Diatoms and coccolithophores play a major role in the regulation of atmospheric carbon dioxide. Biological pump means the carbon uptake by phytoplankton and its export as organic matter to the ocean interior reduces the partial pressure of carbon dioxide in the upper waters and also the diffusive drawdown of atmospheric carbon dioxide.

Phytoplankton are greatly divided into different groups. Diatoms are one of the important classes of phytoplankton, their name is derived from the Greek word meaning 'Cut in half', a reference to their distinctive two-part cell walls made of silica (Virginia and Armbrust, 2009). These are golden brown algae with chloroplasts, single-celled with a glass (silica dioxide) skeleton made of two valves, occur as single cells or in chain-like colonies, store oils in cells, are highly nutritious and can multiply up to 3 times a day. Diatoms, dinoflagellates and blue green algae constitute the dominant phytoplankton group in the tropical coastal waters of Mangaluru in the Arabian Sea (Gowda *et al.*, 2012). It shows integrating response to ambient changes in water quality which may sometimes not be observed by Physico-chemical analysis. Physical and chemical methods allow only instantaneous measurements, therefore restraining the knowledge of water conditions to the moment when the measurements were taken but biological methods give long-term environmental effects. Therefore, physical and chemical methods are complementary to biological methods. So, along with some Physico-chemical parameters biological methods are also influenced to determine the changes in water quality.

The major group of phytoplankton sensitive to the physical and chemical parameters of water such as pH, nutrients, salinity, temperature and water current in which they live (Karthik *et al.*, 2020). Spatial and temporal variations in the bio-chemical composition of phytoplankton and hydrographical parameters have an impact on the distribution and abundance of phytoplankton and thereby it gives an indication about the coastal water quality. Carbon dioxide capture by phytoplankton is a promising technology to solve the environmental problem and also reduce the concentration of greenhouse gasses in the atmosphere in a sustainable manner. The major task of research in this area of carbon sequestration is to identify the algal species and their specific growth conditions, air and water temperature, light intensity, nutrient concentration, photoperiod, aeration and physiochemical characters. The integration of carbon dioxide fixation from flue gas, nutrient removal from effluents and production of biomass from algae may provide a very promising alternative technology to current carbon dioxide capture strategies.

Carbon sequestration in the cochin estuary and the adjacent coastal Arabian Sea was measured by Bhavya *et al.* (2016) using the 13C-$NaHCO_3$ labelling technique and reported that carbon fixation rates in the estuarine water (pre-monsoon) 3.0 to 20.4 μ mol. monsoon 2.55 to 12.4 μ mol. post-monsoon 3.1 to 16.3 μ mol. carbon per hour was significantly higher than the coastal locations (pre-monsoon) 0.02 to 0.1 μ mol. monsoon 0.13 to 3.1 μ mol. post-monsoon 0.007 to 0.032 μ mol. carbon per hour. Carbon dioxide sequestration by the marine planktonic phytoplankton like *Nannochloropsis* spp. and *Isochrysis* spp. as well as macro forms *Gracilaria* spp., *Sargassum* spp. and *Ulva lactuca* was estimated under laboratory conditions by Kaladharan *et al.* (2009) observed that, the green seaweed *Ulva lactuca* registered 100 % utilization of carbon dioxide towards carbon fixation from the ambient water up to 15 mg/l and beyond that, it declined to 60 %.

Biological Carbon Pumping

The biological carbon pump uses the organic carbon sourced from the photosynthesis of phytoplankton, transports and transforms the carbon through the micro-food web and food chain and ultimately deposits it to the sea bed primarily as particulate organic carbon. Therefore, the biological carbon pump is also referred to as the biological deposition pump. Globally, among the carbon captured through photosynthesis, 45% is stored in terrestrial ecosystems as green carbon, and the remaining 55% is stored in the ocean by phytoplankton and coastal blue ecosystems (Nellemann *et al.,* 2009). Analysis of the composition of the particulate fraction in the ocean is required to understand the behaviour and geochemical cycling of both major and trace elements. Phytoplankton has been used as an indicator of environmental and climatic changes. Plankton is a good bio indicator in coastal waters. Plankton is a diverse group of organisms that live in the water column and cannot swim against the current. They reflect the effect of environmental changes in a short time, according to Divya *et al.* (2019) the efficiency of carbon sequestration by the biological pump could decline in the coming decades because respiration tends to increase more with temperature than photosynthesis. Despite these differences in the short-term temperature sensitivities of photosynthesis and respiration, it remains unknown whether the long-term impacts of global warming on the metabolic rates of phytoplankton can be modulated by evolutionary adaptation (Barton *et al.,* 2020).

The organic carbon pump strongly influences the carbon dioxide uptake of the ocean Kaladharan *et al.* (2009). There are indications that the carbon dioxide uptake of the organic carbon pump responds to climate change Libes, (2009) and Hussian *et al.* (2016) observed that the surface ocean warming will strengthen stratification and favour shifts of the phytoplankton community structure towards clades that increase carbon dioxide sequestration by the organic carbon pump. Sarangi *et al.* (2022) enhanced respiration at high temperatures is likely to weaken the organic carbon pump by increasing the respiration of organic matter in the upper water column and lowering the share exported into the deep sea. Basu and Mackey (2018) the ocean is among the largest reservoirs of carbon on Earth, hence, quantifying the efficiency of the biological carbon pump in the face of climate change is a prerequisite to predicting future atmospheric carbon dioxide concentrations (Karl, 2004). It is evident from literature reports that the strength of the biological pump will likely change in the future ocean, in response to a changing climate.

Threats: Activities causing the high rate of loss of coastal blue carbon ecosystems. The main causes of conversion and degradation of blue carbon ecosystems vary around the world but are largely driven by human activities. Common drivers are aquaculture, agriculture, mangrove forest exploitation, terrestrial and marine sources of pollution and industrial and urban coastal development. These impacts are expected to continue and be exacerbated by climate change.

Alternatives: Globally, numerous policies, coastal management strategies, and tools designed for conserving and restoring coastal ecosystems have been developed and implemented. Policies and finance mechanisms being developed for climate change mitigation may offer an additional route for effective coastal management. Blue carbon now offers the possibility to mobilize additional funds and revenue by combining best- practices in coastal management with climate change mitigation goals and needs. Scientists have developed robust methods to reliably measure and monitor the blue carbon stored in the vegetative biomass and soils of mangroves, tidal marshes, and seagrasses. Similarly, effective methods exist to estimate the loss of carbon from these systems if they are degraded or converted. Protocols for measuring blue carbon stored in mangroves have been established for some time and related methods for tidal marshes and seagrass meadows are now becoming standardized

Scope and Potential of Carbon Sequestration by Phytoplankton

Phytoplankton has excellent potential to grow in polluted environments. They remediate a wide range of organic and inorganic nutrients with greater efficiency. Many phytoplankton are reported as accumulators for some specific elements and metals, but on the other hand, they sequester carbon through photosynthesis (Nanabawa *et al.*, 2020). Phytoplankton are the most important primary producers in aquatic ecosystems and they contribute significantly to carbon sequestration, which depends on the photosynthetic process driven by autotrophic organisms. The role of phytoplankton in the absorption of nutrients is reflected by the carbon sequestration potential (Limberger *et al.,* 2019). However, the physiological and ecological characteristics of algae, and the availability and optimal utilization of basic resources such as temperature and nutrients are the main factors controlling phytoplankton growth and production (Purina *et al.,* 2018).

References

Adesalu, T.A. and Nwankwo, D.I., 2010. Euglenoids of Lekki Lagoon: LM and SEM Images of some Taxa. NY. Sci. J., 3(11): 78-85.

Alongi, D.M., 2012. Carbon sequestration in mangrove forests. Carbon management., 3(3): 313-322.

Arunpandi, R. Jyothibabu, L. Jagadeesan, T.R. Gireeshkumar, C. Karnan, S.W.A. and Naqvi., 2017. Noctiluca and copepods grazing on the phytoplankton community in a nutrient-enriched coastal environment along the southwest coast of India Environ. Monit. Assess., 189(7): 1-20.

Babu, A., Varadharajan, D., Vengadesh, P.N., Thilagavathi, B., Manikandarajan, T., Sampathkumar, P. and Balasubramanian, T., 2013. Diversity of Phytoplankton in Different Stations from Muthupettai, South-east Coast of India. J. Mar. Sci. Res., 3(3): 1-11.

Barton, S., Jenkins, J., Buckling, A., Schaum, C.E., Smirnoff, N., Raven, J.A. and Yvon, Durocher, G., 2020. Evolutionary temperature compensation of carbon fixation in marine phytoplankton. Ecol. Lett., 23(4): 722-733.

Basu, S. and Mackey, K.R., 2018. Phytoplankton as key mediators of the biological carbon pump: Their responses to a changing climate. Sustain. Dev., 10(3): 869.

Bhavya, P.S., Kumar, S., Gupta, G.V.M., Sudharma, K.V., Sudheesh, V. and Dhanya, K.R., 2016. Carbon isotopic composition of suspended particulate matter and dissolved inorganic carbon in the Cochin estuary during post-monsoon. Curr. Sci., 111(1-2): 1539-1543.

Chhapgar, B.F. and Kothari, A.S., 2005. Treasures of Indian wildlife. Bombay Natural History Society, 210-212.

Divya, M., Dinesh Kumar, S., Krishnaveni, N. and Santhanam, P., 2019. A Study of Carbon Sequestration by Phytoplankton. J. Plankton Res. Book. Springer, Singapore., pp. 277-284.

Donato, D. C, Kauffman, J.B, Murdiyarso, D., Kurnianto, S., Stidham, M., and Kanninen, M., 2011. Mangroves among the most carbon-rich forests in the tropics. Nat. Geosci., 4 (5): 293-297.

Galloway, J. N., Dentener, F.J., Capone, D.G., Boyer, E.W., Howarth, R.W., Seitzinger, S.P., Asner, G.P., Cleveland, C.C., Green, P.A., Holland, E.A. and Karl, D.M., 2004. Nitrogen cycles: past, present, and future. Biogeochemistry, 70(2): 153-226.

Garcia, N.S., Sexton, J., Riggins, T., Brown, J., Lomas, M.W. and Martiny, A.C., 2018. High variability in cellular stoichiometry of carbon, nitrogen, and phosphorus within classes of marine eukaryotic phytoplankton under sufficient nutrient conditions. Frontiers in microbiology., 9:543.

Gopalakrishnan, R., Sounthararajan, A. and Mohan, M. Tholkapiyan., 2020. The strength and durability of fly ash and quarry dust light weight foam concrete, Materials Today: Proceedings., 4(22): 1117-1124.

Gowda, G., Singh, G., Rajesh, K.M., Mridula, R.M. and Shankar, K.M., 2012. Development of biofilm on artificial substrates in a brackishwater pond near Mangalore, south-west coast of India. J. Aquac. Trop., 27: 13-23.

Halsey, K.H. and Jones, B.M., 2015. Phytoplankton strategies for photosynthetic energy allocation. Annu. Rev. Mar. Sci., 7(1): 265-297.

Hussian, A.E.M., Howayda, H., Toufeek, M.E.F. and Várbíró, G., 2016. Phytoplankton Structure Biochemical Stoichiometry and elemental composition in Lake Nasser Egypt. Int. J. Appl. Environ., 11(1): 211-228.

Irkin, L.C., 2019. Microalgae as a Valuable Additive in Aquaculture. Int. J. Trend Res. Dev., 6(6): 5-8.

Islam, M.R. and Hasan, M., 2016. Climate-induced human displacement: A case study of Cyclone Aila in the south-west coastal region of Bangladesh. Nat. Hazards., 81: 1051-1071.

Kaladharan, P., Veena, S. and Vivekanandan, E., 2009. Carbon sequestration by a few marine algae: observation and projection. J. Mar. Biol. Assoc., 51(1): 107-110.

Karl, D.M. 2004. Nitrogen cycles: past, present, and future. Biogeochemistry., 70(2): 153-226.

Karnan, R. Jyothibabu, N. Arunpandi, R.S. Pandiyarajan, S. Parthasarathi, S. and Santhikrishnan, 2021. Micro plankton size structure induced by a warm-core eddy in the western Bay of Bengal: Role of Trichodesmium abundance. Oceanology., 63(3): 283-300.

Karthik, R., Robin, R.S., Anandavelu, I., Purvaja, R., Singh, G., Mugilarasan, M., Jayalakshmi, T., Samuel, V.D. and Ramesh, R., 2020. Diatom bloom in the Amba River, west coast of India: a nutrient-enriched tropical river-fed estuary. Reg. Stud. Mar. Sci., 35(1): 101244.

Kolmakova, A. A., and KOLMAKOV, V. I., 2019. Amino acid composition of green microalgae and diatoms, cyanobacteria, and zooplankton (Review). Inl. Water Biol., 12: 452- 461.

Libes, S.M., 2009. Introduction to marine biogeochemistry, Academic Press, p. 765.

Limberger, R., Birtel, J., Peter, H., Catalán, N., Da Silva Farias, D., Best, R.j., Brodersen, J., Bürgmann, H. and Matthews, B., 2019. Predator-induced changes in dissolved organic carbon dynamics. Oikos, 128(3): 430-440.

Martiny, A.C., Pham, C.T., Primeau, F.W., Vrugt, J.A., Moore, J.K., Levin, S.A. and Lomas, M.W., 2013. Strong latitudinal patterns in the elemental ratios of marine plankton and organic matter. Nat. Geosci. 6(4): 279-283.

Mee, L., 2012. Between the Devil and the Deep Blue Sea: The coastal zone in an Era of globalisation. Estuar. Coast. Shelf Sci., 96:1-8.

Mishra, P., Begum, M., Gera, A., Kumar, B.C., Deviram, G., Pradhan, U.k., Vashi, A., Bandyopadhyay, D., Naik, S., Panda, U.S. and Murthy, M.R., 2022. Factors responsible for the sudden outburst of Noctiluca scintillans in the Chennai coastal waters, southeast coast of India–a case study. Oceanologia., 64(4): 781-788.

Mohan, G., Anjaneyan, P., Aravind, R. and Raffi, S.M., 2021. An investigation on the phytoplankton dynamics from the coastal waters of New Mangalore Port, Southwest coast of India. Int. J. Ecol. Environ. Sci., 47(1): 61-69.

Nanabawa, U.S., Shrivastava, A., Singh, P. and Mishra, S., 2020. Algae-Based Biofuel Production as a Part of an Industrial Cluster. In Algae and Sustainable Technologies CRC Press., 97-114.

Nellemann, C. Corcoran, E. and EDS, F., 2009. Blue carbon: the role of healthy oceans in binding carbon. A book on Blue carbon, 0122-0123.

Okeke, E.S., Olagbaju, O.A., Okoye, C.O., Addey, C.I., Chukwudozie, K.I., Okoro, J.O., Deme, G.G., Ewusi-Mensah, D., Igun, E., Ejeromedoghene, O. and ODII, E.C., 2022. Microplastic burden in Africa: a review of occurrence, impacts, and sustainability potential of bioplastics. Adv. Chem. Eng., 100402.

Paerl, H.W. and Justic, D., 2013. Estuarine phytoplankton. Estuarine ecology., pp. 85-110.

Parveen, S. and Patidar, S.K., 2022. Revisiting algal lipids and cellular stress-causing strategies for ameliorating the productivity of suitable lipids of microalgae for biofuel applications. Sustain. Energy. Fuels., 6(17): 3907-3925.

Purina, I., Labucis, A., Barda, I., Jurgensone, I. and Aigars, J., 2018. Primary productivity in the Gulf of Riga (Baltic Sea) in relation to phytoplankton species and nutrient variability. Oceanologia, 60(4): 544-552.

Rajkumar, M., Perumal, P., Prabu, V.A., Perumal, N.V. And Rajasekar, K.T. 2009. Phytoplankton diversity in Pichavaram mangrove waters from South-East coast of India. J. Env. Biol., 30: 489-498.

Roy, S., Sathyendranath, S. and Platt, T., 2017. Size-partitioned phytoplankton carbon and carbon-to-chlorophyll ratio from ocean colour by an absorption-based bio-optical algorithm. Remote Sens Environ., 194(1): 177-189.

Santhoshkumar, C. and Perumal, P., 2012. Studies on phytoplankton characteristics in Ayyampattinam coast. India. J. Env. Biol., 33: 585-589.

Sarangi, R.K., Jaiganesh, S.N.N., Patel, B.R., Prajapati, A. and Patel, H.D., 2022. Particulate organic carbon (POC) estimation and algorithm development in northeast Arabian Sea coastal water. Research Square., 3: 2-16.

Sharoni, S. and Halevy, I., 2020. Nutrient ratios in marine particulate organic matter are predicted by the population structure of well-adapted phytoplankton. Science advances., 6(29): 9371.

Sivakumar, R., Sachin, S., Priyadarshini, R. and Ghosh, S., 2022. Sustainable production of eicosapentaenoic acid-rich oil from microalgae: Towards an algal bio refinery. Int. J. Microbial., 132(6): 4170-4185.

Sridhar, R., Thangaradjou, T., Senthil Kumar, S. and Kannan, L., 2006. Water quality and phytoplankton characteristics in the Palk Bay, Southeast coast of India. J. Env. Biol., 27: 561-566.

Sushanth, V.R. and Rajashekhar, M., 2012. Seasonal variation in diatoms in response to physico-chemical characteristics of coastal waters of Uttara Kannada district, West Coast of India. Int. J. Env. Sce., 2(3): 1543-1552.

Swati, S., Panda Dhal, N.K. and Panda, C.R., 2012. Phytoplankton diversity in response to abiotic factors along Orissa coast, Bay of Bengal. An Int. Quartery J. Life Sci.., 2(3): 1818-1832.

Toseland, A.D.S.J., Daines, S.J., Clark, J.R., Kirkham, A., Strauss, J., Uhlig, C., Lenton, T.M., Valentin, K., Pearson, G.A., Moulton, V. and Mock, T., 2013. The impact of temperature on marine phytoplankton resource allocation and metabolism. Nature Climate Change, 3(11): 979-984.

Vasanthi, L.A., Muruganandam, A., Revathi, P., Baskar, B., Jayapriyan, K., Baburajendran, R. and Munuswamy, N., 2014. The application of histo-cytopathological biomarkers in the mud crab Scylla serrata (Forskal) to assess heavy metal toxicity in Pulicat Lake, Chennai. Mar. Pollut. Bull., 81(1): 85–93.

Virginia, E. and Armbrust, E.V. 2009. The life of diatoms in the world's oceans. Nature, 459(7244): 185-192.

Zhang, Y., Zhao, M., CUI, Q., Fan, W., Qi, J., Chen, Y., Zhang, Y., Gao, K., Fan, J., Wang, G. and Yan, C. 2017. Processes of coastal ecosystem carbon sequestration and approaches for increasing carbon sink. China Earth Sci. 60(5): 809-820.

13

Bioremediation: A Green Symphony for Environmental Healing

Kusumlata Goswami and Shubham Disodia

Department of Aquatic Environment Management, College of Fisheries, CCS Haryana Agriculture University, Hisar 125004

Abstract

Environmental pollution, resulting from both human activities and natural factors, is escalating as a consequence of population growth, industrial expansion, and urban development. Bioremediation is a method used to clean up polluted soil and water by using natural helpers like microorganisms, fungi, and plants. Bioremediation is a biological technology which offers cost-effective and low-pollution methods that can be applied directly at contaminated sites for addressing specific types of contamination. This technique uses microbes to break down pollutants and plants to absorb or stabilize harmful substances, especially heavy metals. As pollution becomes a bigger issue, bioremediation offers a promising way to tackle it and help restore damaged ecosystems. Bioremediation can be a more cost-effective solution compared to other technologies for cleaning contaminants, while also minimizing potential risks to human health and the environment. This chapter provides an overview of the current status of bioremediation and highlights its potential areas of application.

Keywords: *Bioremediation, Contamination, Ecosystem, Pollutant*

Introduction

Bioremediation is a natural process that leverages the capabilities of bacteria, fungi, and plants to remove, reduce, degrade, or immobilize environmental pollutants from soil and water. This process plays a crucial role in restoring contaminated sites to a relatively clean, nontoxic state (Kamaludeen *et al.*, 2003; Khalid *et al.*, 2017). As we navigate through an era of heightened environmental awareness, the significance of bioremediation becomes increasingly apparent.

Metabolic processes of various microorganisms enable them to utilize chemical contaminants as an energy source, effectively rendering these pollutants harmless or transforming them into less toxic byproducts (Kamaludeen *et al.*, 2003). This remarkable capability positions bioremediation as a cost-effective and environment friendly solution, gradually gaining traction in environmental cleanup efforts (Ashraf *et al.*, 2019).

Given the urgent need to enhance our understanding of sustainable techniques for remediating contaminated environments, it is vital to explore how biological organisms and plants can absorb and/or convert inorganic contaminants. In this chapter, we will elucidate the concept of bioremediation and detail the various mechanisms underlying different bioremediation strategies applicable to sites contaminated with inorganic pollutants.

Mechanisms of Bioremediation

Bioremediation strategies can be broadly categorized into several types including:

1. **Microbial Bioremediation**: This involves the use of bacteria and fungi to degrade pollutants. Certain strains of microorganisms have evolved to metabolize harmful substances, breaking them down into non-toxic compounds.
2. **Phytoremediation**: This strategy employs plants to absorb, accumulate, and detoxify pollutants from the soil and water. Specific plant species are known for their ability to uptake heavy metals and other inorganic contaminants, making them valuable in restoration efforts.
3. **Bioaugmentation and Biostimulation**: In these approaches, either additional microorganisms are introduced to enhance the existing microbial community (bioaugmentation) or nutrients are added to stimulate the growth of native bacteria (biostimulation), thereby accelerating the degradation process.

Concept of Bioremediation

Bioremediation is defined by the U.S. Environmental Protection Agency (EPA) as “the use of living organisms remove pollutants from soil and water”. This uses organisms such as nonharmful insects to manage agricultural pests or address diseases affecting trees, plants, and garden soil (Cristaldi *et al.*, 2017). The process can involve green plants, which absorb pollutants from the soil or water through their roots, subsequently accumulating these substances in their leaves. Additionally, microorganisms play a crucial role in detoxifying or removing inorganic pollutants from the environment (Khalid *et al.*, 2017). One of the significant advantages of bioremediation is its capacity for permanent in situ remediation, addressing contamination directly at the site rather than simply relocating the problem. Bioremediation is particularly effective for the remediation of heavy metals, metalloids, and other inorganic pollutants in soil and water (Ali *et al.*, 2013; Ashraf *et al.*, 2019). This method has proven to be cost-effective, efficient, novel, and eco-friendly, benefiting from good public acceptance compared to traditional engineering techniques such as excavation, soil incineration, soil washing etc (Ali *et al.*, 2013; Sarwar *et al.*, 2017). The efficiency of bioremediation in removing inorganic pollutants depends on various factors, including the types of plants and microbes involved, as well as the physicochemical properties of the soil and water. The two main categories of bioremediation are in situ and ex situ, both of which involve different methods of application and are suitable for addressing various types of contaminants. In situ bioremediation involves the treatment of contaminated soil, water, or air directly at the site of contamination. This approach does not require the removal of the polluted material, which makes it more cost-effective and environmentally friendly. Ex situ bioremediation involves the removal of contaminated material from its original site to be treated elsewhere. This method allows for more controlled conditions and, in some cases, faster remediation results. Contaminated soil or water is excavated, treated in specialized facilities, and then either returned to the site or disposed of safely after the contaminants have been reduced or eliminated.

Steps of Bioremediation Process

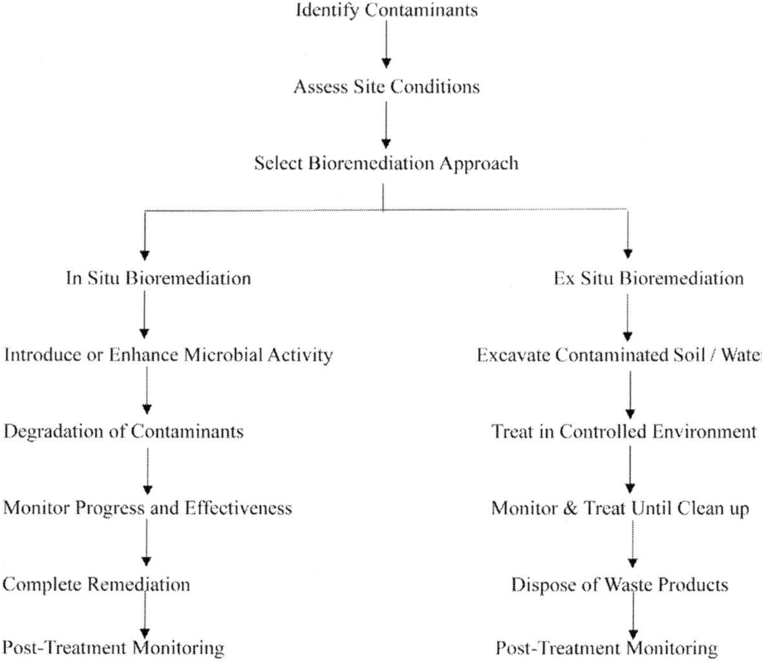

Types of Bioremediations

The term bioremediation encompasses various strategies, primarily focusing on plant-based and microbe-based remediation techniques. These approaches—phytoremediation and microorganism remediation—differ significantly in the processes and mechanisms by which they immobilize, remove, or degrade pollutants (Khalid *et al.*, 2017).

Phytoremediation

Phytoextraction

Phytoextraction is a process that involves the removal of pollutants from soil or water through plant roots. This method enables the translocation and accumulation of these pollutants in the aboveground biomass, specifically in the shoots, which can then be harvested (Ghosh and Singh, 2005; Muthusaravanan *et al.*, 2018). The effective translocation of pollutants is a critical biochemical process, as harvesting root biomass is not practical (Halim *et al.*, 2003; McIntyre, 2003).

Continuous phytoextraction can be achieved by utilizing plants that accumulate high levels of pollutants throughout their life cycle (Sarwar *et al.*, 2017). The phytoextraction process typically involves five major steps:

1. **Pollutant Mobilization in the Rhizosphere**: The process begins with the release of root exudates that mobilize contaminants in the soil.
2. **Pollutant Uptake by Plant Roots**: Pollutants are taken up through the plant's root system.
3. **Translocation Toward Aerial Plant Parts**: The contaminants are transported from the roots to the shoots.
4. **Pollutant Sequestration in Plant Tissues**: The pollutants are stored in the plant's aboveground parts, where they can be harvested.

For optimal phytoextraction, a hyperaccumulator plant should possess several desirable characteristics, including:

- High growth rate
- Substantial aboveground biomass
- Tolerance to high concentrations of inorganic pollutants
- A highly branched root system
- Adaptation to local environmental conditions
- Effective and efficient translocation of pollutants from roots to shoots
- Resistance to pathogens and pests
- Ease of cultivation and harvest

- High accumulation of target pollutants (Ali *et al.*, 2013; Mahar *et al.*, 2016; Khalid *et al.*, 2017; Sarwar *et al.*, 2017).

Pollutant tolerance is particularly crucial, as it allows plant tissues to withstand high concentrations of contaminants with minimal adverse effects on overall plant health. Mechanisms that contribute to this tolerance include cell wall metal binding, active transport of metal ions into vacuoles, chelation of metal ions with proteins and peptides, and complex formation (Memon and Schröder, 2009).

The effectiveness of phytoextraction is influenced by several factors, including the depth of root growth, seasonal weather variations, and climatic conditions (Bhargava *et al.*, 2012). The efficiency of this method can also be enhanced through the application of mobilizing agents such as citric acid, ethylene diamine tetra

acetic acid (EDTA), nitrilotriacetic acid, amino polycarboxylic acids, and ethylene diamine disuccinic acid (Mahar *et al.*, 2016). Phytoextraction is primarily used for the removal of inorganic pollutants from contaminated media, as it offers numerous advantages: it is highly economical, minimally disruptive to the soil and environment, does not require disposal sites, enjoys high public acceptance, and eliminates the need for excavation or transport of contaminated media (Sheoran *et al.*, 2016; Mahar *et al.*, 2016). However, it is not without its challenges. The process relies heavily on specific growing conditions essential for plants, their tolerance to pollutants, and may require a relatively long period to achieve complete site remediation (Khalid *et al.*, 2017). Therefore, the application of phytoextraction should be tailored to the specific context of each case study.

Phytofiltration

Phytofiltration, also known as rhizofiltration, is a bioremediation technique that involves the adsorption or precipitation of pollutants from solutions onto plant roots or their absorption into the roots within the root zone (Khan *et al.*, 2019). This process is facilitated by the synthesis of specific chemicals within the roots, which enhance the adsorption of pollutants. Many plants produce phytochelatins, compounds that increase the binding capacity for contaminants such as metal ions (Singh and Santal, 2015). Phytofiltration can be effectively applied to effluents, contaminated waterways, and groundwater systems. For successful rhizofiltration, it is crucial to understand the speciation of the contaminants and the interactions between all contaminants and nutrients present in the environment.

Ideal Characteristics for Rhizofiltration

An ideal plant for rhizofiltration should possess the following characteristics:

- **Rapidly Growing Roots**: The ability to develop extensive root systems that can effectively absorb contaminants from solutions over extended periods (Dhanam, 2017).
- **Extensive Root Architecture**: Terrestrial plants are often preferred for rhizofiltration due to their fibrous roots, which enable them to draw out contaminants from groundwater and the rhizospheric zone (Pilon-Smits, 2005; Ali *et al.*, 2013; Khan *et al.*, 2019).

By leveraging these characteristics, phytofiltration serves as an effective strategy for remediating contaminated water sources, ultimately contributing to environmental restoration and sustainability.

Phytostabilization

Phytostabilization, also known as phytoimmobilization, is the process of using plants to decrease the mobility and bioavailability of pollutants. This technique aims to prevent contaminants from leaching into groundwater or entering the food chain through various mechanisms, such as adsorption by roots or the formation of insoluble compounds in the root zone (Sarwar *et al.*, 2017; Khan *et al.*, 2019).

Key Characteristics

Phytostabilization can be characterized by two main processes:

1. **Restriction of Pollutant Mobility**: This involves the assimilation and aggregation of pollutants by roots, their adsorption onto root surfaces, or precipitation within the plant's root zone.
2. **Preventing Contaminant Movement**: The deployment of plants helps mitigate the movement of contaminants through wind and water erosion, as well as the drainage and dispersion of contaminated soil (USEPA, 1999).

The ultimate goal of phytostabilization is to stabilize pollutants rather than remove them. By doing so, it diminishes the hazard these contaminants pose to human health and the environment, allowing plants to function similarly to soil amendments.

Limitations and Advantages

While phytostabilization is effective, it is important to note that it is not a permanent solution to contamination. This approach does not reduce the concentration of pollutants in the environment; rather, it helps contain contaminants within the contaminated media, thus reducing their spread (Bolan *et al.*, 2011; Khalid *et al.*, 2017).

One significant advantage of phytostabilization over other phytoremediation techniques is that it reduces the need to treat the aerial parts of plants. Since contaminants are primarily retained in the roots with minimal translocation to the shoots, this approach can be more manageable and efficient. Phytostabilization is particularly effective when rapid immobilization is required to protect groundwater and surface water resources (Jadia and Fulekar, 2009; Gomes *et al.*, 2016). It has shown potential applicability for stabilizing various heavy metals, particularly lead, chromium, and mercury, in the soil, thereby reducing their interaction with surrounding biota (Cunningham and Ow, 1996; Mahar *et al.*, 2016).

Phytovolatilization

Phytovolatilization is a phytoremediation strategy that involves the plant-mediated uptake of contaminants, their transformation into volatile compounds, and the subsequent release of these compounds into the atmosphere. This process occurs either in their original form or as altered compounds due to metabolic activity and transpiration pull (Kumar *et al.*, 2017; Khan *et al.*, 2019).

Transpiration is a key component of this process, involving the evaporation of water vapor from leaf surfaces through stomata into the atmosphere. Certain plant species with extensive root systems possess the ability to uptake and degrade contaminants, aided by specific enzymes or genetic traits (Newman and Reynolds, 2004; Pilon-Smits, 2005; Muthusaravanan *et al.*, 2018). During phytovolatilization, pollutants are absorbed from soil or water, converted into less toxic vapours, and then released into the atmosphere through the plant's transpiration process (Khalid *et al.*, 2017). This technique can be applied to a range of organic pollutants and certain heavy metals, such as arsenic (As), selenium (Se), and mercury (Hg), which exist as gaseous species in the environment (Pajević *et al.*, 2016). However, naturally occurring plants capable of converting metals into volatile forms are quite rare. Consequently, phytovolatilization often relies on genetically modified plants to enhance the ability to volatilize these metals (Khalid *et al.*, 2017).

Limitations and Controversies

Despite its potential, phytovolatilization has significant limitations. The process does not completely eliminate pollutants from the environment; instead, it transfers contaminants from one environmental compartment (soil or water) to another (the atmosphere). This raises concerns, as the contaminants can eventually precipitate with rainfall and re-enter the ecosystem (Ali *et al.*, 2013). Additionally, phytovolatilization is characterized by minimal soil erosion and requires no disposal of contaminated plant biomass, resulting in negligible site disturbance (Rugh *et al.*, 2000; Khalid *et al.*, 2017). However, these attributes contribute to the controversy surrounding phytovolatilization as a remediation technology, as it raises questions about the overall effectiveness and environmental impact of the process (Padmavathiamma and Li, 2007; Gomes *et al.*, 2016).

Phytodegradation

Phytodegradation, also known as phytotransformation, involves the uptake of contaminants and nutrients from water, sediment, or soil, followed by the chemical modification of these contaminants through plant metabolism. This process often results in the inactivation, degradation, or immobilization of pollutants within both the roots and shoots of the plant (Bulak *et al.*, 2014; Gomes *et al.*, 2016). Certain plant species have the ability to degrade absorbed contaminants into less toxic compounds, utilizing their metabolic processes or specific enzymes (Muthusaravanan *et al.*, 2018). As such, phytodegradation serves as a vital metabolic strategy for plants in the detoxification and degradation of environmental pollutants within their tissues (Pajević *et al.*, 2016; Mahar *et al.*, 2016). This approach not only helps in reducing the concentration of harmful substances in the environment but also contributes to the overall health of the ecosystem by transforming pollutants into less harmful forms.

Microorganism Remediation

The principal mechanism of microbial remediation involves immobilizing and reducing the bioavailability of pollutants. While inorganic pollutants such as heavy metals cannot be degraded by microorganisms, they can be transformed into less harmful forms due to changes in their physical and chemical properties (Ashraf *et al.*, 2019).

Mechanisms of Microbial Remediation

In environmental bioremediation, microorganisms can be supported on solid agrowaste, which provides the necessary macro and micronutrients for biofilm formation. This support enhances the metabolic activities of microorganisms, facilitating the solubilization and biodegradation of hydrocarbon contaminants (Mahdi and Aziz, 2017). The interactions between microbes and metal ions occur through various mechanisms, including:

- **Active and Passive Uptake**: Microorganisms can actively transport metal ions or passively absorb them (Gupta and Diwan, 2017).

- **Extracellular Complexation**: Metal ions may bind to extracellular components produced by the microbes.
- **Intracellular Accumulation**: Some metals can be taken up and stored within microbial cells.
- **Oxidation-Reduction Reactions**: Microbial metabolism can facilitate redox reactions that transform pollutants.
- **Precipitation**: Pollutants can precipitate as less soluble forms through microbial activity (Yang *et al.*, 2018; Ashraf *et al.*, 2019).

Among these mechanisms, **bioaccumulation** and **biosorption** are particularly significant.

- **Bioaccumulation** is a two-stage, slow process that involves active metabolic transport within living biomass. It begins with rapid sorption (like biosorption) followed by a slower physiological transport of the sorbate into the cell (Martín-González *et al.*, 2006; Chojnacka, 2010). Effective bioaccumulation requires mechanisms for intracellular binding, such as phytochelatins and metallothioneins, which are proteins synthesized in response to toxic metal ions.
- **Biosorption**, in contrast, occurs passively and is metabolism independent. Both live and dead biomass can bind and concentrate contaminants through various physio-chemical mechanisms (Philippis *et al.*, 2007; Gupta and Diwan, 2017). Pollutants can adhere to polysaccharide slime layers on microbial surfaces via functional groups like carboxyl, amino, phosphate, and sulphate. Additionally, extracellular polymeric substances (EPS)—comprising nucleic acids, proteins, lipids, and complex carbohydrates—play a crucial role in pollutant adsorption.

Microbe-Assisted Phytoremediation

The efficiency of microbial remediation depends on the types of pollutants and microorganisms involved. Microbial activities in the rhizosphere can significantly enhance the effectiveness of phytoremediation processes. Microbe-assisted phytoremediation involves the coupling of contaminant-tolerant plant species with rhizospheric or endophytic microorganisms to improve the remediation of inorganic pollutants (Fester *et al.*, 2014; Gerhardt *et al.*, 2017).

This synergistic approach can enhance remediation through two complementary mechanisms:

1. **Direct Promotion**: Microbes associated with plants can enhance pollutant translocation (facilitating phytoextraction) or reduce the mobility and availability of contaminants in the rhizosphere (phytostabilization).

2. **Indirect Promotion**: Microbes can confer pollutant tolerance to plants and enhance plant biomass production, thus improving the overall removal of pollutants (Gerhardt *et al.*, 2017).

Key features of beneficial rhizospheric microorganisms include:

- Cooperative communication with plant roots to enhance metal uptake.
- Release of compounds that increase pollutant bioavailability.
- Promotion of nutrient and non-essential metal uptake by roots.
- Alteration of pollutant solubility through changes in chemical properties (Doty, 2008; Khalid *et al.*, 2017).

Application and Field Testing

Microorganisms that show promise in laboratory or greenhouse settings must be tested in field conditions, where environmental variables may lead to different outcomes (Burges *et al.*, 2016). Proper application of inoculants is crucial for the success of microbe-assisted phytoremediation. Inoculants can be introduced into contaminated soils through various methods, including seed treatment, foliar sprays, and direct soil inoculation (Bashan *et al.*, 2014; Prasad *et al.*, 2018).

Challenges of Bioremediation

While bioremediation presents a promising solution for the remediation of inorganic pollutant-contaminated media, several challenges must be addressed (Chojnacka, 2010; Mukhopadhyay and Maiti, 2010; Ramamurthy and Memarian, 2012; Ali *et al.*, 2013; Gomes *et al.*, 2016; Mahar *et al.*, 2016; Gerhardt *et al.*, 2017; Ashraf *et al.*, 2019):

1. **Time-Intensive Process**: Bioremediation often requires a prolonged period for effective cleanup, making it less suitable for urgent remediation needs.
2. **Limitations of Hyperaccumulator Plants**: The efficacy of phytoextraction is frequently hindered by the low biomass and slow growth rates of many hyperaccumulator species, which limits their capacity to extract significant amounts of pollutants.
3. **Environmental Conditions**: The success of phytoremediation is heavily influenced by environmental factors, which may not always be favourable for the species employed. This variability can reduce the overall effectiveness of the process.
4. **Multiple Pollutant Contamination**: Sites contaminated with multiple pollutants may necessitate the use of specific plant or microbial species

that are well-adapted or tolerant to the diverse environmental conditions and contaminants present.

5. **Disposal of Polluted Biomass**: Following phytoextraction, proper disposal of the polluted biomass is essential, as it may be classified as hazardous waste, complicating the remediation process.
6. **Limited Bioavailability of Contaminants**: The mobilization of tightly bound fractions of metal ions in soil can be challenging, as their bioavailability is often restricted, making it difficult for plants and microbes to access these contaminants.
7. **Invasive Species Risks**: The introduction of invasive plant species as hyperaccumulators poses a risk to indigenous floral diversity and ecosystem stability, necessitating careful consideration of species selection.
8. **Pollution Level Constraints**: Bioremediation is typically more effective in sites with low to moderate pollution levels, as high levels of contamination can inhibit the growth and effectiveness of the plants and microbes involved.
9. **Stability of Biological Systems**: The exogenous application of microorganisms can disrupt the stability of existing biological treatment systems, potentially leading to unintended ecological consequences.

Addressing these challenges requires careful planning, site assessment, and selection of appropriate bioremediation strategies to ensure effective and sustainable outcomes.

Conclusion and Future Prospects

Bioremediation is gaining traction as a viable method for addressing inorganic pollutant contamination, owing to its numerous advantages, including cost-effectiveness, environmental friendliness, and public acceptance. This chapter has explored various bioremediation strategies and their corresponding mechanisms, paving the way for innovative approaches to pollutant removal from contaminated sites. Despite its promise, bioremediation remains a relatively nascent technology, predominantly in the research phase. Its interdisciplinary nature necessitates expertise in diverse fields such as soil and water chemistry, plant biology, ecology, soil and water microbiology, and environmental engineering. Continued research is essential to deepen our understanding of plant physiology, biochemistry, and contaminant uptake, as well as to assess the potential synergistic effects of multiple contaminants. Looking forward, there is a critical need to identify or engineer suitable plant and microbial systems tailored for effective environmental clean-up. As research progresses, the focus should be on scaling up successful laboratory findings to field applications, ensuring that bioremediation can be implemented effectively in diverse and challenging environments. This holistic approach will be pivotal in harnessing the full potential of bioremediation as a sustainable solution for mitigating inorganic pollution and restoring contaminated ecosystems.

References

Ali, H., Khan, E., Sajad, M.A., 2013. Phytoremediation of heavy metals-concepts and applications. Chemosphere 91, 869-881.

Ashraf, S., Ali, Q., Zahir, Z.A., Ashraf, S., Asghar, H.N., 2019. Phytoremediation: environmentally sustainable way for reclamation of heavy metal polluted soils. Ecotoxicol. Environ. Saf. 174, 714-727.

Bashan, Y., De-Bashan, L.E., Prabhu, S.R., Hernandez, J.P., 2014. Advances in plant growth-promoting bacterial inoculant technology: formula- tions and practical perspectives (1998-2013). Plant Soil 378 (1-2), 1-33.

Bhargava, A., Carmona, F.F., Bhargava, M., Srivastava, S., 2012. Approaches for enhanced phytoextraction of heavy metals. J. Environ. Manag. 105, 103-120.

Bolan, N.S., Park, J.H., Robinson, B., Naidu, R., Huh, K.Y., 2011. Phytostabilization. A green approach to contaminant containment. Adv. Agron. 112, 145-204.

Bulak, P., Walkiewicz, A., Brzeziska, M., 2014. Plant growth regulators-assisted phytoextraction. Biol. Plant. 58, 58), 1-8.

Burges, A., Epelde, L., Benito, G., Artetxe, U., Becerril, J.M., Garbisu, C., 2016. Enhancement of ecosystem services during endophyte-assisted aided phytostabilization of metal contaminated mine soil. Sci. Total Environ. 562, 480-492.

Chojnacka, K., 2010. Biosorption and bioaccumulation-the prospects for practical applications. Environ. Int. 36 (3), 299-307.

Cristaldi, A., Conti, G.O., Jho, E.H., Zuccarello, P., Grasso, A., Copat, C., Ferrante, M., 2017. Phytoremediation of contaminated soils by heavy metals and PAHs. A brief review. Environ. Technol. Innov. 8, 309-326.

Cunningham, S.D., Ow, D.W., 1996. Promises and prospects of phytoremediation. Plant Physiol. 110, 715-719.

Dhanam, S., 2017. Strategies of bioremediation of heavy metal pollutants toward sustainable agriculture. In: Dhanarajan, A. (Ed.), Sustainable Agriculture Towards Food Security. Springer Nature, Singapore, pp. 349-358.

Doty, S.L., 2008. Enhancing phytoremediation through the use of transgenics and endophytes. New Phytol. 179 (2), 318-333.

Fester, T., Giebler, J., Wick, L.Y., Schlosser, D., Kästner, M., 2014. Plant-microbe interactions as drivers of ecosystem functions relevant for the bio- degradation of organic contaminants. Curr. Opin. Biotechnol. 27C (6), 168-175.

Gerhardt, K.E., Gerwing, P.D., Greenberg, B.M., 2017. Opinion: taking phytoremediation from proven technology to accepted practice. Plant Sci. 256, 170-185.

Ghosh, M., Singh, S.P., 2005. A review on phytoremediation of heavy metals and utilization of its byproducts. Appl. Ecol. Environ. Res. 3 (1), 1-18.

Gomes, M.A.C., Hauser-Davis, R.A., De Souza, A.N., Vitória, A.P., 2016. Metal phytoremediation: general strategies, genetically modified plants and applications in metal nanoparticle contamination. Ecotoxicol. Environ. Saf, 134, 133-147.

Gupta, P., Diwan, B., 2017. Bacterial exopolysaccharide mediated heavy metal removal: a review on biosynthesis, mechanism and remediation strategies. Biotechnol. Rep. 13, 58-71.

Halim, M., Conte, P., Piccolo, A., 2003. Potential availability of heavy metals to phytoextraction from contaminated soils induced by exogenous humic substances. Chemosphere 52, 265-275.

Jadia, C.D., Fulekar, M.H., 2009. Phytoremediation of heavy metals: recent techniques. Afr. J. Biotechnol. 8 (8), 921-928.

Kamaludeen, S.P.B., Megharaj, M., Juhasz, A.L., Sethunathan, N., Naidu, R., 2003. Chromium-microorganism interactions in soils: remediation implications. Rev. Environ. Contam. Toxicol. 178, 93-164.

Khalid, S., Shahid, M., Niazi, N.K., Murtaza, B., Bibi, I., Dumat, C., 2017. A comparison of technologies for remediation of heavy metal contami- nated soils. J. Geochem. Explor. 182, 247-268.

Khan, I., Iqbal, M., Shafiq, F., 2019. Phytomanagement of lead-contaminated soils: critical review of new trends and future prospects. Int. J. Environ. Sci. Technol. https://doi.org/10.1007/s13762-019-02431-2.

Kumar, S.S., Kadier, A., Malyan, S.K., Ahmad, A., Bishnoi, N.R., 2017. Phytoremediation and rhizoremediation: uptake, mobilization and se- questration of heavy metals by plants. In: Singh, D., Singh, H., Prabha, R. (Eds.), Plant-Microbe Interactions in Agro-Ecological Perspectives. Springer, Singapore, pp. 367-394.

Mahar, A., Wang, P., Ali, A., Awasthi, M.K., Lahori, A.H., Wang, Q., Li, R., Zhang, Z., 2016. Challenges and opportunities in the phytoremediation of heavy metals contaminated soils: a review, Ecotoxicol. Environ. Saf. 126, 111-121.

Mahdi, A.M.E., Aziz, H., 2017. Hydrocarbon biodegradation using agro-industrial wastes as co-substrates. In: Bhakta, J.N. (Ed.), Handbook of Research on Inventive Bioremediation Techniques. IGI Global Book Series Advances in Environmental Engineering and Green Technologies, IGI Global, USA, pp. 1635-1665.

Martín-González, A., Díaz, S., Borniquel, S., Gallego, A., Gutiérrez, J.C., 2006. Cytotoxicity and bioaccumulation of heavy metals by ciliated pro- tozoa isolated from urban wastewater treatment plants. Res. Microbiol. 157 (2), 108-118.

McIntyre, T., 2003. Phytoremediation of heavy metals from soils. Adv. Biochem. Eng. Biotechnol. 78, 97-123.

Memon, A.R., Schröder, P., 2009. Implications of metal accumulation mechanisms to phytoremediation. Environ. Sci. Pollut. Res. 16, 162-175.

Mukhopadhyay, S., Maiti, S.K., 2010. Phytoremediation of metal enriched mine waste: a review. Glob. J. Environ. Res. 4 (3), 135-150.

Muthusaravanan, S., Sivarajasekar, N., Vivek, J.S., Paramasivan, T., Naushad, M.U., Prakashmaran, J., *et al.*, 2018. Phytoremediation of heavy metals: mechanisms, methods and enhancements. Environ. Chem. Lett. 16, 1339-1359.

Newman, L.A., Reynolds, C.M., 2004. Phytodegradation of organic compounds. Curr. Opin. Biotechnol. 15, 225-230.

Padmavathiamma, P.K., Li, L.Y., 2007. Phytoremediation technology: hyper-accumulation metals in plants. Water Air Soil Pollut. 184 (1-4), 105-126.

Pajević, S., Borišev, M., Nikolić, N., Arsenov, D.D., Orlović, S., Župunski, M., 2016. Phytoextraction of heavy metals by fast-growing trees: a review. In: Ansari, A., Gill, S., Gill, R., Lanza, G., Newman, L. (Eds.), Phytoremediation. Springer, Cham, pp. 29-64.

Philippis, R.D., Paperi, R., Sili, C., 2007. Heavy metal sorption by released polysaccharides and whole cultures of two exopolysaccharide-producing cyanobacteria. Biodegradation 18 (2), 181-187.

Pilon-Smits, E., 2005. Phytoremediation. Annu. Rev. Plant Biol. 56, 15-39. Prasad, M.N.V., Freitas, H.M., 2003. Metal hyperaccumulation in plants-biodiversity prospecting for phytoremediation technology. Electron. J. Biotechnol. 6, 285-321.

Prasad, M.N.V., Favas, P.J., Maiti, S.K., 2018. Biogeotechnologies for Mine Site Rehabilitation. Elsevier, USA.

Ramamurthy, A.S., Memarian, R., 2012. Phytoremediation of mixed soil contaminants. Water Air Soil Pollut. 223 (2), 511-518.

Rugh, C.L., Bizily, S.P., Meagher, R.B., 2000. Phytoremediation of environmental mercury pollution. In: Raskin, I., Ensley, B.D. (Eds.), Phytoremediation of Toxic Metals Using Plants to Clean Up the Environment. Wiley, New York, pp. 193-229.

Sarwar, N., Imran, M., Shaheen, M.R., Ishaque, W., Kamran, M.A., Matloob, A., Rehim, A., Hussain, S., 2017. Phytoremediation strategies for soils contaminated with heavy metals: modifications and future perspectives. Chemosphere 171, 710-721.

Sheoran, V., Sheoran, A.S., Poonia, P., 2016. Factors affecting phytoextraction: a review. Pedosphere 26 (2), 148-166.

Singh, N.P., Santal, A.R., 2015. Phytoremediation of heavy metals: the use of green approaches to clean the environment. In: Ansari, A., Gill, S., Gill, R., Lanza, G., Newman, L. (Eds.), Phytoremediation. Springer, Cham, pp. 115-129.

USEPA, 1999. Report on Bioavailability of Chemical Wastes With Respect to the Potential for Soil Remediation. T28006: QT-DC-99-003260.

Yang, Z., Shi, W., Yang, W., Liang, L., Yao, W., Chai, L., Gao, S., Liao, Q., 2018. Combination of bioleaching by gross bacterial biosurfactants and flocculation: a potential remediation for the heavy metal contaminated soils. Chemosphere 206, 83-91.

14

Sponge City: A Solution to Urban Flooding in Climate Change Prone Areas

***Kurapati Nagendrasai*[1]*, Kishore Kumar Krishnani*[2] *and Guntapalli Sravani*[3]**

[1]*Aquatic Environmental Management, AEHM Division, ICAR- Central Institute of Fisheries Education, Mumbai, Maharashtra, 400061, India*

[2]*ICAR-Indian Institute of Agricultural Biotechnology, Ranchi-834003, Jharkhand, India*

[3]*Aquaculture Division, ICAR-Central Institute of Fisheries Education, Mumbai Maharashtra-400061, India*

Abstract

Climate change and the rapid progress of urbanization lead to abiotic stresses of frequent flooding events and pollution seriously threatening the sustainable development of modern cities. The construction of Sponge City can effectively alleviate these problems and improve a city's resilience to environmental changes and coping with floods making them more sustainable. China was the first country to build Sponge Cities for urban flood control, rainwater harvest, ecological restoration of urban land, and water quality improvement. Many Indian cities also face the risk of flooding during monsoon and this risk is going to increase year after year in the whole nation. India is getting bruised both ways- lack of water & excess flood. So, India too needs a mission that mitigates flood risk and provides a pathway to water security. The most promising solution across the world at this time is the idea to adopt Sponge City construction. Incorporating the Sponge City idea under the area of smart city project would save countless lives and lead to the implementation of sustainable cities. The expense of making this a reality may cost some money, but these projects will provide enormous benefits. Sponge City design thus would act as a defence mechanism during rainstorms, supporting greater absorption and water retention. Thus, it leads as a pathway for sustainable Indian cities.

Keywords: *Flood control, sponge city, resilience, urban flooding, climate change.*

Introduction

Floods are the most common and costly natural catastrophic event occurring around the world, wreaking havoc on both lives and the economy. Urbanization, high-intensity human activities and fast population expansion have all contributed to significant global environmental concerns in recent years. Among these issues was the impact of urban stormwater runoff on the urban environment. (Dingkun *et al.,* 2020) Extreme urban flooding disasters are caused by the combined effect of rainfall, the degree of urbanization, and the poor capacity of drainage systems. As a result, improved urban rainfall management is required in the context of global climate change and urbanization development.

The sponge city concept, which was first developed in China, is an urban resilience-building measure. This majorly focuses on water management in cities to mitigate the effects of natural disasters and other changes. The objective of the initiative is that the inflows from rainfall events are absorbed, infiltrated, stored, and cleaned by way of technology and natural methods. 44% of global disasters are linked to flooding, extreme weather patterns are likely to increase with climate change, and nature-based solutions to climate change can be 50 per cent more cost-effective. As climate change leads to more extreme weather events like heavy rainfall and droughts, it has become crucial to integrate floodwater management systems in cities. The primary objective of sponge cities is not only to prevent urban flooding but also to improve the ecology and biodiversity and address water resource shortages. Sponge cities are urban areas designed with a permeable construction model that allows for the absorption and storage of rainwater.

Transforming Cities into Sponge Cities

To transform cities into sponge cities, additional green spaces like parks, trees, and natural drainage systems need to be integrated. Increasing the presence of parks, greenery, and drainage systems enhances a city's "absorbency" and resilience to floods. Ground surfaces that mimic sponges can absorb rainwater, which is then naturally filtered by the soil and can recharge urban aquifers. The collected water can be extracted through urban or peri-urban wells, treated, and reused for the city's water supply. Digital mapping tools can assist cities in identifying suitable spaces for rainwater harvesting, ponds, and inner-city gardens, as well as understanding the associated risks of not implementing these measures.

The Sponge City performs the same function as a sponge, effectively adapting to environmental changes and handling natural calamities. (Saho *et al.*, 2020; Liang *et al.*, 2020) Water can be absorbed in case of heavy rainfall, and the stored water can be released when needed in times of drought. As a result, Sponge City projects can contribute to managing/mitigating floods while also effectively collecting rainwater for future use, meeting the objectives of green infrastructure and sustainable development. (Zhang and Zhang 2019)

Core Notions Involved in Developing Sponge City

Sponge City has four basic concepts. The ***first*** is to improve the city's surface ability to absorb and retain precipitation to supply water and mitigate stormwater runoff. The ***second*** concept is about managing water ecology through water self-purification systems and providing ecological waterfront design. The ***third*** focuses on using green infrastructure to purify/filter, restore/repair, modify and reuse stormwater. (Nguyen *et al.,* 2019; Nguyen *et al.,* 2020) ***Fourthly***, permeable pavements are being used in urban road construction.

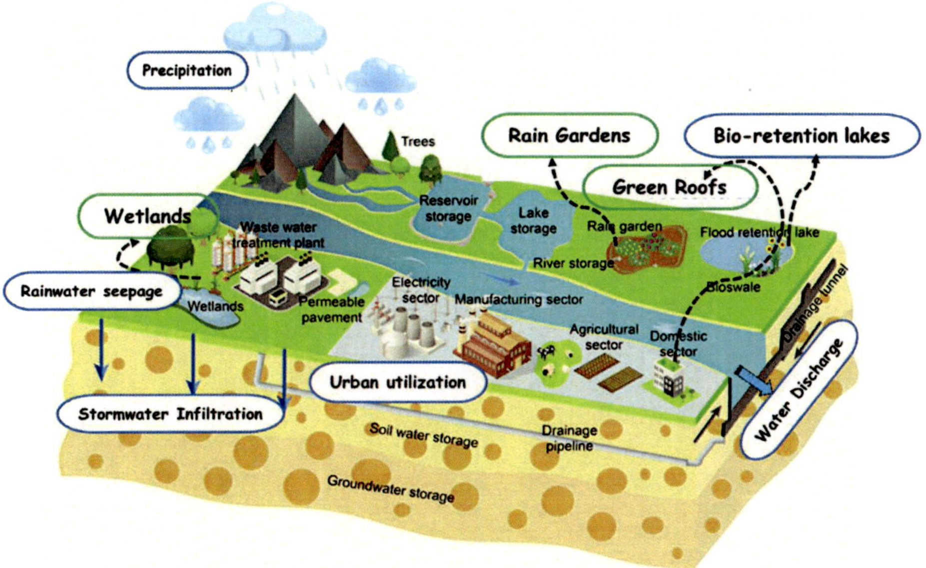

Fig 1: Schematic representation of a Sponge City

Resourcing rainwater: Rainwater harvesting is the gathering and storage of rainwater from rooftops and other building surfaces for household purposes and drought relief. (He *et al.,* 2019) Rainwater harvesting is now focused solely on restoring and recycling stormwater, with little consideration given to its additional advantages. To properly resource rainwater, it is required to first understand the area's hydrological characteristics, which include water surface runoff, discharge, speed, flow time, size and peak time. This aids in the improvement of the link between natural water networks and urban drainage systems, hence controlling urban floods. (Zevenbergen and Pathirana

2018) A ductile iron pipe system, which includes a pipe trench and ductile iron pipes utilized for rainwater collection. This serves as a rainwater storage space beneath the road. Water that reaches non-polluted surfaces, such as roof surfaces, can be discharged straight into this storage space. Polluted rainwater is treated before being released into the rainwater storage system.

Ecological water management: The Sponge City idea ensures that the aquatic environment is environmentally regenerated via the use of a self-purification system and waterfront design. The development of healthy water landscapes is also an important aspect of ecological water management. Many variables impact water self-purification, including the soil, plants, microbes, and the physical and biological chemical processes of flowing water. Ecological shoreline design is another method of ecological water management. (Jiang *et al.,* 2018) Natural and artificial ecological waterfronts will assist in defending against riverbank erosion in the creation of a Sponge City.

Green infrastructure: Green infrastructure has been created as a solution for environmental protection and the sustainability of urban areas. Sponge City is being built with green infrastructures and nature-based solutions such as detention basins, filter drains, infiltration systems, filter steeps, wetlands and swales. (Sun *et al.,* 2020) Green roofs and bio-retention are the two main forms of green infrastructure used in Sponge City.

Green roofs: Green roofs are often referred to as living roofs and rooftop gardens. Green roofs are constructed by covering them with vegetation or plants. They have several advantages, including mitigating the urban heat island effect, reducing stormwater runoff, increasing air and water quality, reducing energy utilization and promoting wildlife habitat and plant life. Green roofs are made up of five major layers: a waterproof membrane, a layer of drainage system, a filter membrane, soil and plants. (Zhang 2017)

Bio-retention: Bio-retention systems are sometimes commonly referred to as bio-filters or rain gardens. A bio-retention system typically consists of five components: a drainage layer, a transition layer, a submerged zone, filter media and a detention layer. These aid in the filtration of contaminated runoff and the removal of toxins through biological processes using active living plants and sandy loam layers. Bio-retention systems are regarded as one of the most feasible methods for sustainable rainwater management. As little sponges for cities, these systems have been a popular technique for controlling urban floods. (Jia *et al.,* 2022)

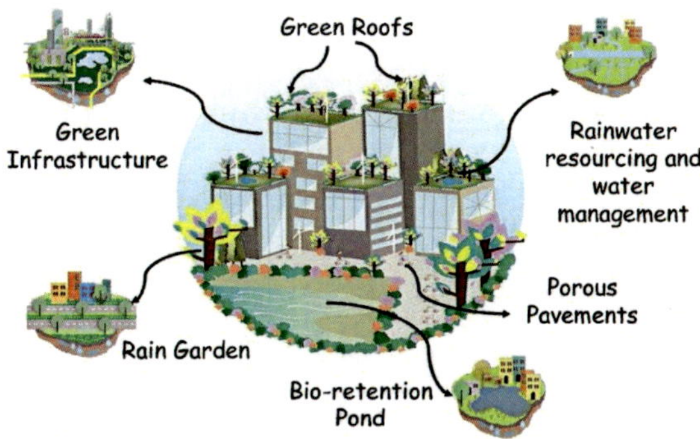

Fig 2: Essential individual elements constituting a Sponge City

Rain gardens and bioswales: Building rain gardens and bioswales is an effective method of retaining water in cities. A rain garden is a depression in the ground planted with aboriginal plants that collect rainwater. House spouts may be diverted into a rain garden rather than a sewer using this configuration, reducing sewage overflows during heavy downpours. A bioswale is a bigger, more designed or engineered version of a rain garden. Water can be temporarily stored and slowly drained away in this manner.

Urban permeable pavement: Permeable pavement is a Sponge City technology that builds ground pavement with permeable materials. (Fig:3) Permeable pavement aids in precipitation infiltration, groundwater purification for urban supplies, water runoff reduction, cooling, humidification, noise reduction, and environmental and ecological soil regeneration. Roads, parking lots and urban site pathways are all good examples of urban pavement. Permeable concrete, as the primary material used in the construction of sponge cities, has the potential to solve flooding problems due to its continuous voids created by gap grading the aggregate and eliminating the use of fine aggregate, allowing water and air to easily pass from the surface to the underground. (Shen *et al.,* 2020)

Fig 3: Commonly used permeable pavement materials in Sponge City

Path of Sponge City Construction

Sponge city construction, as a methodical undertaking, follows the following procedures: ***First*** and foremost, the city's native environment must be preserved. While undergoing urban building and development processes, the preservation of the original rivers, lakes, wetlands, ponds, ditches, and other natural places must be taken care of, and their protection must be maximized, as well as their maintenance operations.

Second, ecological restoration and repair work must be done regularly. Ecological measures must be employed for the recovery and restoration of damaged waterways and other natural habitats. A specific amount of ecological space must constantly be preserved, and water ecological infrastructure must also be constructed.

Third, the notion of low-impact development must be put into action. The intensity of urban growth must be moderated to avoid the destruction of the original urban water ecological environment. Rivers, lakes, and ditches must be correctly excavated, the water area must be extended, rainfall collection, infiltration, and purification must be fostered, and the size and quality of the urban "sponge" must thus be enhanced.

Fourth, the use and management of urban rain and flood resources must be significantly improved. Surface runoff caused by urban rains must be controlled properly. Permeable areas must be given for simple infiltration, and sufficient drainage must be supplied. A flood disaster risk management system is based on the idea of guaranteeing safety. To ensure safety, a flood disaster risk management system, personal hedging, disaster prevention and mitigation for structures, and other adaptive measures must be implemented. (Zhang *et al.,* 2018)

Top gains of Sponge Cities

Rainwater may be naturally absorbed, retained and reused for irrigation, residential usage, and other requirements during a drought by employing sponge city tactics. China intends to catch and utilize 70% of rainfall in its major centers by implementing sponge city methods. Other countries might benefit from this creative initiative. (Su *et al.,* 2019) Following the innovation principle, a sponge city may address water scarcity problems rather than generate them. Instead of diverting rainfall, Sponge City keeps it for usage inside its borders. It is a resource that may be used to supplement our water supply. In the long run, sponge cities will cut carbon emissions and aid in the battle against climate change. Reducing the quantity of stone and concrete used in urban infrastructure automatically reduces the amount of cement CO_2

emissions. (Zhou *et al.,* 2018) Cement contributes around 8% of all man-made greenhouse gas emissions, so any reduction is beneficial.

A sponge city, when correctly implemented, may lessen the frequency and severity of floods while also improving water quality. Stormwater capture aids in the prevention of beach contamination. Green space, for example, may increase quality of life, improve air quality, and minimize urban heat islands. Cities are generally hotter than their surroundings. Sponge city designs try to mitigate excess heat by using natural elements. Tree planting is a highly efficient method of enhancing the microclimate of a city. Water may be utilized to maintain green spaces, provide an outdoor area for residents and workers, and even be used to grow food. This, in turn, can increase the richness of urban ecosystems by offering new homes for a greater range of creatures. Grass swales, artificial ponds, wetlands, and rain gardens all contribute to urban development's aesthetic attractiveness and soothing impact. Sponge cities also bring aesthetic splendour to cities through the use of grass swales, wetlands, artificial ponds, and rain gardens. (Leng *et al.,* 2020)

The Union government's 'sponge cities' mission will cover 15 urban local bodies of Andhra Pradesh, including Visakhapatnam, Rajahmundry, Kakinada, Guntur, Kadapa, Eluru, Bhimavaram, Ongole, Adoni, etc. Under the initiative, plans will be prepared for the rejuvenation of 64 water bodies in these 15 cities/towns. Sponge Cities' mission is promoted by the Union government by incentivizing the cities to conduct surveys of the water.

Challenges in Sponge Cities

As the notion of a sponge city grows in popularity, the technologies used to build it are being investigated and acknowledged by governments all over the world. However, it faces severe implementation and scalability hurdles. There are technological, physical, and financial hurdles, as well as legal and regulatory challenges, public acceptance issues, and interagency or data-sharing challenges (Fig: 4). Given the technological obstacles, an inadequate research basis will limit the project's positive impact needlessly. In addition, present practices show a tendency to employ one comparable model for every section of the country, which cannot be acceptable because local conditions differ in various locations. As a result, sponge city plans should be designed after a thorough assessment of local conditions and potential. Inadequate information and guidance, as well as a lack of design standards and regulations, as well as insufficient education and training in its ideas, may result in poor planning and implementation, leading to additional major maintenance issues in the future. Furthermore, sponge city tactics need more frequent and ongoing upkeep. There are also certain maintenance issues, such as the fact that some sponge city measures are placed on private land, making it impossible for state organizations to guarantee that regular maintenance is carried out. Some sponge city measures may not be appropriate or cannot be implemented in some regions owing to the physical qualities of the land, climate, or other considerations. Traditional urban stormwater systems transport stormwater through underground pipes, however in sponge city practice, stormwater is infiltrated into the earth or held onsite. This necessitates the acquisition of extra land. Financial difficulties are also a significant influence.

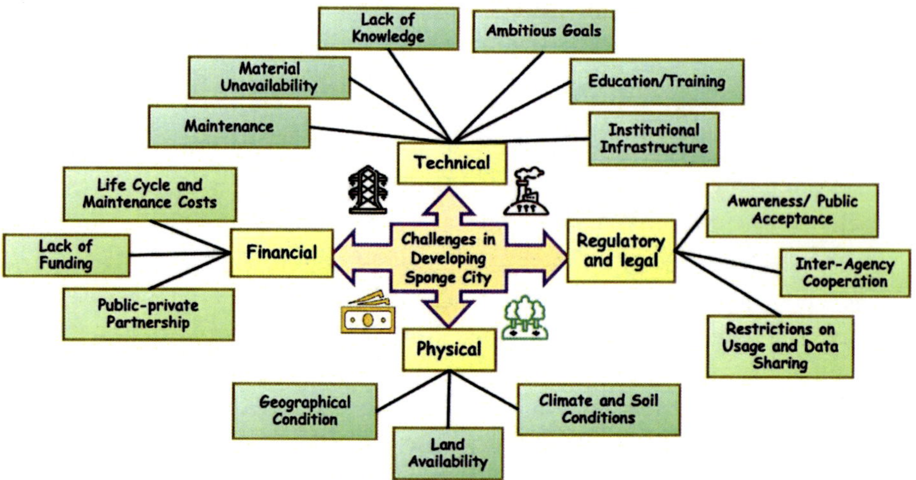

Fig 4: Challenges in developing Sponge City

Currently, financing for initial construction is granted, but funding for future operation and maintenance is not adequately handled. Also, because the project's duration is unpredictable, the life-cycle benefits, including environmental, ecological, and social advantages, cannot be accurately estimated. Because there is no clear image of the future rewards, both public and private businesses will face increased financial risk. (Li *et al.,* 2018) Building codes, plumbing and health standards, constraints regarding street width, drainage laws and parking spots, and restrictions on the use of recovered rainwater are all tough factors to consider. Public Acceptance Obstacles include Its success might also be hampered by public opinion and rejection of its interest. There are additional challenges with inter-agency collaboration and data sharing, i.e., a lack of inter-agency cooperation leads to issues with data and information sharing. It stifles research and development. In certain cases, time-consuming efforts were aimed towards collecting the same data or information. (Li *et al.,* 2017)

Need for Sponge Cities in India

India is suffering from both a scarcity of water and an overabundance of floodwaters. Incorporating the Sponge City concept into a smart city project would save countless lives while also paving the way for the adoption of sustainable communities. The cost of making this a reality may be significant, but the rewards of these undertakings will be huge. Countries throughout the world are dealing with climate change by implementing innovative and sustainable solutions.

NITI Aayog, in its Composite Water Management Index (CWMI) Report, stated in detail about the water crisis in our country. As per the report, Indian cities are grappling with water scarcity due to overpopulation and urbanization. According to the NITI Aayog assessment, several cities in India, including Chennai, Delhi, and Bangalore, would run out of groundwater. According to India Meteorological Department statistics, the South-West monsoon during June-August 2023 was below normal in 42% of the districts. Rainfall in the country was 32% below average in August, and 62% below normal in the southern states. This year's rainfall in India was the lowest in the past 122 years - that is, since 1901. With just approximately a month till the end of the Southwest monsoon, the lower rainfall would not only have a severe impact on agriculture, but it may also result in significant water shortages across the country. In India, the net quantity of water that may be consumed in a year is projected to be 1,121 billion cubic meters (bcm). However, figures from the Ministry of Water Resources suggest that overall water consumption will be 1,093 billion cubic meters in 2025 and 1,447 billion cubic meters in 2050. This suggests that India would face a severe water deficit within the next ten years. According to the Central Ground Water Board, Punjab's groundwater in the first 100 meters would be depleted by 2029.

In contrast, it is also obvious that many of these towns and cities face significant urban flooding problems every year as a result of heavy rainfall. Over the last decade, India has had several massive floods in cities such as Hyderabad, Kerala, Mumbai, Chennai, Kashmir, and others. We need to put this extra rainfall into a water storage system so that the water issue may be addressed while also controlling flooding. The current infrastructure will be upgraded along the lines of a sponge city to accommodate the effects of excessive rains and surface runoffs. During rainstorms, the Sponge City architecture would operate as a defense mechanism, allowing for higher absorption and water retention. As a result, it serves as a model for sustainable Indian cities.

Conclusion

Heavy rainfall is not the only cause of urban flooding. Rainfall, urbanization, inadequate drainage infrastructure, and the high intensity of human activity all contribute to an urban flooding disaster. Because global climate change and urbanization are causing major issues, improving urban rainfall and flood control should be prioritized. To address the harmful effects of urbanization, several cities are reverting to their pre-urbanization state. Bringing in more wildlife or constructing around existing green space will solve many of the difficulties if done right. A sponge city may be able to minimize urban water logging but not eliminate it, and the effect of sponge city development is dependent on its infiltration and storage capacity. Even if there are financial constraints, the sponge idea should be adopted as soon as possible in nations that are rapidly developing, particularly developing countries like India. Porous pavements and natural solutions can be used in tandem to address flooding issues. Urban wetlands as well as forests are being restored in cities like Bangalore. Similarly, sponge city solutions should be widely applied in India to preserve water and eliminate urban floods. The sponge city building can have a good impact on urban water resource management, urban flood control, ecological construction, and land usage.

References

Dingkun, Y.I.N., Zhengxia, C.H.E.N., Qian, L.I., Haifeng, J.I.A., Zhengquan, L.I.U. and Lei, S.H.E.N., 2020. Influence of rainfall characteristics on runoff control of a sponge reconstructed community in a rainy city. Journal of Tsinghua University (Science and Technology), 61(1), pp.50-56.

Femin, M.I. and George, E.H., 2022. Review on sponge city construction for flood management. Sustainability, Agri, Food and Environmental Research, 10. https://doi.org/10.7770/safer-V10N1-art2537

Guan, X., Wang, J. and Xiao, F., 2021. Sponge City strategy and application of pavement materials in Sponge City. Journal of Cleaner Production, 303, p.127022. https://doi.org/10.1016/j.jclepro.2021.127022

He, B.J., Zhu, J., Zhao, D.X., Gou, Z.H., Qi, J.D. and Wang, J., 2019. Co-benefits approach: Opportunities for implementing sponge city and urban heat island mitigation. Land use policy, 86, pp.147-157. https://doi.org/10.1016/j.landusepol.2019.05.003

Jia, H., Hu, J., Huang, T., Chen, A.S. and Ma, Y., 2022. Urban Runoff Control and Sponge City Construction. Water, 14(12), p.1910. https://doi.org/10.3390/w14121910

Jiang, Y., Zevenbergen, C. and Ma, Y., 2018. Urban pluvial flooding and stormwater management: A contemporary review of China's challenges and "sponge cities" strategy. Environmental science & policy, 80, pp.132-143. https://doi.org/10.1016/j.envsci.2017.11.016

Köster, S., 2021. How the Sponge City becomes a supplementary water supply infrastructure. Water-Energy Nexus, 4, pp.35-40. https://doi.org/10.1016/j.wen.2021.02.002

Liang, C., Zhang, X., Xu, J., Pan, G. and Wang, Y., 2020. An integrated framework to select resilient and sustainable sponge city design schemes for robust decision-making. Ecological Indicators, 119, p.106810. https://doi.org/10.1016/j.ecolind.2020.106810

Leng, L., Mao, X., Jia, H., Xu, T., Chen, A.S., Yin, D. and Fu, G., 2020. Performance assessment of coupled green-grey-blue systems for Sponge City construction. Science of the Total Environment, 728, p.138608. https://doi.org/10.1016/j.scitotenv.2020.138608

Li, H., Ding, L., Ren, M., Li, C. and Wang, H., 2017. Sponge City construction in China: A survey of the challenges and opportunities. Water, 9(9), p.594. https://doi.org/10.3390/w9090594

Li, Z., Xu, S. and Yao, L., 2018. A systematic literature mining of sponge city: trends, foci and challenges standing ahead. Sustainability, 10(4), p.1182. https://doi.org/10.3390/su1004118

Nguyen, T.T., Ngo, H.H., Guo, W. and Wang, X.C., 2020. A new model framework for sponge city implementation: Emerging challenges and future developments. Journal of environmental management, 253, p.109689. https://doi.org/10.1016/j.jenvman.2019.109689

Nguyen, T.T., Ngo, H.H., Guo, W., Wang, X.C., Ren, N., Li, G., Ding, J. and Liang, H., 2019. Implementation of a specific urban water management-Sponge City. Science of the Total Environment, 652, pp.147-162. https://doi.org/10.1016/j.scitotenv.2018.10.168

Shao, W., Zhou, J., Liu, J., Xie, P., Yang, Z. and Li, W., 2020, August. Characteristics and effectiveness of water management methods for 'sponge cities' in China. In Proceedings of the Institution of Civil Engineers-Water Management (Vol. 173, No. 4, pp. 163-171). Thomas Telford Ltd. https://doi.org/10.1680/jwama.19.00016

Shen, W., Liu, Y., Wu, M., Zhang, D., Du, X., Zhao, D., Xu, G., Zhang, B. and Xiong, X., 2020. Ecological carbonated steel slag pervious concrete prepared as a key material of sponge city. Journal of cleaner production, 256, p.120244. https://doi.org/10.1016/j.jclepro.2020.120244

Su, D., Zhang, Q.H., Ngo, H.H., Dzakpasu, M., Guo, W.S. and Wang, X.C., 2019. Development of a water cycle management approach to Sponge City construction in Xi'an, China. Science of the Total Environment, 685, pp.490-496. https://doi.org/10.1016/j.scitotenv.2019.05.382

Sun, Y., Deng, L., Pan, S.Y., Chiang, P.C., Sable, S.S. and Shah, K.J., 2020. Integration of green and gray infrastructures for sponge city: Water and energy nexus. Water-Energy Nexus, 3, pp.29-40. https://doi.org/10.1016/j.wen.2020.03.003

Zevenbergen, C., Fu, D. and Pathirana, A., 2018. Transitioning to sponge cities: Challenges and opportunities to address urban water problems in China. Water, 10(9), p.1230. https://doi.org/10.3390/w10091230

Zhang, S., Li, Y., Ma, M., Song, T. and Song, R., 2018. Storm water management and flood control in sponge city construction of Beijing. Water, 10(8), p.1040. https://doi.org/10.3390/w10081040

Zhang, Y.S., 2017. Sponge City theory and its application in landscape. World Construction, 6(1), pp.29-33. https://doi.org/10.18686/wc.v6i1.84

Zhang, Z. and Zhang, L., 2019, November. Development of Sponge Cities and Construction Strategies of Different Climate Zones in China. In International Symposium on Advancement of Construction Management and Real Estate (pp. 255-269). Singapore: Springer Singapore. https://doi.org/10.1007/978-981-15-8892-1_18

Zhou, J., Liu, J., Shao, W., Yu, Y., Zhang, K., Wang, Y. and Mei, C., 2018. Effective evaluation of infiltration and storage measures in sponge city construction: a case study of Fenghuang City. Water, 10(7), p.937. https://doi.org/10.3390/w10070937

15

Conservation of Aquatic Ecosystem An Ecosystem-Based Approach

Rinkesh Nemichand Wanjari[1], Zaheer Abass[2], Hudisa Banoo[2] and Dechen Chuskit[2]

[1]*Department of Fisheries Resource Management, School of Fisheries, Centurion University of Technology and Management, Paralakhemundi, Odisha-761211, India*

[2]*Division of Fisheries Resource Management (FRM), Faculty of Fisheries, SKUAST-Kashmir, Rangil, Ganderbal, (J&K), India-190006*

Abstract

Aquatic ecosystems comprising freshwater, coastal, and marine environments-are vital for biodiversity, food security, and economic well-being, particularly in developing countries. Nevertheless, human activities increasingly threaten these ecosystems, including overfishing, pollution, habitat destruction and climate change. This chapter examines the conservation of these ecosystems through an ecosystem-based management (EBM) approach, which is crucial for enhancing ecosystem resilience and preserving biodiversity. EBM provides a holistic framework incorporating ecological, social, and economic dimensions, moving beyond traditional management strategies focusing on individual policies or species. By considering the entire social-ecological system, EBM fosters sustainable fisheries and aquaculture while ensuring the long-term availability of ecosystem services. The chapter underscores the significance of policy integration, stakeholder involvement, and innovative solutions to safeguard and sustain aquatic ecosystems amid persistent environmental challenges.

Keywords: *Aquatic ecosystems, EBM, Biodiversity, Sustainability, Climate change*

Introduction

Aquatic ecosystems span diverse habitats, from freshwater springs, rivers, lakes, and wetlands on continents and islands to estuaries, coastal shallows, reefs, and the open seas (Arthington *et al.*, 2016). These environments are home to an astonishing variety of fish—over 30,000 known species (Nelson *et al.*, 2016)—exhibiting remarkable diversity in size, form, physiology, habitat preferences, diets, and life-history strategies (Helfman *et al.*, 2009). The conservation of aquatic ecosystems is a critical issue in today's rapidly changing world, where fisheries and aquaculture play a vital role in global food systems. These sectors are essential for food security, nutrition, and livelihoods, particularly in developing countries. Fisheries provide about 20% of animal protein to 3.3 billion people worldwide and are a key source of micronutrients (Action, 2020). Moreover, over 600 million people are employed in fisheries and aquaculture, with many involved in small-scale artisanal operations deeply tied to coastal communities (FAO, 2022). Aquaculture, in particular, has emerged as a crucial player in addressing global food security, contributing 59% of worldwide fish production, with 130.9 million tons produced in 2022, valued at USD 312.8 billion (SOFIA, 2024). The global consumption of aquatic food has more than doubled since 1961, reflecting a significant rise in demand (Dongyu, 2022). This trend underscores the urgent need for sustainable aquaculture practices to ensure that healthy water systems can continue providing nutritious food. However, the sustainability of these aquatic ecosystems is under severe threat. Freshwater biodiversity is critically endangered, facing escalating pressures globally (Higgins *et al.*, 2005). Climate change exacerbates these threats, disrupting ecosystems and jeopardizing food security and economic stability, particularly in countries heavily reliant on natural resources (Allison *et al.*, 2009). As climate change impacts the availability and distribution of fisheries, concerns about the long-term viability of these sectors intensify, making it imperative to take urgent action to ensure resilient food systems and protect livelihoods (Barange *et al.*, 2018). Conventional approaches to managing aquatic resources typically focused on individual policies or flagship species, have often failed to meet policy goals (Long *et al.*, 2015). In contrast, Ecosystem-Based Management (EBM) offers a more holistic approach, emphasizing the interconnections within ecosystems. EBM is increasingly seen as an essential tool for modern decision-making (Börgstrom *et al.*, 2015; Cormier *et al.*, 2017). Unlike traditional methods, which often assume that policy failures result from a lack of understanding of ecological processes, EBM takes a broader view that incorporates ecological, social, and economic factors (Ruckelshaus *et al.*, 2009). This comprehensive perspective is crucial, given that aquatic ecosystems, including freshwater, marine environments, and coastal, are rich in biodiversity and provide critical ecosystem services (Bennett *et al.*, 2015). Unfortunately, human activities—such as pollution, habitat destruction, invasive species, overfishing, and climate change—are putting many of these ecosystems at risk of irreversible damage (Maximillian *et al.*, 2019; Elisha and Felix, 2021). An integrated approach is necessary to ensure the long-term sustainable management of aquatic ecosystems.

Ecosystem-based management (EBM) is a promising framework that combines policy objectives, scientific knowledge, tools, and stakeholder engagement to support sustainability. EBM's collaborative approach aims to restore and enhance ecosystem resilience, ensuring the continued provision of ecosystem services while protecting biodiversity (Gómez *et al.*, 2016a; 2016b; Langhans *et al.*, 2019). As we move forward, it is increasingly clear that the outdated dichotomy between economic improvement and environmental protection no longer applies. Instead, a 21st century approach that combines traditional methods with ecosystem-based strategies offers cost-effective solutions for water management (Vörösmarty *et al.*, 2018). This shift is essential for safeguarding the health and functionality of aquatic ecosystems in the face of growing global challenges. This chapter examines the conservation of aquatic ecosystems using an ecosystem-based management (EBM) approach, focusing on restoring ecosystem resilience, sustaining biodiversity, and ensuring the sustainable provision of ecosystem services. It also addresses the integration of ecological, social, and economic factors in managing the challenges posed by human activities and climate change.

Critical Components of Aquatic Ecosystem Conservation

Biodiversity Conservation

Biodiversity conservation aims to protect all life forms on Earth and ensure the health and functionality of natural ecosystems. This involves the preservation, maintenance, sustainable use, recovery, and enhancement of biological diversity (Chivian, 2002). Conservation encompasses the sustainable management of resources, balancing protection and use, while preservation specifically refers to keeping ecosystems intact and unaltered. Sustainable development plays a vital role in biodiversity conservation, aiming to fulfill current needs without jeopardizing the ability of future generations to meet their own. It fosters equity within the present generation and future generations (Maggio, 1996). Achieving a balance among environmental health, development, and societal needs is essential for effective biodiversity conservation and requires robust enforcement and implementation of environmental policies and conventions (Perrings and Gadgil, 2003; Ukpoju *et al.*, 2023). Despite freshwater ecosystems, such as rivers, lakes, and wetlands, accounting for only 0.01% of the Earth's water, they support a disproportionately large share of global biodiversity (McAllister *et al.*, 1997). Freshwater fish alone represent about one-quarter of all living vertebrate species, with approximately 44,000 scientifically recognized species of freshwater organisms. Alarmingly, freshwater biodiversity is often more at risk than terrestrial biodiversity. According to the IUCN Red List 2000, around 20% of amphibians and

30% of fish (primarily freshwater) were classified as threatened. Freshwater biodiversity faces numerous threats, including the direct impacts of dams, the introduction of non-native species, overfishing, pollution, stream alterations, and water extraction and diversion. Moreover, indirect effects from land-use activities such as logging, agriculture, industry, urban development, and mining further endanger these ecosystems (Prasad *et al.*, 2002). To safeguard freshwater biodiversity, effective conservation strategies must be developed and implemented. The aquatic conservation strategy focuses on safeguarding and preserving the ecological integrity of watersheds and marine ecosystems (Ramachandra *et al.*, 2002).

1. To safeguard aquatic systems and the species, populations, and communities uniquely adapted to them, it is crucial to preserve the distribution, diversity, and complexity of watershed and landscape-scale features (Geist and Hawkins, 2016). This approach ensures that these vital ecosystems continue to thrive.
2. Maintaining and preserving water quality is crucial to support healthy riparian, aquatic, and wetland ecosystems (Ramachandra *et al.*, 2005). Water quality should remain within a range that upholds these systems' biological, physical, and chemical integrity, promoting the organisms' survival, growth, reproduction, and migration within aquatic and riparian communities (Swanson *et al.*, 2017).
3. Maintaining and conserving plant communities' species composition and structural diversity in riparian zones and wetlands is essential. This ensures effective thermal regulation during summer and winter, enhances nutrient filtering, and manages surface and bank erosion and channel migration (Naiman and Decamps, 1997). Furthermore, preserving these communities provides adequate amounts and distributions of coarse woody debris, vital for sustaining physical complexity and stability within these ecosystems (Gurnell *et al.*, 1995).
4. Maintaining and conserving spatial and temporal connectivity within and between watersheds is vital. This includes lateral, longitudinal, and drainage network connections, such as floodplains, wetlands, upslope areas, and headwater tributaries (Wohl, 2017). These connections must provide chemically and physically unobstructed pathways to essential habitats that meet the life-history requirements of aquatic and riparian-dependent species.
5. Long-term fish conservation will require a blend of targeted management actions. This approach should include conservation reserves that safeguard habitats rich in species, essential resources,

areas of unique species evolution, and regions with a high concentration of threatened endemic species. For freshwater-protected areas to be practical, they need control over critical elements such as upstream drainage networks, adjacent land, riparian zones, and downstream areas (Dudgeon *et al.*, 2006). Connectivity among habitats and a diversity of habitat patches should also be maintained. Advancements in systematic river conservation planning now incorporate methods that address connectivity along multiple axes—longitudinal, lateral (between river and floodplain), vertical (surface and groundwater), and temporal—while also recognizing threats that could undermine biodiversity (Linke *et al.*, 2015), including climate change impacts (Pittock *et al.*, 2008).

6. Conservation and management of aquatic ecosystems necessitate collaborative research that integrates natural, social, and interdisciplinary studies. This approach is vital for understanding key components such as water quality monitoring, socio-economic dependencies, biodiversity, and related activities, all essential for developing long-term conservation strategies (Kiran and Ramachandra, 1999). To achieve this, it is essential to have multidisciplinary-trained professionals who can promote awareness of the ecosystem's significance in local schools, colleges, and research institutions. They can enhance public understanding of aquatic ecosystem restoration, its goals, and effective methods by initiating educational programs.

State Fish: A Strategic Approach to Aquatic Conservation in India

In 2006, the ICAR-National Bureau of Fish Genetic Resources (ICAR-NBFGR) in Lucknow, Uttar Pradesh, introduced the "State Fish" concept, compiling a list of 16 Indian states to promote the conservation of fish germplasm resources. The initiative Prof. W. S. Lakra proposed encourages states to adopt a specific native fish species as their official "state fish," akin to other state symbols like animals and trees (Sarkar *et al.*, 2024). This approach fosters collaboration among critical stakeholders, including local communities and governmental bodies, to develop targeted conservation strategies that enhance the selected species. By engaging communities in fish germplasm conservation, the initiative promotes biodiversity and supports sustainable fisheries development across the states (Goswami *et al.*, 2023; Lakra *et al.*, 2010). Figure 1 depicts the state fish of various Indian states.

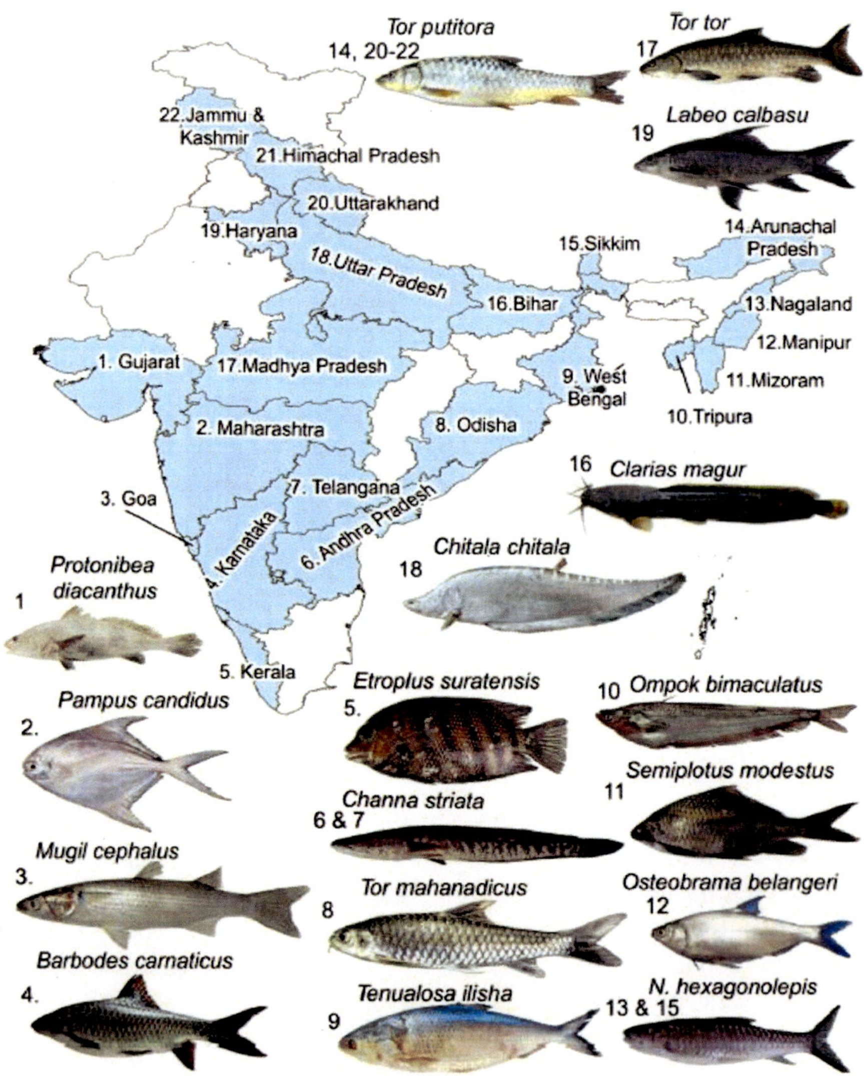

Figure 1. State Fish as Symbols of Biodiversity: India's Approach to Aquatic Conservation (Sarkar *et al.*, 2024)

Restoration and Management

Human activities and stressors that threaten freshwater and marine fish populations are projected to become more pervasive, intense, and harmful unless countered through prevention, enhanced management practices—such as sustainable fisheries—and restoration or adaptation efforts. In freshwater ecosystems, decades of research and practical experience provide valuable insights into integrated catchment management, habitat restoration techniques (Roni *et al.*, 2008), dam management and removal strategies (Olden, 2016), the implementation of environmental flows (Arthington, 2012), and the restoration of riparian and floodplain processes (Kingsford, 2015). Fish populations can be rehabilitated through appropriate regulations, such as catch-and-release practices, no-take zones in critical breeding and recruitment areas, and even managed relocation and reintroduction efforts (Cooke and Schramm, 2007). Recovery initiatives can target individual species or encompass entire river basin ecosystems. Many studies integrate essential knowledge of fish biology with focused threat assessments to develop recovery plans for specific species. Approximately 35,000 advanced fingerlings of rohu, catla, mrigal and calbasu were released into the Ganga River. In addition to ranching efforts, around 100 fish were tagged with external floy tags to track their migration pathways, facilitating management methods to conserve natural fish genetic resources. A mass Dolphin Awareness Program was also organized to promote the conservation of the country's National Aquatic Animal.

Ratification of Environmental Policies

Policymakers should prioritize fundamental research conducted by institutes in conjunction with "infield" experiments that apply practical methodologies from upstream to downstream. This integrated approach will provide accurate, in-situ assessments of pollution impacts, ecological conditions, and gaps in restoration strategies (Srivastava, 2007). Particular attention should be given to rivers crossing interstate and international boundaries, necessitating coordinated management of river regulation, water allocation, and irrigation practices. These restoration efforts are often politically influenced by socioeconomic disputes among riparian states or countries (Ramana, 1992). Effective policy management requires committees to align river restoration objectives with long-term research to assess ecological impacts accurately. This integration of science and policy is essential for informed decision-making. Given the dynamic nature of ecosystems and societal changes, management strategies should be adaptable and resilient to ongoing uncertainties (Green *et al.*, 2015; Abell *et al.*, 2017). With climate change increasingly affecting the ability of many organisms to adapt and survive, policies must incorporate relocation strategies for selected species to suitable habitats, thereby enhancing their survival and persistence (Dudgeon, 2006). Policymakers should collaborate with researchers to ensure species selection minimizes adverse effects on native populations. Furthermore, policy guidelines should be periodically updated to reflect new scientific findings and socio-economic considerations (Reid *et al.*, 2019).

Environment Awareness and Public Participation

A key challenge in restoration and conservation efforts lies in effectively prioritizing sites and objectives. Achieving successful restoration on national and international scales requires the development of integrated strategies that address multiple, region-specific goals simultaneously (Abell *et al.*, 2019; Stefanidis *et al.*, 2019). A holistic approach is essential to enhance management strategies globally, focusing on population-scale assessments, resource demands, political and institutional frameworks, technical capacities, and practical awareness and education policies. In the current era of global warming, water quality and quantity pressures have intensified, necessitating proactive conservation actions through collaborative efforts among all stakeholders, including policymakers, resource users, ecosystem managers, citizens, NGOs, and government entities. Various initiatives, such as implementing regulatory instruments, providing fiscal incentives (e.g., agri-environment schemes, green climate funds), and creating market opportunities for investments in nature-based solutions and restoration projects with financial returns, are vital. Additionally, societal actions, including awareness programs, educational initiatives, and volunteer participation in conservation activities, play a crucial role in addressing emerging threats to freshwater ecosystems (Birkhofer *et al.*, 2015; Wohl *et al.*, 2015; Szałkiewicz *et al.*, 2020; Cantonati *et al.*, 2020).

Integrated Aquatic Ecosystem Management

Effective integrated aquatic ecosystem management requires comprehensive study, a solid understanding, and proactive management of interconnected water systems. This approach includes the interactions between groundwater, surface water, and return flows, considering both water quantity and quality alongside biological components and examining relationships between upstream and downstream areas. These water systems should be analyzed within the larger environmental context, considering socio-economic factors, opportunities, and the local political and cultural landscape (Langhans *et al.*, 2019). Water resources must be viewed as social, environmental, and economic assets, with each aspect playing a vital role in the political conversation. This dialogue should address the interests of local communities, their water-dependent livelihoods, and the health of their water environments. Inclusive participation from users and managers at all levels is essential to provide diverse insights and foster community investment. The primary goal of integrated aquatic system management is to sustain the multifunctional use of water systems over time. Meeting the essential water needs of both people and ecosystems should be prioritized, with a strong emphasis on protecting key ecological and physical processes. It is also critical to consider the impacts on downstream water bodies, such as seas, lakes, deltas, and coastal areas, ensuring they receive full and thoughtful attention. Critical factors for sustainable management must be emphasized in this framework (Ramachandra *et al.*, 2005).

1. A systems approach emphasizes individual components and their connections, recognizing that any disturbance within one part of the system can impact others. These effects may vary: some may be indirect and mitigated by the system's natural resilience, while others may have a direct, significant impact that can amplify as it moves through the system. Although systems analysis is valuable, avoiding overly complex models and analyses that hinder practical understanding and application is essential. The focus should remain on developing insights that can be effectively translated into actionable knowledge.
2. Integrating water and environmental management is essential and has broad, strong support. Integrated aquatic ecosystem management can be further enhanced by incorporating environmental impact assessments (EIAs), water resource modelling, and land-use planning. Adopting a catchment or watershed approach means that water management should go hand in hand with managing interconnected natural resources, including soil, forests, air, and biotic life. This holistic strategy ensures that interdependencies among resources are respected and managed sustainably.
3. Decentralization should be prioritized to make river basin management accessible to individual citizens and responsive to local conditions and preferences. Localized management allows for flexible approaches that reflect community needs. At the same time, even tasks with a broader scope can be effectively decentralized through cooperation among local governments (such as panchayats within a river basin) or under the guidance of a higher-level authority. This process should be transparent, phased, and planned to ensure clarity and effective implementation.

Threats to Aquatic Ecosystems: Addressing Conservation Challenges

Aquatic ecosystems worldwide face severe threats from numerous human-induced pressures, such as altered water flows, barriers to migration, overfishing, changes to riparian zones and floodplains, climate change, and the spread of invasive species (Bănăduc *et al.*, 2022; Ogidi and Akpan, 2022). These factors have placed riverine fish among the most endangered species globally. This special issue addresses these challenges, focusing on vulnerable species and high-risk river systems while exploring various conservation and restoration strategies for endangered fish populations and their habitats (Olden *et al.*, 2010; Evans *et al.*, 2016). Critical approaches include creating aquatic protected areas, stakeholder co-management, enhanced flow management, habitat connectivity restoration, and invasive species control. The contributions discuss both resident and migratory fish species recognized as endangered by national and international bodies, as well as those identified as at-risk by local organizations. Case studies from multiple countries illustrate integrated scientific and practical approaches to conserving aquatic biodiversity and promoting sustainable ecosystem management (Cooke *et al.*, 2012). Figure 2 illustrates the major threats to aquatic biodiversity.

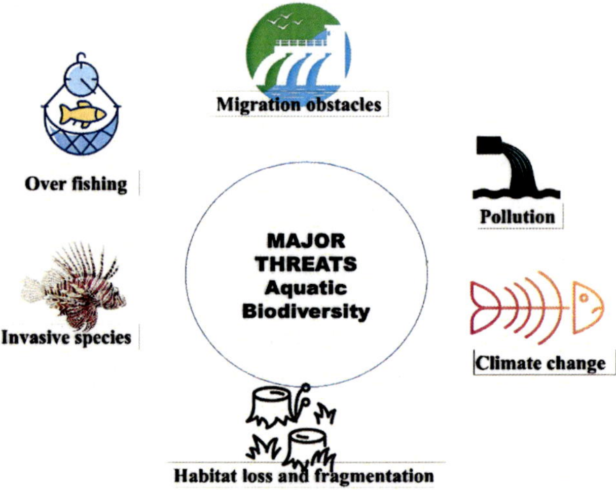

Figure 2. Major threats to aquatic biodiversity

Ecosystem-based Conservation Strategies

Protecting Biodiversity Hotspots and Keystone Species

Protecting biodiversity hotspots and keystone species is essential for maintaining ecosystem health and resilience (Kumar *et al.*, 2023; Jiang *et al.*, 2024). Biodiversity hotspots in regions with high levels of unique species diversity and significant habitat loss are critical for conservation as they support many species found nowhere else on Earth (Mittermeier *et al.*, 2011; Marchese, 2015). Keystone species, meanwhile, play a pivotal role in their ecosystems, influencing the structure and function of their environments disproportionately to their abundance. The loss of keystone species can trigger cascading effects that disrupt ecosystem balance, affecting various other species and ecological processes. Conservation efforts focused on these hotspots and keystone species can preserve the unique biodiversity within these regions and the ecological services and stability they provide. We can safeguard ecosystems that sustain life, contribute to climate resilience, and support human well-being by prioritizing protection and restoration in these areas (Johnson *et al.*, 2017).

Habitat Restoration and Ecological Connectivity

Habitat restoration and ecological connectivity are crucial for supporting biodiversity and ecosystem resilience (Miller *et al.*, 2017). Restoring degraded habitats revitalizes essential ecological functions, enabling native species to thrive and reestablishing natural processes like nutrient cycling and water purification (Nellemann and Corcoran, 2010). Ecological connectivity, the network of natural corridors linking habitats, ensures species can move, interact, and adapt across landscapes. This is especially vital for migratory species and those with large territorial ranges, as it allows them to access breeding, feeding, and sheltering areas, which can improve their survival rates. Connectivity also bolsters genetic diversity by linking fragmented habitats, helping populations adapt to environmental changes, including those driven by climate change. Habitat restoration and connectivity form a foundation for sustainable conservation, enabling ecosystems to function as interconnected, resilient landscapes supporting wildlife and human needs (Gexist and Hawkins, 2016).

Community-driven Conservation Efforts

Community-driven conservation efforts are essential for sustainable ecosystem management, as they engage local people in protecting the natural resources they rely on (Ruiz-Mallén *et al.*, 2015; Armitage *et al.*, 2020). These initiatives harness local communities' unique knowledge and cultural values, making conservation strategies more adaptable and relevant to the specific environmental and social context. When communities lead conservation efforts, they are more likely to support and sustain these initiatives over time, as they see direct benefits such as improved livelihoods, increased food security, and enhanced resilience to environmental changes (Munang *et al.*, 2011). Community-driven efforts also foster a sense of stewardship and shared responsibility, where locals become active guardians of biodiversity, wildlife, and natural resources. Conservation programs can achieve long-term, meaningful impacts that balance ecological health with community welfare by incorporating community voices and prioritizing inclusive, collaborative decision-making (Zhang *et al.*, 2023).

Conservation Reserves

Long-term fish conservation will require a blend of targeted management actions. This approach should include conservation reserves that safeguard habitats rich in species, essential resources, areas of unique species evolution, and regions with a high concentration of threatened endemic species. For freshwater-protected areas to be practical, they need control over critical elements such as upstream drainage networks, adjacent land, riparian zones, and downstream areas (Dudgeon *et al.*, 2006). Connectivity among habitats and a diversity of habitat patches should also be maintained. Advancements in systematic river conservation planning now incorporate methods that address connectivity along multiple axes—longitudinal, lateral (between river and floodplain), vertical (surface and groundwater), and temporal—while also recognizing threats that could undermine biodiversity (Linke *et al.*, 2015), including climate change impacts (Pittock *et al.*, 2008). Conservation planning and legislation must also consider the socio-economic context, balancing biodiversity protection with the challenges posed by land use, river infrastructure, human activity, and climate change. For instance, species-rich river basins, such as the Mekong, Congo, and Amazon, which face pressures from numerous new hydropower projects, urgently require strategic conservation planning (Frederico *et al.*, 2016; Winemiller *et al.*, 2016).

Conclusion

Aquatic ecosystems are vital for supporting biodiversity and human livelihoods, yet they are increasingly vulnerable to human-induced pressures and climate change. Effective conservation and restoration strategies, such as ecosystem-based management, offer a comprehensive framework to address these challenges by integrating ecological, social, and economic factors. Protecting biodiversity, restoring habitats, maintaining environmental connectivity, and involving local communities in conservation efforts are essential to ensuring aquatic ecosystems' long-term health and functionality. Policymakers should work closely with scientists to create adaptive management strategies that are guided by current research and socio-economic factors, ensuring the protection of these vital ecosystems. Through concerted efforts, it is possible to sustain the multifunctional use of aquatic systems, ensuring that they continue to provide essential services for both current and future generations.

References

Abell, R., Lehner, B., Thieme, M. and Linke, S., 2017. Looking beyond the fenceline: Assessing protection gaps for the world's rivers. Conservation Letters, 10(4), pp.384-394.

Action, S. I. (2020). World fisheries and aquaculture. Food and Agriculture Organization, 2020, 1-244.

Allison, E. H., Perry, A. L., Badjeck, M. C., Neil Adger, W., Brown, K., Conway, D., Halls, A.S., Pilling, G.M., Reynolds, J.D., Andrew, N.L., & Dulvy, N. K. (2009). Vulnerability of national economies to the impacts of climate change on fisheries. Fish and fisheries, 10(2), 173-196.

Armitage, D., Mbatha, P., Muhl, E.K., Rice, W. and Sowman, M., 2020. Governance principles for community centered conservation in the post 2020 global biodiversity framework. Conservation Science and Practice, 2(2), p.e160.

Arthington, A., 2012. Environmental flows: saving rivers in the third millennium (Vol. 4). Univ of California Press.

Arthington, A.H., Dulvy, N.K., Gladstone, W. and Winfield, I.J., 2016. Fish conservation in freshwater and marine realms: status, threats and management. Aquatic Conservation: Marine and Freshwater Ecosystems, 26(5), pp.838-857.

Bănăduc, Doru, Vladica Simić, Kevin Cianfaglione, Sophia Barinova, Sergey Afanasyev, Ahmet Öktener, Grant McCall, Snežana Simić, and Angela Curtean-Bănăduc. "Freshwater as a sustainable resource and generator of secondary resources in the 21st century: Stressors, threats, risks, management and protection strategies, and conservation approaches." International Journal of Environmental Research and Public Health 19, no. 24 (2022): 16570.

Barange, M., Bahri, T., Beveridge, M. C., Cochrane, K. L., Funge-Smith, S., & Poulain, F. (2018). Impacts of climate change on fisheries and aquaculture. United Nations' Food and Agriculture Organization, 12(4), 628-635.

Bennett, E.M., Cramer, W., Begossi, A., Cundill, G., Díaz, S., Egoh, B.N., Geijzendorffer, I.R., Krug, C.B., Lavorel, S., Lazos, E. and Lebel, L., 2015. Linking biodiversity, ecosystem services, and human well-being: three challenges for designing research for sustainability. Current opinion in environmental sustainability, 14, pp.76-85.

Birkhofer, K., Diehl, E., Andersson, J., Ekroos, J., Früh-Müller, A., Machnikowski, F., Mader, V.L., Nilsson, L., Sasaki, K., Rundlöf, M. and Wolters, V., 2015. Ecosystem services—current challenges and opportunities for ecological research. Frontiers in Ecology and Evolution, 2, p.87.

Borgström, S., Bodin, Ö., Sandström, A., & Crona, B. (2015). Developing an analytical framework for assessing progress toward ecosystem-based management. Ambio, 44(3), 357–369.

Cantonati, M., Poikane, S., Pringle, C.M., Stevens, L.E., Turak, E., Heino, J., Richardson, J.S., Bolpagni, R., Borrini, A., Cid, N. and Čtvrtlíková, M., 2020. Characteristics, main impacts, and stewardship of natural and artificial freshwater environments: Consequences for biodiversity conservation. Water, 12(1), p.260.

Chivian, E., 2002. Biodiversity: its importance to human health. Center for Health and the Global Environment, Harvard Medical School, Cambridge, MA, 23.

Cooke, S. J., & Schramm, H. L. (2007). Catch-and-release science and its application to conservation and management of recreational fisheries. Fisheries Management and Ecology, 14(2), 73-79.

Cooke, S. J., Paukert, C., & Hogan, Z. (2012). Endangered river fish: factors hindering conservation and restoration. Endangered species research, 17(2), 179-191.

Cormier, R., Kelble, C. R., Anderson, M. R., Allen, J. I., Grehan, A., & Gregersen, O. (2017). Moving from ecosystem-based policy objectives to operational implementation of ecosystem based management measures. ICES Journal of Marine Science, 74, 406–413.

Dongyu, Q. (2024). 2024 The State of World Fisheries and Aquaculture-Blue Transformation in Action. The State of World Fisheries and Aquaculture, R1-232.

Dudgeon, D., & Smith, R. E. (2006). Exotic species, fisheries and conservation of freshwater biodiversity in tropical Asia: the case of the Sepik River, Papua New Guinea. Aquatic Conservation: Marine and Freshwater Ecosystems, 16(2), 203-215.

Elisha, O.D. and Felix, M.J., 2021. Destruction of coastal ecosystems and the vicious cycle of poverty in Niger Delta Region. Journal of Global Agriculture and Ecology, 11(2), pp.7-24.

Evans, D.M., Che-Castaldo, J.P., Crouse, D., Davis, F.W., Epanchin-Niell, R., Flather, C.H., Frohlich, R.K., Goble, D.D., Li, Y.W., Male, T.D. and Master, L.L., 2016. Species recovery in the United States: increasing the effectiveness of the Endangered Species Act. Issues in Ecology, 2016(20).

FAO (2022). "Fishery and Aquaculture Statistics," in Global Fisheries Commodities Production and Trade 1976-2018 (FishstatJ) (Fishery and Aquaculture Statistics. Global Fisheries commodities production and trade, Rome 1976–2018 (FishstatJ)

Geist, J. and Hawkins, S.J., 2016. Habitat recovery and restoration in aquatic ecosystems: current progress and future challenges. Aquatic Conservation: Marine and Freshwater Ecosystems, 26(5), pp.942-962.

Gómez, C. M., Delacámara, G., Arevalo-Torres, J., Barbiere, J., Barbosa, A.L., & Iglesias-Campos, A. (2016b). The AQUACROSS innovative concept. Deliverable 3.1, European Union's Horizon 2020 Framework Programme for Research and Innovation Grant Agreement No. 642317.

Gómez, C. M., Delacámara, G., Arévalo-Torres, J., Barbière, J., Barbosa, A. L., Boteler, B., Culhane, F., Daam, M., Gosselin, M.-P., Hein, T., Iglesias-Campos, A., Jähnig, S., Lago, M., Langhans, S., Martínez-López, J., Nogueira, A., Lillebø, A., O'Higgins, T., Piet, G., Pletterbauer, F., Pusch, M., Reichert, P., Robinson, L., Rouillard, J., & Schlüter, M. (2016a). The AQUACROSS innovative concept. Deliverable 3.1. AQUACROSS, European Union's Horizon 2020 Framework Programme for Research and Innovation Grant Agreement No. 642317. Technical Report (February 19th, 2016). European Union (H2020 FP Grant Agreement)-AQUACROSS.

Goswami, P., & Singha, S. (2023). Fish diversity of the Beki River, Assam, India: Present status and conservation needs. Ecocycles, 9(2), 1-6.

Green, P.A., Vörösmarty, C.J., Harrison, I., Farrell, T., Sáenz, L. and Fekete, B.M., 2015. Freshwater ecosystem services supporting humans: Pivoting from water crisis to water solutions. Global Environmental Change, 34, pp.108-118

Gurnell, A.M., Gregory, K.J. and Petts, G.E., 1995. The role of coarse woody debris in forest aquatic habitats: implications for management. Aquatic conservation: marine and freshwater ecosystems, 5(2), pp.143-166.

Helfman, G.S., Collette, B.B., Facey, D.E. and Bowen, B.W., 2009. The diversity of fishes: biology, evolution, and ecology. John Wiley & Sons.

Higgins, J.V., Bryer, M.T., Khoury, M.L. and Fitzhugh, T.W., 2005. A freshwater classification approach for biodiversity conservation planning. Conservation Biology, 19(2), pp.432-445.

Jiang, F., Song, P., Gu, H., Zhang, J., Xu, B., Li, B., Liang, C., Gao, H., Cai, Z., Zhang, M. and Zhang, T., 2024. New shortcut for boundary delimitation and functional zoning of national parks based on keystone species in China: A case study of kunlun mountains national park. Ecological Indicators, 159, p.111675.

Johnson, S. A., Ober, H. K., & Adams, D. C. (2017). Are keystone species effective umbrellas for habitat conservation? A spatially explicit approach. Journal for Nature Conservation, 37, 47-55.

Kingsford, R. T., Mac Nally, R., King, A., Walker, K. F., Bino, G., Thompson, R. & Humphries, P. (2015). A commentary on 'Long-term ecological trends of flow-dependent ecosystems in a major regulated river basin', by Matthew J. Colloff, Peter Caley, Neil Saintilan, Carmel A. Pollino and Neville D. Crossman. Marine and Freshwater Research, 66(11), 970-980.

Kiran, R., & Ramachandra, T. V. (1999). Status of wetlands in Bangalore and its conservation aspects. ENVIS Journal of Human Settlements, 16-24.

Kumhar, G.L., VijayKumar, R. and Kumar, V., 2023. Biodiversity Conservation in the Face of Climate Change. Environment and Climate Change, p.171.

Langhans, S. D., Jähnig, S. C., Lago, M., Schmidt-Kloiber, A., & Hein, T. (2019). The potential of ecosystem-based management to integrate biodiversity conservation and ecosystem service provision in aquatic ecosystems. Science of the Total Environment, 672, 1017–1020.

Linke, S., & Bruckmeier, K. (2015). Co-management in fisheries–experiences and changing approaches in Europe. Ocean & Coastal Management, 104, 170-181.

Long, R. D., Charles, A., & Stephenson, R. L. (2015). Key principles of marine ecosystem-based management. Marine Policy, 57, 53–60.

Maggio, G.F., 1996. Inter/intra-generational equity: current applications under international law for promoting the sustainable development of natural resources. Buff. Envtl. LJ, 4, p.161.

Marchese, C., 2015. Biodiversity hotspots: A shortcut for a more complicated concept. Global Ecology and Conservation, 3, pp.297-309.

Maximillian, J., Brusseau, M.L., Glenn, E.P. and Matthias, A.A., 2019. Pollution and environmental perturbations in the global system. In Environmental and pollution science (pp. 457-476). Academic Press.

McAllister, D.E., Hamilton, A.L. and Harvey, B., 1997. Global freshwater biodiversity: striving for the integrity of freshwater ecosystems. Sea wind: bulletin of Ocean Voice International; 11 (3).

Miller, B.P., Sinclair, E.A., Menz, M.H., Elliott, C.P., Bunn, E., Commander, L.E., Dalziell, E., David, E., Davis, B., Erickson, T.E. and Golos, P.J., 2017. A framework for the practical science necessary to restore sustainable, resilient, and biodiverse ecosystems. Restoration Ecology, 25(4), pp.605-617.

Mittermeier, R.A., Turner, W.R., Larsen, F.W., Brooks, T.M. and Gascon, C., 2011. Global biodiversity conservation: the critical role of hotspots. In Biodiversity hotspots: distribution and protection of conservation priority areas (pp. 3-22). Berlin, Heidelberg: Springer Berlin Heidelberg.

Munang, R.T., Thiaw, I. and Rivington, M., 2011. Ecosystem management: Tomorrow's approach to enhancing food security under a changing climate. Sustainability, 3(7), pp.937-954.

Naiman, R.J. and Decamps, H., 1997. The ecology of interfaces: riparian zones. Annual review of Ecology and Systematics, 28(1), pp.621-658.

Nellemann, C. and Corcoran, E. eds., 2010. Dead planet, living planet: Biodiversity and ecosystem restoration for sustainable development: A rapid response assessment. UNEP/Earthprint.

Nelson, J.S., Grande, T.C. and Wilson, M.V., 2016. Fishes of the World. John Wiley & Sons.

Ogidi, O.I. and Akpan, U.M., 2022. Aquatic biodiversity loss: impacts of pollution and anthropogenic activities and strategies for conservation. In Biodiversity in Africa: potentials, threats and conservation (pp. 421-448). Singapore: Springer Nature Singapore.

Olden, J. D. (2016). Challenges and opportunities for fish conservation in dam-impacted waters. Conservation of freshwater fishes, 107-148.

Olden, J.D., Kennard, M.J., Leprieur, F., Tedesco, P.A., Winemiller, K.O. and García-Berthou, E., 2010. Conservation biogeography of freshwater fishes: recent progress and future challenges. Diversity and Distributions, 16(3), pp.496-513.

Perrings, C. and Gadgil, M., 2003. Conserving biodiversity: Reconciling local and global public benefits. Providing global public goods: Managing globalization, 4, pp.532-555.

Pittock J, Hansen LJ, Abell R. 2008. Running dry: freshwater biodiversity, protected areas and climate change. Tropical Conservancy Biodiversity 9: 30–38.

Prasad, S.N., Ramachandra, T.V., Ahalya, N., Sengupta, T., Alok Kumar, Tiwari, A.K., Vijayan V.S. and Lalitha Vijayan, 2002. Conservation of wetlands of India – a review, Tropical Ecology, 43 (1): 173-186

Ramachandra, T. V., Ahalya, N., & Murthy, R. (2005). Aquatic ecosystems: conservation, restoration and management. Aquatic ecosystems-Conservation, restoration and management, 26-50.

Ramachandra, T.V., Kiran R and Ahalya N. 2002, Status, Conservation and Management of Wetlands, Allied Publishers Pvt Ltd, Bangalore.

Ramana, M.V.V., 1992. Inter-state river water disputes in India. Stosius Incorporated/Advent Books Division.

Reid, A.J., Carlson, A.K., Creed, I.F., Eliason, E.J., Gell, P.A., Johnson, P.T., Kidd, K.A., MacCormack, T.J., Olden, J.D., Ormerod, S.J. and Smol, J.P., 2019. Emerging threats and persistent conservation challenges for freshwater biodiversity. Biological Reviews, 94(3), pp.849-873.

Roni, P., Hanson, K., & Beechie, T. (2008). Global review of the physical and biological effectiveness of stream habitat rehabilitation techniques. North American Journal of Fisheries Management, 28(3), 856-890.

Ruckelshaus, M., Essington, T., & Levin, P. (2009). Puget Sound, Washington, USA. In K. M. H. Leslie (Ed.), Ecosystem-based management for the oceans (pp. 201–226). Washington, DC: Island Press

Ruiz-Mallén, I., Schunko, C., Corbera, E., Rös, M. and Reyes-García, V., 2015. Meanings, drivers, and motivations for community-based conservation in Latin America. Ecology and Society, 20(3).

Sarkar, U. K., Kathirvelpandian, A., Kantharajan, G., Tyagi, L. K., & Lakra, W. S. (2024). The concept of "state fishes of India": Current status, knowledge gaps, and strategic plans for conservation and sustainable utilization. Journal of Fish Biology.

SOFIA, 2024. The State of World Fisheries and Aquaculture 2024: Blue Transformation in Action. Food and Agriculture Organisation of the United Nations.

Srivastava, V.K., 2007. River ecology in India: present status and future research strategy for management and conservation. Proceedings-Indian National Science Academy, 73(4), p.255.

Stefanidis, K., Sarika, M. and Papastegiadou, E., 2019. Exploring environmental predictors of aquatic macrophytes in water-dependent Natura 2000 sites of high conservation value: Results from a long-term study of macrophytes in Greek lakes. Aquatic Conservation: Marine and Freshwater Ecosystems, 29(7), pp.1133-1148.

Swanson, S., Kozlowski, D., Hall, R., Heggem, D. and Lin, J., 2017. Riparian proper functioning condition assessment to improve watershed management for water quality. Journal of Soil and Water Conservation, 72(2), pp.168-182.

Szałkiewicz, E., Sucholas, J. and Grygoruk, M., 2020. Feeding the Future with the Past: Incorporating Local Ecological Knowledge in River Restoration. Resources, 9(4), p.47.

Ukpoju, E.A., Abatan, A., Obaedo, B.O., Balogun, O.D. and Adegbite, A.O., 2023. Assessing the Effectiveness of Biodiversity Conservation Strategies in Nigeria: A Comprehensive Review. International Journal of Applied Research in Social Sciences, 5(10), pp.577-598.

Vörösmarty, C.J., Osuna, V.R., Cak, A.D., Bhaduri, A., Bunn, S.E., Corsi, F., Gastelumendi, J., Green, P., Harrison, I., Lawford, R. and Marcotullio, P.J., 2018. Ecosystem-based water security and the Sustainable Development Goals (SDGs). Ecohydrology & Hydrobiology, 18(4), pp.317-333.

Wohl, E., 2017. Connectivity in rivers. Progress in Physical Geography, 41(3), pp.345-362.

Wohl, E., Lane, S.N. and Wilcox, A.C., 2015. The science and practice of river restoration. Water Resources Research, 51(8), pp.5974-5997.

Zhang, W., ElDidi, H., Masuda, Y. J., Meinzen-Dick, R. S., Swallow, K. A., Ringler, C. & Aldous, A. (2023). Community-based conservation of freshwater resources: learning from a critical review of the literature and case studies. Society & Natural Resources, 36(6), 733-754.

Index

A

Abiotic:48, 159, 181, 231, 247
Algae: 11, 25, 56, 66, 69, 77, 111, 120, 123, 156, 158, 169, 206, 210, 222, 224, 226, 227, 229, 230, 231
Algaecide: 206
Ambio: 271
Antibiofilm: 114
Antibiotic: 114, 115, 117
Antibiotics: 56, 114, 115, 131, 160, 161, 177, 180, 182, 193, 194
Antimicrobial: 61, 62, 63, 115, 117, 130, 132, 186, 210
Aquaculture: 1, 2, 4, 5, 6, 7, 8, 9, 15, 16, 17, 18, 33, 35, 36, 39, 40, 41, 42, 43, 44, 45, 46, 47, 48, 49, 50, 67, 68, 73, 109, 115, 116, 121, 122, 125, 127, 128, 129, 130, 131, 132, 135, 148, 159, 173, 194, 197, 203, 212, 213, 221, 228, 230, 259, 260, 270, 271, 272, 274
Aquatic: 1, 2, 3, 4, 5, 6, 7, 8, 9, 10, 11, 12, 13, 14, 15, 16, 17, 18, 19, 20, 22, 24, 26, 27, 28, 31, 32, 33, 34, 35, 36, 39, 40, 42, 44, 46, 48, 49, 50, 51, 52, 53, 54, 55, 56, 57, 58, 59, 60, 61, 62, 63, 64, 65, 66, 67, 68, 69, 70, 71, 72, 73, 74, 75, 76, 78, 80, 82, 84, 86, 88, 90, 92, 94, 96, 98, 100, 102, 104, 106, 109, 110, 111, 112, 113, 114, 115, 116, 117, 118, 119, 120, 121, 122, 123, 124, 125, 126, 127, 128, 129, 130, 131, 132, 133, 134, 135, 136, 137, 138, 139, 140, 141, 142, 143, 144, 146, 147, 148, 149, 150, 151, 152, 153, 154, 155, 156, 157, 158, 159, 160, 161, 162, 163, 164, 165, 166, 167, 168, 169, 170, 171, 172, 173, 174, 175, 176, 177, 179, 180, 181, 182, 183, 184, 185, 186, 187, 188, 189, 190, 191, 192, 193, 194, 195, 196, 197, 198, 199, 200, 201, 202, 203, 204, 205, 206, 207, 208, 209, 210, 211, 212, 213, 214, 215, 216, 217, 218, 219, 220, 222, 224, 226, 228, 229, 230, 232, 233, 234, 236, 238, 240, 242, 244, 246, 248, 250, 252, 254, 256, 259, 260, 261, 262, 263, 264, 265, 266, 267, 268, 269, 270, 271, 272, 273, 274, 275
Autotrophic: 219, 229

B

Bioaccessibility: 177
Bioaccumulate: 65, 68, 77, 136, 160, 161, 179, 180, 204
Bioaccumulation: 51, 59, 64, 65, 68, 73, 74, 112, 134, 153, 155, 157, 160, 161, 162, 163, 165, 172, 173, 174, 175, 176, 177, 208, 209, 213, 214, 215, 217, 241, 244, 245
Bioaccumulative: 183
Bioactive: 218
Bioadsorption: 195
Bioadvantaged: 103
Bioanal: 201
Bioanalytical: 129, 149, 200
Bioattenuation: 111
Bioaugmentation: 111, 171, 234
Bioavailability: 127, 153, 164, 165, 172, 173, 176, 177, 205, 238, 240, 242, 243, 246
Bioavailable: 206, 209
Biobased: 102
Biocatalysts: 28, 97, 122
Biochar: 46, 106, 190, 195, 197
Biochem: 245
Biochemical: 58, 60, 70, 77,115, 124, 154, 185, 191, 205, 208, 213, 218, 225, 230, 236
Biochemistry: 18, 65, 74, 129, 180, 199, 213, 215, 216, 218, 243
Biocide: 196
Biocompatibility: 90, 92
Biocompatible: 92
Bioconcentration: 65, 153, 161, 163, 164, 165, 166, 172, 173, 174, 175, 176, 177
Biodegradability: 90, 93, 100, 198
Biodegradable: 1, 14, 15, 16, 17, 18, 77, 90, 96, 100, 129, 145, 146, 147, 191, 192
Biodegradation: 14, 25, 70, 81, 177, 190, 191, 192, 198, 240, 245
Biodigesters: 36

Biodiversity: 1, 2, 6, 7, 13, 15, 18, 27, 32, 33, 40, 46, 49, 51, 70, 113, 135, 136, 154, 159, 160, 166, 172, 188, 203, 205, 207, 209, 211, 213, 215, 217, 220, 245, 248, 259, 260, 261, 262, 263, 264, 268, 269, 270, 271, 272, 273, 274
Bioenergy: 109, 129
Bioengineering: 196, 197, 199
Bioethanol: 125, 129
Biofertilisation: 112
Biofertilization: 109, 112, 113
Biofertilizers: 112, 113, 122, 129, 131
Biofilms: 109, 113, 114, 115, 119, 120, 128, 129, 130, 131
Biofloc: 4, 8, 16, 17, 18, 109, 115, 128, 129
Biofuels: 5, 35, 52, 110, 122, 125, 231
Biogas: 33
Biogenic: 78, 119, 199, 222
Biogeochemical: 64, 109, 110, 127, 159, 223, 226
Biogeochemistry: 230
Bioindicators: 109, 122, 131, 132, 213
Bioleaching: 126, 129, 130, 131, 246
Biological: 1, 10, 11, 12, 15, 16, 18, 19, 20, 25, 35, 48, 49, 64, 66, 76, 77, 80, 90, 91, 97, 111, 112, 113, 116, 121, 126, 130, 132, 145, 153, 154, 156, 160, 163, 164, 169, 171, 172, 177, 182, 186, 187, 188, 191, 192, 193, 194, 195, 198, 199, 202, 214, 216, 217, 219, 221, 223, 224, 226, 227, 228, 229, 233, 234, 243, 250, 261, 262, 266, 274
Biology: 17, 18, 37, 109, 125, 128, 181, 199, 201, 213, 214, 215, 216, 217, 243, 265, 272, 274
Biomagnification: 51, 65, 66, 153, 154, 155, 162, 163, 172, 173, 174, 176, 177
Biomarkers: 129, 168, 173, 175, 213, 217, 232
Biomass: 4, 34, 90, 115, 122, 131, 132, 190, 207, 211, 214, 219, 220, 221, 224, 227, 228, 236, 240, 241, 242, 243
Biomedical: 79, 120, 128
Biomolecular: 103, 198
Biomonitoring: 130, 177, 196
Bionanocomposite: 105
Biopolymer: 91, 93, 103
Biopolymers: 90, 91, 102, 104, 105
Bioprocesses: 109, 111, 113, 115, 117, 119, 121, 123, 125, 126, 127, 129, 130, 131, 132
Bioprocessing: 121, 128
Bioreactor: 85, 190, 197, 201
Bioreactors: 84, 115, 145, 191, 198, 199
Biorefineries: 122, 129, 131
Biorefinery: 102, 129, 131
Bioremediation: 6, 13, 16, 18, 25, 28, 103, 109, 110, 111, 112, 113, 115, 126, 127, 130, 131, 132, 168, 169, 171, 173, 196, 233, 234, 235, 236, 237, 239, 240, 241, 242, 243, 244, 245
Biosensors: 72, 109, 122, 124, 128, 129
Biosolids: 143, 144, 148, 195
Biosorption: 28, 91, 112, 160, 241, 244
Biostimulation: 111, 234
Biosurfactant: 127, 129
Biosurfactants: 126, 129, 246
Biota: 183, 205, 207, 215, 239
Biotechnological: 125, 129, 130, 132
Biotechnology: 33, 109, 118, 122, 125, 126, 129, 130, 131, 132, 150, 174, 196, 197, 199, 216, 247
Biotic: 131, 181, 267
Biotransformation: 167, 168, 173, 174, 175, 177

C

Climate: 1, 2, 3, 15, 16, 17, 21, 24, 39, 40, 41, 42, 43, 44, 45, 47, 48, 49, 50, 59, 196, 219, 220, 222, 228, 229, 230, 231, 247, 248, 249, 251, 252, 253, 254, 255, 256, 257, 259, 260, 261, 263, 265, 266, 267, 268, 269, 270, 271, 273, 274
Coastal: 2, 3, 6, 10, 13, 14, 17, 40, 42, 43, 44, 48, 50, 63, 123, 124, 139, 142, 148, 151, 158, 159, 174, 183, 205, 207, 215, 219, 220, 221, 222, 223, 225, 226, 227, 228, 229, 230, 231, 232, 259, 260, 267, 271, 273
Conservation: 5, 7, 13, 15, 16, 17, 18, 28, 113, 122, 129, 151, 163, 259, 260, 261, 262, 263, 264, 265, 266, 267, 268, 269, 270, 271, 272, 273, 274, 275

D

Degradation: 1, 4, 20, 25, 26, 27, 28, 29, 59, 60, 64, 78, 86, 94, 98, 103, 104, 109, 112, 119, 128, 138, 142, 148, 150, 152, 157, 193, 196, 197, 199, 202, 221, 222, 228, 234, 240, 244

Desalination: 1, 10, 15, 16, 85, 102, 103, 104, 105, 106, 196
Disinfection: 11, 21, 22, 76, 89, 103, 180, 182

E

Ecosystem: 1, 2, 3, 4, 6, 7, 8, 10, 12, 13, 14, 15, 16, 17, 18, 20, 22, 24, 26, 28, 29, 31, 32, 33, 34, 35, 36, 40, 42, 44, 46, 48, 49, 50, 51, 52, 54, 56, 58, 60, 62, 64, 66, 68, 70, 72, 74, 76, 78, 80, 82, 84, 86, 88, 90, 92, 94, 96, 98, 100, 102, 104, 106, 110, 111, 112, 114, 116, 118, 119, 120, 122, 123, 124, 126, 128, 130, 132, 134, 135, 136, 138, 140, 142, 144, 146, 148, 150, 152, 153, 154, 155, 156, 157, 158, 160, 162, 163, 164, 166, 168, 170, 172, 174, 175, 176, 177, 180, 182, 184, 186, 188, 190, 192, 194, 196, 198, 200, 202, 203, 204, 205, 206, 207, 208, 210, 212, 214, 216, 218, 220, 221, 222, 224, 225, 226, 228, 230, 232, 233, 234, 236, 238, 240, 242, 243, 244, 246, 248, 250, 252, 254, 256, 259, 260, 261, 262, 263, 264, 265, 266, 267, 268, 269, 270, 271, 272, 273, 274, 275
Energy: 1, 2, 8, 9, 10, 11, 15, 16, 22, 24, 27, 28, 36, 49, 78, 86, 89, 94, 96, 99, 100, 101, 102, 106, 110, 118, 121, 125, 126, 128, 129, 159, 193, 194, 196, 199, 222, 223, 230, 231, 233, 250, 256, 257
Eutrophication: 5, 113, 159, 166

F

Filtration: 1, 2, 3, 6, 9, 10, 11, 12, 15, 19, 21, 22, 26, 76, 83, 84, 85, 89, 91, 92, 101, 103, 104, 116, 133, 137, 142, 143, 145, 147, 151, 181, 188, 189, 211, 250
Fisheries: 1, 15, 17, 18, 19, 36, 37, 39, 40, 41, 42, 43, 45, 48, 49, 50, 51, 75, 109, 128, 130, 133, 153, 179, 203, 212, 213, 219, 220, 221, 222, 223, 224, 233, 247, 259, 260, 263, 264, 270, 271, 272, 273, 274
Freshwater: 5, 10, 17, 20, 31, 32, 64, 68, 72, 84, 100, 113, 116, 129, 133, 136, 147, 170, 173, 176, 197, 199, 203, 207, 209, 213, 215, 216, 217, 219, 222, 225, 259, 260, 261, 262, 263, 264, 266, 270, 271, 272, 273, 274, 275

G

Green: 1, 2, 3, 4, 5, 7, 9, 11, 12, 13, 15, 16, 17, 28, 50, 96, 97, 102, 103, 105, 106, 131, 146, 151, 206, 207, 213, 216, 224, 226, 227, 230, 233, 234, 235, 237, 239, 241, 243, 244, 245, 246, 248, 249, 250, 253, 256, 257, 265, 266, 272, 274
Greenhouse: 45, 46, 49, 50, 121, 220, 227, 242, 253
Groundwater: 32, 59, 76, 93, 104, 151, 159, 179, 180, 183, 184, 195, 197, 198, 200, 201, 202, 237, 238, 239, 251, 255, 263, 266, 270

H

Habitat: 1, 2, 3, 4, 12, 13, 15, 18, 68, 157, 222, 250, 259, 260, 263, 265, 268, 269, 270, 272, 274
Habitats: 1, 2, 3, 4, 14, 20, 27, 51, 71, 78, 110, 122, 128, 133, 142, 145, 147, 148, 157, 159, 166, 176, 179, 220, 221, 222, 224, 252, 259, 262, 263, 265, 267, 269, 270, 272

I

Infiltration: 58, 68, 250, 251, 252, 256, 257
Invasive: 1, 3, 18, 40, 121, 136, 152, 243, 260, 267, 268

M

Marine: 3, 13, 14, 15, 18, 20, 32, 40, 43, 44, 48, 49, 50, 64, 111, 121, 124, 125, 128, 129, 130, 133, 134, 135, 136, 139, 140, 142, 145, 147, 148, 149, 150, 159, 163, 165, 166, 169, 173, 174, 175, 176, 177, 185, 187, 188, 195, 206, 207, 209, 210, 213, 214, 215, 217, 218, 219, 221, 222, 223, 225, 226, 227, 228, 229, 230, 231, 259, 260, 262, 264, 271, 272, 273, 274
Membrane: 10, 11, 16, 19, 24, 28, 66, 75, 76, 77, 79, 81, 83, 84, 85, 86, 87, 88, 89, 90, 91, 92, 93, 94, 95, 96, 97, 99, 100, 101, 102, 103, 104, 105, 106, 107, 145, 160, 177, 188, 189, 190, 191, 194, 196, 197, 198, 199, 201, 202, 205, 208, 217, 250
Membranes: 9, 10, 21, 23, 24, 28, 29, 76, 77, 84, 85, 86, 87, 90, 91, 92, 97, 98, 99, 101, 102, 103, 104, 105, 106, 163, 165, 189, 190, 195, 197, 198, 201, 206

Microalgae: 5, 16, 46, 122, 155, 168, 170, 171, 175, 221, 222, 230, 231
Microbial: 4, 10, 17, 25, 27, 28, 29, 58, 61, 68, 69, 72, 90, 109, 110, 111, 112, 113, 114, 115, 116, 117, 118, 119, 120, 121, 122, 123, 124, 125, 126, 127, 128, 129, 130, 131, 132, 166, 169, 175, 187, 196, 219, 231, 234, 240, 241, 242, 243
Microbiology: 17, 29, 128, 129, 130, 132, 151, 174, 200, 230, 243
Microbiome: 117, 173, 193, 200
Microbiota: 116, 117, 125, 130, 168, 176, 218
Microfiltration: 21, 76, 85, 86, 89, 94, 103, 105, 106, 189, 198, 200, 202
Microplastics: 3, 21, 27, 115, 121, 133, 134, 135, 136, 137, 138, 139, 140, 141, 142, 143, 144, 145, 146, 147, 148, 149, 150, 151, 152, 153, 154, 155, 157, 170, 172, 184, 185, 195, 198, 200
Monitoring: 1, 6, 7, 9, 14, 15, 16, 17, 21, 24, 29, 44, 47, 50, 59, 112, 119, 120, 123, 124, 128, 129, 130, 131, 136, 138, 139, 151, 155, 171, 172, 181, 186, 200, 212, 263

N

Nanofiltration: 9, 23, 28, 76, 85, 86, 89, 90, 94, 103, 189, 190, 194, 195, 196, 198, 201, 202
Nanoplastics: 134, 148, 198
Nanosolar: 39, 40
Nanotechnology: 1, 9, 15, 16, 19, 23, 103, 104, 105, 128
Nutrients: 5, 6, 7, 12, 27, 31, 33, 34, 35, 109, 110, 111, 112, 113, 115, 116, 120, 126, 158, 159, 215, 222, 223, 225, 226, 229, 234, 237, 240

O

Oxidation: 19, 23, 25, 26, 28, 84, 93, 94, 95, 102, 104, 129, 167, 182, 189, 192, 197, 198, 199, 201, 211, 241
Oxygen: 5, 8, 9, 23, 26, 66, 67, 95, 119, 125, 136, 145, 157, 158, 159, 160, 191, 193, 206, 207, 209, 223

P

Phycoremediation: 170, 177
Phytodegradation: 13, 240, 245
Phytoremediation: 13, 15, 17, 171, 172, 176, 212, 216, 234, 236, 239, 241, 242, 244, 245, 246
Plastics: 3, 14, 110, 133, 134, 135, 138, 142, 144, 145, 146, 147, 157, 177, 183, 195
Pollution: 1, 2, 3, 4, 6, 12, 14, 15, 16, 18, 19, 20, 21, 22, 23, 24, 25, 27, 28, 29, 31, 35, 40, 58, 59, 68, 70, 71, 73, 74, 75, 76, 77, 78, 79, 80, 99, 102, 107, 110, 111, 112, 114, 119, 122, 123, 131, 132, 133, 134, 135, 136, 139, 140, 141, 142, 143, 144, 145, 146, 148, 149, 150, 151, 155, 157, 158, 159, 168, 169, 171, 173, 174, 175, 176, 179, 185, 186, 193, 195, 196, 197, 198, 199, 201, 203, 204, 205, 207, 208, 209, 210, 211, 212, 213, 214, 215, 216, 217, 228, 233, 243, 244, 246, 247, 259, 260, 262, 265, 273
Probiotics: 109, 116, 117, 122, 128, 130, 132

R

Rainwater: 247, 248, 249, 250, 251, 252, 254
Reclamation: 85, 198, 244
Recycling: 5, 6, 31, 97, 99, 115, 116, 129, 145, 146, 150, 249
Remediation: 1, 13, 15, 16, 17, 21, 27, 29, 93, 94, 102, 104, 111, 112, 113, 115, 118, 123, 127, 129, 131, 151, 177, 193, 201, 202, 211, 212, 234, 235, 236, 237, 240, 241, 242, 243, 244, 245, 246
Renewable: 2, 8, 10, 24, 35, 90, 96, 100, 110, 126, 128
Resilience: 6, 7, 14, 39, 40, 45, 47, 48, 49, 50, 115, 116, 125, 160, 247, 248, 259, 261, 267, 268, 269
Restoration: 2, 6, 13, 15, 17, 111, 112, 118, 119, 234, 238, 247, 252, 263, 264, 265, 266, 267, 268, 269, 270, 271, 272, 273, 274, 275
Rhizofiltration: 13, 172, 237, 238
Rhizoremediation: 245

S

Seawater: 10, 76, 101, 134, 150, 213
Sedimentation: 4, 21, 22, 27, 157, 221
Smart: 7, 8, 15, 16, 17, 19, 23, 247, 255
Smartphone: 139, 149, 150
Solar: 8, 9, 10, 11, 15, 16, 18, 24, 28, 94, 96, 105, 198, 200

Species: 1, 2, 3, 4, 6, 7, 12, 18, 20, 25, 26, 33, 40, 41, 43, 44, 45, 46, 48, 51, 55, 60, 61, 62, 63, 66, 67, 68, 95, 109, 115, 116, 121, 123, 124, 125, 135, 136, 137, 152, 154, 155, 156, 157, 158, 159, 160, 161, 162, 165, 166, 167, 168, 169, 170, 174, 176, 188, 193, 206, 207, 208, 209, 210, 213, 214, 217, 221, 222, 223, 224, 227, 231, 234, 239, 240, 241, 242, 243, 259, 260, 261, 262, 263, 265, 267, 268, 269, 270, 271, 272

Stormwater: 133, 143, 150, 152, 186, 248, 249, 250, 253, 254, 256

Sustainability: 1, 2, 3, 4, 5, 6, 7, 8, 9, 11, 13, 15, 16, 17, 18, 20, 21, 27, 28, 31, 35, 39, 40, 42, 47, 48, 49, 75, 112, 116, 121, 125, 127, 129, 130, 146, 150, 154, 172, 201, 231, 238, 250, 256, 257, 259, 260, 261, 271, 273

T

Technology: 1, 2, 3, 4, 5, 7, 8, 9, 10, 11, 12, 13, 15, 16, 17, 18, 23, 24, 25, 27, 29, 39, 40, 41, 49, 73, 76, 82, 83, 84, 93, 94, 99, 100, 101, 102, 103, 104, 106, 109, 115, 123, 127, 128, 129, 130, 131, 132, 144, 148, 149, 151, 174, 175, 177, 189, 190, 194, 195, 196, 197, 198, 201, 202, 216, 217, 227, 233, 240, 243, 244, 245, 248, 251, 256, 259

Temperature: 5, 8, 23, 24, 29, 41, 42, 49, 59, 66, 84, 92, 96, 106, 120, 126, 141, 142, 157, 159, 199, 205, 215, 223, 224, 225, 226, 227, 228, 229, 231

Treatment: 1, 9, 11, 12, 15, 16, 18, 20, 21, 22, 23, 25, 26, 27, 28, 29, 69, 75, 76, 82, 83, 84, 85, 89, 91, 93, 94, 98, 99, 100, 102, 103, 104, 105, 106, 109, 114, 115, 119, 121, 122, 125, 126, 127, 132, 133, 137, 142, 143, 145, 148, 151, 153, 155, 158, 170, 172, 174, 176, 179, 180, 181, 182, 185, 186, 188, 189, 190, 191, 192, 193, 194, 195, 196, 197, 198, 199, 200, 201, 202, 211, 235, 242, 243, 245

Trophic: 6, 16, 17, 45, 46, 64, 65, 66, 157, 159, 161, 162, 163, 170, 173, 177, 205

U

Ultrafiltration: 21, 76, 83, 84, 85, 86, 89, 90, 92, 94, 104, 105, 106, 189, 190, 194, 195, 196, 197, 198, 201

W

Wastewater: 1, 3, 10, 11, 12, 13, 15, 16, 18, 20, 23, 25, 26, 27, 28, 29, 46, 68, 75, 76, 78, 79, 83, 86, 89, 90, 91, 93, 96, 98, 99, 100, 102, 103, 104, 105, 106, 109, 114, 115, 121, 122, 125, 126, 127, 132, 133, 135, 141, 142, 143, 145, 150, 151, 153, 155, 158, 170, 171, 176, 177, 179, 181, 182, 183, 185, 186, 188, 189, 190, 191, 192, 193, 194, 195, 196, 197, 198, 199, 200, 201, 202, 215, 216, 245

Water: 1, 2, 3, 4, 5, 6, 7, 8, 9, 10, 11, 12, 13, 14, 15, 16, 18, 19, 20, 21, 22, 23, 24, 25, 26, 27, 28, 29, 33, 34, 35, 36, 40, 41, 42, 44, 45, 46, 48, 49, 50, 51, 54, 59, 60, 64, 65, 66, 67, 68, 70, 71, 72, 73, 75, 76, 77, 78, 79, 80, 81, 82, 83, 85, 86, 87, 89, 91, 92, 93, 94, 95, 97, 99, 100, 101, 102, 103, 104, 105, 106, 107, 109, 110, 113, 115, 116, 118, 119, 120, 121, 122, 123, 124, 125, 128, 130, 131, 132, 133, 134, 135, 136, 137, 138, 139, 140, 141, 142, 143, 144, 145, 147, 148, 149, 150, 151, 152, 153, 154, 155, 156, 157, 158, 159, 160, 162, 163, 164, 165, 169, 171, 172, 174, 177, 179, 180, 181, 182, 183, 184, 185, 188, 189, 190, 191, 192, 193, 194, 195, 196, 197, 198, 199, 200, 201, 202, 203, 204, 206, 209, 212, 213, 214, 215, 216, 220, 221, 222, 223, 224, 226, 227, 228, 230, 231, 233, 234, 235, 236, 238, 239, 240, 243, 245, 246, 247, 248, 249, 250, 251, 252, 253, 255, 256, 257, 260, 261, 262, 263, 265, 266, 267, 269, 271, 272, 274, 275

Wetlands: 1, 12, 15, 16, 18, 28, 32, 121, 193, 194, 196, 197, 201, 212, 213, 220, 221, 250, 252, 253, 256, 259, 261, 262, 272, 274

X

Xenobiot: 73

Xenobiotic: 167, 168

Xenobiotics: 167, 168